ANNALS OF THE NEW YORK ACADEMY OF SCIENCES

Volume 982

EDITORIAL STAFF

Managing Editor
JUSTINE CULLINAN

Associate Editors
JOYCE HITCHCOCK
STEFAN MALMOLI
RICHARD STIEFEL

The New York Academy of Sciences
2 East 63rd Street
New York, New York 10021

THE NEW YORK ACADEMY OF SCIENCES
(Founded in 1817)

BOARD OF GOVERNORS, September 2002 – September 2003

TORSTEN N. WIESEL, *Chairman of the Board*
JOHN T. MORGAN, *Treasurer*
ELLIS RUBINSTEIN, *Chief Executive Officer* [ex officio]

Honorary Life Governors
WILLIAM T. GOLDEN   JOSHUA LEDERBERG

Governors

	ELEANOR BAUM	KAREN E. BURKE	
	LAWRENCE B. BUTTENWIESER	PRAVEEN CHAUDHARI	
BRIAN FERGUSON	GERALD FISCHBACH		JOHN H. GIBBONS
MICHAEL GOLDEN	RONALD L. GRAHAM		MARNIE IMHOFF
JACQUELINE LEO	BRUCE McEWEN		PAUL MARKS
RONAY MENSCHEL	JOHN F. NIBLACK		SANDRA PANEM
PETER RINGROSE	JOHN J. ROCHE		LEE G. VANCE
	DEBORAH WILEY		

HELENE L. KAPLAN, *Counsel* [ex officio]

CARCINOGENESIS BIOASSAYS AND PROTECTING PUBLIC HEALTH

COMMEMORATING THE LIFEWORK OF
CESARE MALTONI AND COLLEAGUES

ANNALS OF THE NEW YORK ACADEMY OF SCIENCES
Volume 982

CARCINOGENESIS BIOASSAYS AND PROTECTING PUBLIC HEALTH

COMMEMORATING THE LIFEWORK OF CESARE MALTONI AND COLLEAGUES

Edited by Myron A. Mehlman, Eula Bingham, Philip J. Landrigan, Morando Soffritti, Fiorella Belpoggi, and Ronald L. Melnick

The New York Academy of Sciences
New York, New York
2002

Copyright © 2002 by the New York Academy of Sciences. All rights reserved. Under the provisions of the United States Copyright Act of 1976, individual readers of the Annals *are permitted to make fair use of the material in them for teaching or research. Permission is granted to quote from the* Annals *provided that the customary acknowledgment is made of the source. Material in the* Annals *may be republished only by permission of the Academy. Address inquiries to the Permissions Department (permissions@nyas.org) at the New York Academy of Sciences.*

Copying fees: *For each copy of an article made beyond the free copying permitted under Section 107 or 108 of the 1976 Copyright Act, a fee should be paid through the Copyright Clearance Center, Inc., 222 Rosewood Drive, Danvers, MA 01923 (www.copyright.com).*

⊚ *The paper used in this publication meets the minimum requirements of the American National Standard for Information Sciences—Permanence of Paper for Printed Library Materials, ANSI Z39.48-1984.*

Library of Congress Cataloging-in-Publication Data

Carcinogenesis bioassays and protecting public health: commemorating the lifework of Cesare Maltoni and colleagues / edited by Myron A. Mehlman, Eula Bingham, Philip J. Landrigan, ... [*et al.*]
   p.; cm. — (Annals of the New York Academy of Sciences, ISSN 0077-8923; v. 982)
  The result of a conference held on April 29–30, 2002, in New York.
  Includes bibliographical references and index.
   ISBN 1-57331-406-4 (cloth: alk. paper) — ISBN 1-57331-407-2 (paper: alk. paper)
  1. Carcinogenicity testing—Congresses. 2. Maltoni, Cesare—Congresses.
   [DNLM: 1. Carcinogenicity Tests—Congresses. 2. Biological
Assay—Congresses. 3. Carcinogens, Environmental—analysis—Congresses.
4. Organic Chemicals—toxicity—Congresses. 5. Public Health—Congresses.
QZ 202 C26442 2002] I. Maltoni, Cesare. II. Mehlman, Myron A. III. Bingham, Eula.
IV. Landrigan, Philip J. V. Series.
Q11.N5 vol. 982
[RC268.65]
500 s—dc21
[616.99/40

                                                      2002153477

GYAT/B-M Press
Printed in the United States of America
ISBN 1-57331-406-4 (cloth)
ISBN 1-57331-407-2 (paper)
ISSN 0077-8923

ANNALS OF THE NEW YORK ACADEMY OF SCIENCES

Volume 982
December 2002

CARCINOGENESIS BIOASSAYS AND PROTECTING PUBLIC HEALTH

COMMEMORATING THE LIFEWORK OF CESARE MALTONI AND COLLEAGUES

Editors
MYRON A. MEHLMAN, EULA BINGHAM, PHILIP J. LANDRIGAN,
MORANDO SOFFRITTI, FIORELLA BELPOGGI, AND RONALD L. MELNICK

This volume is the result of a conference entitled **Carcinogenesis Bioassays and Protecting Public Health: Commemorating the Lifework of Cesare Maltoni and Colleagues** sponsored by the New York Academy of Sciences in collaboration with the Collegium Ramazzini and the National Institute of Environmental Health Sciences and held on April 29–30, 2002 in New York, New York.

CONTENTS

Introduction. *By* MYRON A. MEHLMAN	ix
Remembering Professor Cesare Maltoni. *By* MYRON A. MEHLMAN	1
Ramazzini Foundation Cancer Program: History and Major Projects, Life-Span Carcinogenicity Bioassay Design, Chemicals Studied, and Results. *By* MORANDO SOFFRITTI, FIORELLA BELPOGGI, FRANCO MINARDI, AND CESARE MALTONI	26
Results of Long-Term Experimental Studies on the Carcinogenicity of Methyl Alcohol and Ethyl Alcohol in Rats. *By* MORANDO SOFFRITTI, FIORELLA BELPOGGI, DANIELA CEVOLANI, MARINA GUARINO, MICHELA PADOVANI, AND CESARE MALTONI	46
Results of Long-Term Carcinogenicity Bioassays on *Tert*-Amyl-Methyl-Ether (TAME) and Di-Isopropyl-Ether (DIPE) in Rats. *By* FIORELLA BELPOGGI, MORANDO SOFFRITTI, FRANCO MINARDI, LUCIANO BUA, ELISA CATTIN, AND CESARE MALTONI	70
Results of Long-Term Experimental Studies on the Carcinogenicity of Formaldehyde and Acetaldehyde in Rats. *By* MORANDO SOFFRITTI, FIORELLA BELPOGGI, LUCA LAMBERTINI, MICHELINA LAURIOLA, MICHELA PADOVANI, AND CESARE MALTONI	87

Results of Long-Term Carcinogenicity Bioassay on Vinyl Acetate Monomer in Sprague-Dawley Rats. *By* FRANCO MINARDI, FIORELLA BELPOGGI, MORANDO SOFFRITTI, ADRIANO CILIBERTI, MICHELINA LAURIOLA, ELISA CATTIN, AND CESARE MALTONI 106

Results of Long-Term Experimental Studies of Carcinogenicity of Ethylene-bis-Dithiocarbamate (Mancozeb) in Rats. *By* FIORELLA BELPOGGI, MORANDO SOFFRITTI, MARINA GUARINO, LUCA LAMBERTINI, DANIELA CEVOLANI, AND CESARE MALTONI 123

Carcinogenic Effects of Benzene: Cesare Maltoni's Contributions. *By* MYRON A. MEHLMAN 137

Carcinogenicity of Methyl-Tertiary Butyl Ether in Gasoline. *By* MYRON A. MEHLMAN 149

Asbestos Fibers Contributing to the Induction of Human Malignant Mesothelioma. *By* YASUNOSUKE SUZUKI AND STEVEN R. YUEN 160

Carcinogenicity and Mechanistic Insights on the Behavior of Epoxides and Epoxide-Forming Chemicals. *By* RONALD L. MELNICK 177

Primary Prevention Protects Public Health. *By* LORENZO TOMATIS 190

The National Toxicology Program Rodent Bioassay: Designs, Interpretations, and Scientific Contributions. *By* JOHN R. BUCHER 198

Chemicals Studied and Evaluated in Long-Term Carcinogenesis Bioassays by Both the Ramazzini Foundation and the National Toxicology Program: In Tribute to Cesare Maltoni and David Rall. *By* JAMES HUFF 208

Index of Contributors 231

Financial assistance was received from:
Major funding
- COLLEGIUM RAMAZZINI/RAMAZZINI FOUNDATION
- NATIONAL TOXICOLOGY PROGRAM/NATIONAL INSTITUTE OF ENVIRONMENTAL HEALTH SCIENCES— NATIONAL INSTITUTES OF HEALTH

Supporters
- AGENCY FOR TOXIC SUBSTANCES AND DISEASE REGISTRY
- THE MUSHETT FAMILY FOUNDATION
- NATIONAL CANCER INSTITUTE—NATIONAL INSTITUTES OF HEALTH

The New York Academy of Sciences believes it has a responsibility to provide an open forum for discussion of scientific questions. The positions taken by the participants in the reported conferences are their own and not necessarily those of the Academy. The Academy has no intent to influence legislation by providing such forums.

Cesare Maltoni

Introduction

Participants were welcomed by Professor Philip J. Landrigan, Chairman of the Department of Community and Preventive Medicine of the Mount Sinai School of Medicine and President of the Collegium Ramazzini, and Dr. Rashid Shaikh of the New York Academy of Sciences. Giorgio Setti, Deputy Mayor of Carpi, Italy, then presented an overview of the history of Carpi and the contributions of Bernardino Ramazzini and their relevance into the twenty-first century. Myron Mehlman presented the history of Maltoni's life, his scientific accomplishments, and his legacy.

Professor Cesare Maltoni (November 17, 1930–January 21, 2002) was a world-renowned leader in the research of causation of cancers by industrial chemicals in the workplace and environment. Born in Faenza, Italy, Maltoni conducted long-term carcinogenicity studies in animals on nearly 200 chemicals. He was the first to demonstrate that vinyl chloride is a carcinogenic agent that produces angiosarcomas of the liver and other tumors in experimental animals; similar tumors subsequently were found to occur among persons working with vinyl chloride in industry. Maltoni was the first to demonstrate and report that benzene was a multipotential carcinogen that caused carcinomas of the zymbal gland, oral and nasal cavities, the skin, forestomach, mammary glands, liver, and the hemolymphoreticular systems—for example, leukemias.

The scientific studies presented in this volume on carcinogenesis bioassays of chemicals and products demonstrate that when carcinogenic studies are properly designed and carried out, they can provide protection to the public and the environment. These studies confirm that animal studies remain the gold standard in predicting carcinogenicity of chemicals to humans and that studies conducted with high levels of exposure can be extrapolated to very low-level human exposure over time.

Soffritti and associates describe the research program that began 30 years ago at the Ramazzini Foundation (RF). Carcinogenicity bioassays and epidemiological studies were carried out. Carcinogenicity bioassays were conducted on 200 separate substances using 148,000 animals. In general, animals were exposed to the test substance for a finite period—for example, 1–2 years—and then observed for tumor occurrence until their spontaneous death.

Of the agents and compounds tested, 47 showed clear evidence of carcinogenicity. New or updated reports on agents found to be carcinogenic included in this volume are aldehydes (formaldehyde and acetaldehyde), alcohols, vinyl acetate monomer, the fungicide ethylene-bis-dithio-carbamate (Mancozeb), and oxygenated fuel additives—for example, methyl tertiary butyl ether (MTBE), ethyl tertiary butyl ether (ETBE), tert-amyl-methyl ether (TAME), and di-isopropyl ether (DIPE). As in previous reports on other carcinogens such as xylene, vinyl chloride, vinylidene chloride, acrylonitrile, vinyl acetate, styrene oxide, plastic polymers, chrysotile, asbestos, and more, the majority of tumors occurred after the usual study period of 2 years (104 weeks). The experience of the Ramazzini Foundation shows the importance of conducting the experiments for the lifetime of the animal because, had this not been done, many compounds would have wrongly been reported as noncarcinogenic. The occurrence of tumors later in life in rodents is analogous to that in humans.

Mehlman reviews animal studies and epidemiological data on the carcinogenicity of benzene. Benzene has been shown to cause tumors of hematopoietic and lymphoreticular systems and of several other tissues and organs in experimental animals. In humans, benzene exposure is associated with increases in all types of leukemia as well as multiple myeloma and lymphoma and other hemolymphoreticular tumors. Epidemiological studies show that there is no safe level of benzene since hemolymphoreticular malignancies have been produced in humans exposed to low as well as high levels of benzene. Over the years, permissible exposure levels of benzene have decreased from 1000 ppm in 1900 to the present occupational standard set by OSHA of 1 ppm, owing to increasing proof of carcinogenicity of benzene in both animal bioassays and humans exposed to benzene. Current occupational benzene standards are not protective of human health.

Mehlman reviews animal carcinogenicity studies of the gasoline additive methyl tertiary butyl ether (MTBE), some of which are included in this volume (Belpoggi *et al.*), that show MTBE causes tumors of various organs in various rodent strains and makes the argument for classification of MTBE as a human carcinogen. He also describes neurological, respiratory, and allergic symptoms in groups of individuals exposed to MTBE both during its production and in the use of fuel containing MTBE. Such reports are based on symptom surveys taken when MTBE was present in gasoline and contrasted to symptoms reported when MTBE was removed. The time-related association of symptoms present and the presence of MTBE in gasoline or in contaminated drinking water are strong evidence that neurological, respiratory, and allergic symptoms reported by exposed persons are due to this gasoline additive. Based on the animal bioassay data showing carcinogenicity of MTBE, humans exposed to this substance must be considered to be at risk of developing tumors as a result of exposure to MTBE.

Suzuki and Yuen report on their examination of fibers from tissues of 168 cases of human mesothelioma using high-resolution analytical microscopy. It is generally accepted that there is no threshold for asbestos-related diseases and malignant mesothelioma at any level of fibers above zero. OSHA's current standard for asbestos analysis uses light microscopy (phase microscopy) and counts only those fibers that are larger than 5 µm with an aspect ratio larger than 3:1 and assumes that fibers shorter than 5 µm are not carcinogenic. For this reason, many investigators have neglected to count short fibers (<5 µm)—for example, the Arthur D. Little report on DumDum. The study of Suzuki and Yuen shows that the OSHA standard for identifying and characterizing carcinogenicity of asbestos fibers is not adequate since the majority of asbestos fibers included in these tumors are not counted. These authors show that only 4% of all asbestos fibers detected in lung and mesothelial tissue from mesothelioma patients fit Stanton's criteria, which is the major basis of the OSHA standards. The majority of asbestos fibers from lung and mesothelial tissue in these patients were less than 5 µm in length (89.5%). The study also showed that only 2.3% (247/10,575) of fibers detected in both lung and mesothelial tissues fit Stanton's hypothetical dimensions (>8 µm in length and >0.25 µm in width). They also conclude that short, thin asbestos fibers are carcinogenic because they are the principal type of fibers identified in human cancerous tissue and that chrysotile is the most common asbestos type in mesothelial tissue. Suzuki and Yuen conclude that asbestos fiber analysis must include short, thin asbestos fiber types; and that both lung and mesothelial tissues must be studied to determine the types of asbestos fibers associated with the induction of human malignant mesothelioma. Their results support the induction of human malignant mesothelioma by chrysotile.

Melnick reviews the carcinogenicity of epoxides and epoxide-forming chemicals, as well as toxicokinetic issues that have an impact on target organ doses and toxicodynamic issues concerning the effects of epoxides at target sites. Epoxides and epoxide-forming chemicals have been shown to be carcinogenic in experimental studies conducted by the RF (vinyl chloride, acrylonitrile, styrene, styrene oxide, and benzene) and by the National Toxicology Program (NTP) (ethylene oxide, 1,3-butadiene, isoprene, chloroprene, glycidol, acrylonitrile, and benzene). These laboratory animal studies prompted additional investigations of the potential carcinogenicity of other epoxide-forming chemicals (e.g., vinyl bromide, vinyl fluoride), as well as extensive mechanistic studies on epoxide carcinogenesis. In addition to data from epidemiological studies that have been performed on several of these chemicals, animal data and properly used mechanistic information are important for evaluating human cancer risk.

Tomatis describes reasons why animal bioassays are necessary and effective in predicting the risk of carcinogenicity in humans. While mechanisms are useful and may help to more accurately characterize and quantify risk, the more recent use of mechanistic studies to evaluate evidence of carcinogenicity has not necessarily been oriented toward protection of human health. In the absence of absolute certainty, which cannot be attained, it is reasonable to consider animal bioassays as the gold standard in predicting carcinogenicity.

Bucher describes details of the NTP animal study protocol, methods used to determine chemicals and doses to be tested, results of chemicals studied, and chemical groups under study now. The NTP has produced 505 technical reports presenting the results of 2-year rodent cancer studies on discrete substances, mixtures, or physical agents such as industrial chemicals, drugs, pesticides, and common contaminants in air and water, to metals, hormones, natural toxins, and selected physical agents such as electromagnetic fields. An additional 70 studies are either completed or ongoing. Recently, many unregulated herbal medicines and dietary supplements, a number of drinking water disinfection by-products, some DNA-based therapeutics, dioxins and major dioxin-like chemicals, estrogenic agents, and one antiandrogen have been entered in the study.

Bucher also discusses study design and statistical methods used by the NTP and the Ramazzini Foundation and how various approaches can be complementary. He stresses that the animal bioassay has been increasingly used to address fundamental issues in toxicology and carcinogenesis.

Huff compared 14 chemicals that were studied by the NTP in the United States and by the RF for Cancer Research in Italy. Eleven of the 14 chemicals studied in both laboratories gave results that were concordant. The chemicals for which results differed were vinylidene chloride, xylene, and toluene. Maltoni's studies showed that these chemicals were carcinogenic, while NTP studies did not detect carcinogenicity. Huff shows that the main difference between the RF and NTP studies was in the duration of the experiments. Had the NTP experiments been extended for the same period of time as the RF studies—for example, for the lifetime of the animal—they most likely would have produced the same findings of carcinogenicity of these chemicals as was found by the RF.

In summary, the conference shows that long-term carcinogenesis bioassays that are carefully designed and carried out do identify carcinogenic potential to humans from exposure to environmental, industrial, and occupational agents, chemicals, and products. These experimental bioassays continue to be the gold standard for providing chemical carcinogenesis information that serves to protect workers and the public.

ACKNOWLEDGMENTS

I would like to personally thank staff members of the New York Academy of Sciences—in particular, Rashid Shaikh, Renée Wilkerson-Brown, Barbara M. Goldman, Joyce Hitchcock, and Richard Stiefel for their invaluable help in organizing, conducting, and publishing the results of this meaningful conference.

I also want to thank all participants and others who helped with suggestions of topics and speakers, review of the manuscripts submitted, and general support. In particular, I would like to thank Dr. James Huff for his presentation and manuscript, both of which describe so well the history of the Ramazzini Foundation and the National Toxicology Program and describe the outstanding contributions of the late David P. Rall, to whom this volume is also dedicated. The reader is referred to his manuscript, which closes this volume, on the contributions of his mentor and friend, Dr. David Rall.

Special thanks go to my wife, Karyl Norcross Mehlman, for her tenacious help in developing, organizing, and overseeing all parts of the conference and the publication of its proceedings.

—Myron A. Mehlman
Department of Environmental Medicine
The Mount Sinai Medical Center
New York, New York

Remembering Professor Cesare Maltoni

MYRON A. MEHLMAN

Department of Environmental Medicine, The Mount Sinai Medical Center, New York, New York 10029, USA

ABSTRACT: Professor Cesare Maltoni, a renowned leader in the research of the hazards of carcinogens in the workplace, died on January 22, 2001 at the age of 70. Born in Faenza (Ravenna), Italy on November 17, 1930, he received his M.D. degree from the University of Bologna in 1954–1955. He was Director of the Institute of Oncology of Bologna (1964 to 1997), Director of the Bologna Centre for the Prevention and Detection of Tumours and Oncological Research (1966 to 1989), and Scientific Director, European Foundation of Oncology and Environmental Sciences "B. Ramazzini" from 1993 until he died. Maltoni conducted long-term carcinogenic studies on some 200 agents. He was the first to demonstrate that vinyl chloride is a carcinogen that produces, among other tumors, angiosarcoma of the liver. He was the first to show that benzene is a powerful multipotential carcinogen. Maltoni authored more than 700 original scientific publications, books, and proceedings. He was editor and coeditor of many journals. Among his many awards were the Stokinger Award, American Conference of Governmental Industrial Hygienists (ACGIH), Kansas City, 1995; International Award "B. Ramazzini" of the Collegium Ramazzini, Washington, 1995; International I.J. Selikoff Memorial Award, Washington, 1995; and the Sigillum Magnum of the University of Bologna, 1997. Many fellow Ramazzinians and his coworkers wish to be considered part of the group of, as he once wrote, his "always family friends" and to remember happy moments when we were together. A man of great stature and many contributions, Cesare Maltoni will never be forgotten.

KEYWORDS: Maltoni; Collegium Ramazzini; vinyl chloride; benzene; carcinogenicity; aromatic hydrocarbons

INTRODUCTION

Professor Cesare Maltoni (FIG. 1) was a world-renowned leader in the research of causation of cancer by industrial carcinogens in the workplace and environment. He conducted long-term chronic carcinogenicity studies on nearly 200 chemicals. He was the first to demonstrate that vinyl chloride is a carcinogenic agent producing angiosarcoma and other cancers of the liver. He was also the first to demonstrate that benzene is a powerful, multipotential carcinogen.

Professor Maltoni died on January 22, 2001 at the age of 70 at his home in Bologna, Italy. He was a member and officer of many national and international sci-

---

Address for correspondence: Myron A. Mehlman, Ph.D., 7 Bouvant Dr., Princeton, NJ 08540. Voice: 609-683-4750; fax: 609-683-0838.
mehlman@rcn.com

FIGURE 1. Cesare Maltoni shortly before his death on January 22, 2001.

FIGURE 2. Cesare Maltoni's parents, Maria Porisini Maltoni and Carlo Maltoni. (Taken from old family photographs.)

entific organizations and advisory boards and received world-wide recognition for his many accomplishments. Professor Maltoni, along with his associates, published more than 700 scientific articles. He was a person of great stature and accomplishment and will never be forgotten.

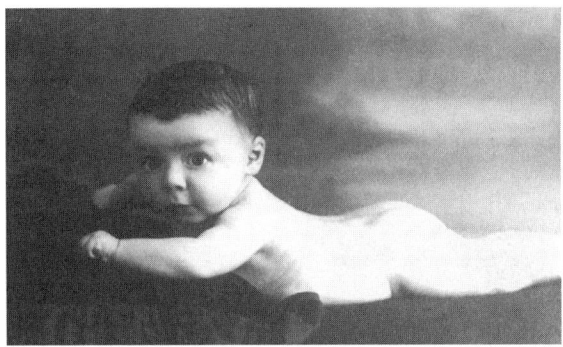

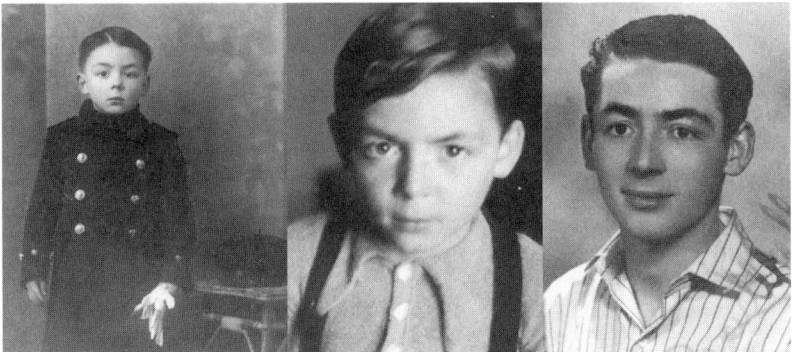

FIGURE 3. Family photographs of Cesare Maltoni during his early life.

BACKGROUND AND HISTORY

Cesare Maltoni was born in Faenza (Ravenna), Italy on November 17, 1930, to Maria Porisini Maltoni and Carlo Maltoni. Photographs of Maltoni's parents and of his early life are shown in FIGS. 2 and 3. Maltoni graduated from the University of Bologna in 1955, where he received his medical degree. He and his advisor, Professor Favilli, are shown in FIGURE 4 (upper left) at the time of awarding of the M.D. degree. Maltoni's experimental thesis was entitled "Changes in the Dermis during Experimental Skin Carcinogenesis." Maltoni at the time of his cousin's graduation is shown in Figure 4 (upper right). Maltoni's parents died when he was very young, and he was cared for all of his life by his aunt, Afra Maltoni (FIG. 4, bottom) of Faenza, Italy.

FIGURE 4. Cesare Maltoni and Professor Favilli, his advisor, at the time of awarding of the M.D. degree from the University of Bologna in 1955 *(upper left)*. Maltoni at the time of his cousin's graduation *(upper right)*. Afra Maltoni, Maltoni's aunt, who raised him and was a tower of strength throughout his life. Afra Maltoni resides in Faenza, Italy.

CAREER HIGHLIGHTS

Throughout his professional life, Maltoni received many professional and scientific appointments, honors and held many positions. He served as:

- Secretary General of the Collegium Ramazzini, an international academy consisting of 180 members who are experts on the relationship between cancer development, environment, and work and health, since its foundation in 1982;
- Past President (1981–1986) and Honorary President (1986–2001) of the Italian Society for Tumor Prevention, Diagnosis and Therapy;
- Past Chairman of the International Committee for Human Tumour Investigation;
- Member of the Italian National Board of Health (1984–1989);
- Member of the National Commission for Mutagenesis, Carcinogenesis and Teratogenesis of the Italian Institute of Public Health (now National Commission of Toxicology) from 1975–1989;
- Member of the Académie Internationale de Lutuèce;
- Member of the Académie Européenne des Sciences des Arts et des Lettres;
- Member of the National Technical Scientific Commission for Biosafety for evaluation of potential health risk from genetically modified organisms, Italian Ministry of the Environment in 2000;
- National Chairman of the Environmental and Occupational Carcinogenesis Committee set up in 2000 by the Italian League for the Fight against Cancer.

For his work, Cesare Maltoni received many awards. But a few are:

- the Golden Medal "Faentino Lontano" of the town of Faenza (1974);
- the International Award for Cancer Prevention of the Italian League for the Fight against Cancer of Latina (1994);
- the Stokinger Award of the American Conference of Governmental Industrial Hygienists (ACGIH) (1995);
- the International Award "B. Ramazzini" of the Collegium Ramazzini (1995);
- the International I.J. Selikoff Memorial Award (1995);
- the Sigillum Magnum of the University of Bologna (1997);
- the Golden Medal of the Italian Society of Tumours (SIT), Prevention, Diagnosis, Therapy (2000).

Professor Maltoni's legacy and major scientific contributions include:

- establishment of the relevance of connective tissue changes in the carcinogenesis and natural history of the formation of tumors and metastases;
- pioneering of the use of animal models for the identification of environmental carcinogens, the assessment of carcinogenic risks, and the evaluation of tumor chemoprotective compounds;

- documentation of the multipotential nature of a large number of carcinogenic agents;
- establishment of the carcinogenicity of several important industrial agents including vinyl chloride, vinylidene chloride, acrylonitrile, trichloroethylene, formaldehyde, gasolines, gasoline aromatics, and MTBE;
- documentation of high oncogenic risk due to exposure to asbestos used in the railroad and in the sugar industries;
- establishment of the Nominative Bologna Registry of Mortality with particular regard to cancer;
- direction of screening of 270,000 women for early diagnosis of uterine cervical cancer and 125,000 women for early diagnosis of breast cancer;
- creation of the first Italian hospice for terminal cancer patients.

OCCUPATIONAL CARCINOGENESIS

In 1973, Professor Maltoni convened a major symposium in Bologna, Italy entitled "Advances in Tumour Prevention, Detection and Characterization." Renzo Dal Zotto assisted him with organizational aspects of this symposium. FIGURE 5 shows Cesare Maltoni and Lanfranco Turci, President of the Emilia Romagna Region, welcome distinguished scientists from all over the world who participated in this major symposium. The proceedings of this conference were published in 1974 in *Excerpta Medica* (FIG. 6).

Maltoni was known for his meticulous and carefully documented experiments. He studied 198 chemicals and agents and conducted 394 separate experiments using 138,281 animals. Of the 135 agents studied, 68.9% were found to be carcinogenic, 5.92% showed borderline carcinogenicity, and 25.18% were found to be non-carcinogenic in the animals tested.

Professor Maltoni authored and co-authored more than 700 original scientific publications, numerous books, and proceedings in the national and international literature, and was editor or co-editor of many journals. Some of his early colleagues and associates were M. Soffritti, F. Belpoggi, R. Dal Zotto (FIG. 7, top), M.G. Perino (FIG. 7, lower left), and life-long friend and associate, artist Gianna Boschi (FIG. 7, lower right), F. Minardi, D. Caretti, C. Pinto, B. Conti, G. Lefemine, A. Masina, A. Ciliberti, V. Patella, G. Paladini, U. Veronesi, I.J. Selikoff, M.A. Mehlman, H. Popper, and many others.

Maltoni's studies resulted in scientific publications of major significance, some of which included studies of the carcinogenesis of vinyl chloride[1] and of vinylidene chloride,[2] benzene,[3–5] and trichloroethylene.[6]

VINYL CHLORIDE CARCINOGENESIS

Maltoni's study of the carcinogenesis of vinyl chloride[1] began in 1971 and lasted until 1983. It included more than 7,000 animals, the majority of which were submitted to chronic treatment. All were kept under observation until spontaneous death.

FIGURE 5. Maltoni and Lanfranco Turci, President of Emilia Romagna Region, welcome distinguished scientists who participated in a major symposium entitled "Advances in Tumour Prevention, Detection and Characterization" held in Bologna, Italy, in 1973.

Preprint of the International Congress Series No. 322

Advances in tumour prevention, detection and characterization.

Vol. 2: Cancer detection and prevention

Proceedings of the Second International Symposium on Cancer Detection and Prevention

Bologna, April 9 - 12, 1073.

1974, Excerpta Medica, Amsterdam

FIGURE 6. Proceedings of the 1973 conference were published in *Excerpta Medica*.

FIGURE 7. Maltoni's early associates and supporters: Fiorella Belpoggi, Renzo Dal Zotto, and Morando Soffritti (*top panel*), Giorgio Perino (*lower left*) and life-long friend, artist Gianna Boschi (*lower right*).

More than 200,000 histological slides were examined. To our knowledge, these studies are the largest ever performed in carcinogenesis in one laboratory on a single industrial compound. This project produced a good deal of important information. It:

- clearly demonstrated that vinyl chloride is a carcinogen that produces a variety of tumors in different tissues and organs;
- indicated that liver angiosarcoma is the most specific marker of vinyl chloride-related tumors;
- promoted the epidemiological investigations that led to the discovery of liver angiosarcoma in vinyl chloride-exposed workers;
- produced dose–response data, thereby providing the scientific basis for regulatory measures; and
- provided the scientific proofs that long-term carcinogenicity bioassays can predict oncogenic risk for humans and can be the basis for quantitative risk assessment.

Maltoni's studies on vinyl chloride:

- led to the control of vinyl chloride exposure;
- promoted carcinogenicity studies of numerous chemically correlated compounds to assess their potential risk;
- stimulated national and international regulations for the control of environmental toxic agents including the passing of legislation requiring pre-production testing;
- constituted a model for research protocols in the field of occupational and environmental carcinogens including a large part of those contained in the Good Laboratory Practice Acts for long-term carcinogenicity bioassays;
- demonstrated that experimental findings can be extrapolated to human pathology in the area of potential carcinogens;
- contributed to basic knowledge of factors and mechanisms of carcinogenesis; and
- demonstrated that studies in the field of environmental and occupational carcinogenesis must no longer be considered retrograde ancillary research but that such studies must be an important component in the decision-making process.

BENZENE: A MULTIPOTENTIAL CARCINOGEN

Maltoni's study of the carcinogenicity of benzene[3–5,7] showed benzene to be a multipotential carcinogen in experimental animals. It is important to note that experimental data on carcinogenicity of benzene prior to 1976 were few and insubstantial. IARC Monograph No. 7 (1974) concluded that "the data reported do not permit the conclusion that carcinogenic activity has been demonstrated."

In 1976, Maltoni started a long-term bioassay study on benzene administered by gavage to Sprague-Dawley rats at the Institute of Oncology of Bologna. There were two main aims of that experiment: (1) to contribute to knowledge of oncogenic potential of benzene; and (2) to test parallelism of effects produced by various agents in humans and in experimental animals because a lack of correlation between effects of arsenic on humans and animals had long produced skepticism on the predictivity of experimental bioassays.

Maltoni[4] later wrote that "...carcinogenic activity [of benzene] had not been demonstrated [previously] because of the inadequacy of the experiments." Maltoni's experiments showed that benzene caused carcinomas of the zymbal gland, the oral and nasal cavities, the skin, the forestomach, mammary glands, liver, and hemo-lymphoreticular system in the form of leukemias.

TRICHLOROETHYLENE CARCINOGENESIS

Maltoni's studies on the carcinogenesis of trichloroethylene[6] (TCE) began in 1976. These studies employed 3,948 animals of different species and strains. In all experiments but one, animals were submitted to TCE by inhalation. All animals were kept under observation until spontaneous death. To our knowledge, this experimental project is the largest performed on TCE carcinogenicity. This project provided the following information on TCE:

- TCE is carcinogenic in two different strains of mice.
- TCE causes tumors in rats.
- Renal lesions and tumors in rats are the most specific marker of TCE-correlated pathology.

Under the experimental conditions, the evidence of TCE (without epoxide stabilizer) carcinogenicity is based on the following results:

Following long-term inhalation exposure:
- TCE caused an increase in hepatomas in male Swiss and B6C3F1 mice.
- TCE caused lung tumors in male Swiss and female B6C3F1 mice.
- A dose-related increase in Leydig (interstitial) cell tumors of the testes was observed in treated Sprague-Dawley rats.
- There is some evidence that TCE produced an increase in the incidence of leukemias (mainly immunoblastic lymphosarcomas) that was not dose-related.
- TCE induced the onset of a few characteristic kidney tubular adenomas in males at the highest dose tested.

Following ingestion exposure:
- A slight borderline increase in leukemia was observed in male rats.

SIGNIFICANCE OF MALTONI'S PROTOCOLS

FIGURE 8 illustrates a concept of which Maltoni was particularly fond: "Is the glass half empty or half full?" Maltoni's experimental protocols were designed to answer that question.

Typically, industry conducted or sponsored long-term oncogenic studies that were generally designed *not* to completely answer the question of whether the glass was half full or half empty. In order to introduce dispute as to the question of carcinogenicity of a compound, the study design consistently called for animals to be given at least one very high dose of the compound being tested, often 8,000 ppm, which is a magnitude greater than to which humans would be exposed. This provided an argument that humans are never exposed to these types of doses, and so results could not be applied to humans.

Legator[8] demonstrated that for benzene, vinyl chloride, and butadiene, high doses greatly underestimated the carcinogenic risk to humans. In addition, in the U.S. EPA ED_{01} study, Staffa and Mehlman[9] published a major study using 24,192 animals. The study showed that there is no safe level for carcinogens and that no dose above zero was without risk. These studies refuted the fallacious argument, which is still today being used by industry health professionals and their consultants to defend potentially carcinogenic products and to avoid reaching an answer.

The second technique generally supported by industry was to limit the time of animal exposure to potential carcinogens to two years or less. By imposing this limitation, the real risk of carcinogenicity of a compound could be substantially underestimated and, in some cases, totally missed. In FIGURE 9, the effect of extending the length of the experiment to more fully determine carcinogenicity during the life span of the animal shows the carcinogenic effect of xylene revealed in the period of time after the usual 2-year cutoff period.

The third method used by some manufacturers of potentially carcinogenic products was to attack the messenger. Maltoni's meticulously conducted experiments and experimental goals made such attacks frivolous. In particular, the duration of Maltoni's experiments more closely paralleled human experience in that tumors often develop not in the prime or life, but later, as the individual ages. This medical concept cannot be challenged. Maltoni's goals in studying the carcinogenicity of chemicals were:

- to identify new potential carcinogenic materials;
- to evaluate the effects of technological modifications of already demonstrated carcinogenic materials with the aim of lowering their risk;
- to assess, in quantitative terms, the relative carcinogenic risk of different materials with particular regard to alternative choices, and to guide the formulation of future materials and choices;
- to help to predict the target organs;
- to define the carcinogenic potential of the physical and chemical properties of the test compounds;
- to determine the role of different biological and experimental factors affecting the neoplastic response and, consequently, to shed light on the mechanism of the possible oncogenic effects of compounds found or suspected to be carcinogenic;
- to help trace the natural history of tumors induced by the test compound;
- to conduct experiments at low levels of exposure; and
- to carry out experiments for extended periods of time, often as long as 30–36 months—basically for the lifetime of the animal.

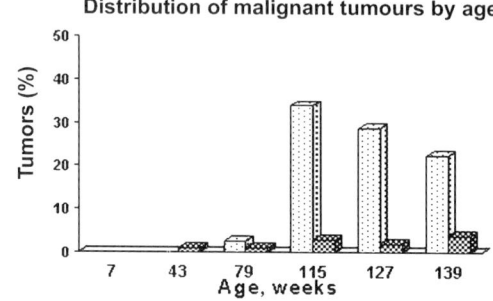

FIGURE 8. Maltoni's approach to research questions was always to be complete and explore all possibilities, thus his favorite quotation, "Is the glass half empty or half full?"

FIGURE 9. Life-span studies of the appearance of malignant tumors in animals exposed to xylene four days per week from age 7–104 weeks. All animals alive at 139 weeks were sacrificed at that time. Detection of malignant tumors after the usual cut-off point of 104 weeks (2 years) shows the importance of continuation of the experiments for more than 2 years because the majority of tumors develop in treated animals, but not in controls, after this point.

FIGURE 10. First meeting of Fellows of the Collegium Ramazzini in 1983. Fellows present are as follows: Seated: Cesare Maltoni, Werther Cigarini, Mayor of Carpi, and Irving J. Selikoff. Standing: Muzaffer Aksoy, Guiseppe Paladini, Jorge Chiriboga, Vito Foà, Giovangiacomo Giacomo, Jerry Stara, Myron Mehlman, Bo Holmberg, Sheldon Samuels, John Harington, Ruth Lilis, Pericle DiPietro, and Elihu Richter.

FIGURE 11. Professor Irving J. Selikoff, First President of the Collegium Ramazzini.

FIGURE 12. Professor Cesare Maltoni, First General Secretary of the Collegium Ramazzini.

FIGURE 13. Professor Norton Nelson.

FIGURE 14. Morando Soffritti, General Secretary, and Philip Landrigan, President, of the Collegium Ramazzini.

FIGURE 15. David P. Rall (deceased), former Treasurer; Eula Bingham, Past President; Myron Mehlman, North American Secretariat and former Treasurer of the the Collegium Ramazzini.

FIGURE 16. Daniel Teitelbaum, Treasurer, and Charles Xintaras, Webmaster, of the Collegium Ramazzini.

FIGURE 17. Members of the Executive Council of the Collegium Ramazzini elected in 2001. *Top*: Ellen Silbergeld, Fiorella Belpoggi; *middle*: Vito Foà, Arthur Frank; *bottom*: Carlos Santos-Burgoa, and Elihu Richter.

FIGURE 18. Early supporters of the Collegium Ramazzini: Major Bergianti (*upper left*), Senator Luigi Orlandi (*upper right*), Ramazzini Fellows (*lower left*), Lanfranco Turci, Past President, Emilia Romagna Region (*lower right*).

FIGURE 19. *Top*: Marja Sorsa, Anders Englund; *middle*: Bernard Goldstein, Bo Holmberg; *bottom*: Gordon Atherley, James Dodge.

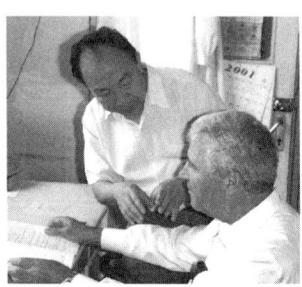

FIGURE 20. Fellows of the Collegium Ramazzini. *Top*: Mostafa H. Mostafa and Songnian Yin with Myron Mehlman; *middle*: Muzaffer Aksoy with Maths Berlin, Tor Norseth; *bottom*: Anthony Mazzochi, Nikolai Izmerov.

FIGURE 21. Fellows of the Collegium Ramazzini. *Top*: William Nicholson, Morris Greenberg; *middle*: Melissa McDiarmid, James Huff; *bottom*: Michael Gochfeld, John Bailar, III.

FIGURE 22. Ramazzini Fellows. *Top*: John Andrews, Jr., Mary Wolff; *middle*: Christopher de Rosa, Richard Wedeen; *bottom*: Henry Falk, Nachman Brautbar.

MEHLMAN: REMEMBERING CESARE MALTONI

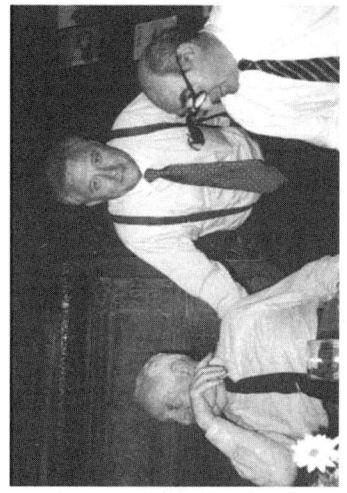

FIGURE 23. Ramazzini Fellows: *Upper left*: Myron Mehlman, Nicolai Izmarov, Philip Landrigan; *upper right*: Irving Selikoff, Myron Mehlman, Cesare Maltoni; *lower left*: Irving Selikoff and Arthur Ashford; *lower right*: Maths Berlin and David Hoel.

FIGURE 24. Fellows and friends of the Collegium Ramazzini. *Top*: Paul Lioy, Nachman Brautbar, Elihu Richter, Marina Thorborg. *Bottom*: Karyl Norcross Mehlman, Marina Thorborg, Carol Rice, Eula Bingham.

FIGURE 25. Meeting of Drs. Karyl and Myron Mehlman, Giordani Corradini, Morando Soffritti, and Mayor Demos Malavasi in March 2001, at which time the continued commitment of the City of Carpi to the Collegium Ramazzini was reaffirmed.

THE COLLEGIUM RAMAZZINI

For the major part of his life, Professor Maltoni was dedicated to the philosophy of Bernardino Ramazzini, a philosophy that spawned the Collegium Ramazzini. Bernardino Ramazzini[10] was an 18th century physician and scholar who realized the link between occupational exposure to dangerous chemicals and conditions and adverse effects on workers' health. The Collegium Ramazzini was organized as a non-profit organization dedicated to detecting adverse working conditions and chemical exposures, improving working conditions, and minimizing the danger of such exposure to workers. Along with giants in epidemiology, environmental health, and occupational health—including Norton Nelson, Irving Selikoff, David P. Rall, Eula Bingham, Philip J. Landrigan, Richard Lemen, Massimo Crespi, Vito Foà, Joseph Fraumeni, Bernard Goldstein, K. Tsichiya, Bernard Weinstein, Arthur Upton, D. Bates, Lars Freiberg, William Nicholson, Tor Norseth, Maths Berlin, Anders Englund, Barry Johnson, and others—Maltoni devoted much of his life to the protection of workers. In this light, Selikoff, Maltoni, Mehlman, and Samuels founded the Collegium Ramazzini, an organization of international health professionals dedicated to improving environmental and occupational health. During his lifetime, Cesare Maltoni worked, traveled, and participated in development of policy with Fellows of the Collegium from 35 countries who were leaders in the fields of occupational and environmental health, labor, and public policy. The first meeting of the fledgling Collegium was held in Carpi, Italy in 1983. A photograph of many of the initial Fellows of the Collegium Ramazzini is shown as FIGURE 10. The City of Carpi, home of Bernardino Ramazzini, has served as the annual meeting place of Fellows of the Collegium Ramazzini for nearly 20 years and will do so for years to come, largely due to Professor Maltoni's efforts and stature in the scientific community.

History and memories of meetings of the Collegium Ramazzini are immortalized in photographs taken in Italy, the USA, Sweden, Japan, India, Russia, and many other countries (FIGS. 11–25). Maltoni was present at almost all of the times these photographs were taken. These photographs are a tribute to Dr. Maltoni's memory and to the close friends and colleagues who worked both directly and indirectly to promote public health and safety in the workplace and the environment. Castle Bentivoglio in Bologna, Italy, where Maltoni's research continues today, is shown in FIGURE 26.

CONCLUSION

Professor Maltoni was a true friend of working men and women. The effects of Professor Maltoni's lifetime work and achievements place him in a very small and select group of world scholars like Ramazzini and Hippocrates. Not only was Cesare Maltoni one of the greatest experimental oncologists, but he was also a physician who made important medical and oncological research contributions. And most importantly, just as Professor Irving J. Selikoff, who was a close colleague and personal friend of Professor Maltoni, he was a teacher who practiced Ramazzini's philosophy that "It is better to prevent than to cure."

Gentleness, kind-heartedness, helpfulness in the cause of the underdog: these were human qualities that are associated with him. Yet he could be firm, not to say

FIGURE 26. Research begun by C. Maltoni continues at Castle Bentivoglio in Bologna, Italy.

scathing, in the face of arrogance, posing, opportunism, or off-handedness. He was a man of finesse and culture, a lover of art, literature, and poetry. He seemed to possess a key to these, based on an uncommon respect for fine detail.

To his close co-workers, he leaves work, actions to be taken, projects, and culture, but above all a spirit that he doggedly sought to instill: a sense of mission, the determination to carry it out, and "the habit of telling the truth, come what may."

Many of us fellow Ramazzinians, his co-workers, wish to be considered as a part of the group of, as he once wrote, his "always family friends" and to remember happy moments when we were together. We will remember Professor Maltoni's devotion of his life, his energy, and his love of saving lives for the betterment of mankind. A man of great stature and many contributions, Cesare Maltoni will never be forgotten.

REFERENCES

1. MALTONI, C. *et al.* 1984. Experimental research on vinyl chloride carcinogenesis. *In* Archives of Research on Industrial Carcinogenesis, Vol. II. C. Maltoni & M.A. Mehlman, Eds. Princeton Scientific Publishing Corp. Princeton, NJ.
2. MALTONI, C. *et al.* 1985. Experimental research on vinylidine chloride carcinogenesis. *In* Archives of Research on Industrial Carcinogenesis, Vol III. C. Maltoni & M.A. Mehlman, Eds. Princeton Scientific Publishing Corp. Princeton, NJ.
3. MALTONI, C. & C. SCARNATO. 1977. Le prime prove sperimentali del l'azione carcerogena del benzene. Gli Ospedali della Vita **4:** 111–113.

4. MALTONI, C. et al. 1983. Benzene: a multipotential carcinogen. Results of long term bioassays performed at the Bologna Institute of Oncology. Am. J. Indust. Med. **4:** 589–630.
5. MALTONI, C. et al. 1989. Benzene: an experimental multipotential carcinogen. Results of the long-term bioassays performed at the Bologna Institute of Oncology. Environ. Health Persp. **82:** 109–124.
6. MALTONI, C. et al. 1986. Experimental research on trichloroethylene carcinogenesis. *In* Archives of Research on Industrial Carcinogenesis, Vol V. C. Maltoni & M.A. Mehlman, Eds. Princeton Scientific Publishing Corp. Princeton, NJ.
7. MALTONI, C. 1982. Benzene: a multipotential carcinogen. Acta Oncologica **3:** 1–4.
8. LEGATOR, M. 1997. Underestimating risk using data derived from mechanistic or animal bioassay data: reply to the Bond *et al.* letter to the editor. J. Clean Technol. Environ. Toxicol. Occup. Med. **6:** 215–220.
9. STAFFA, J.A. & M.A. MEHLMAN. 1979. Innovations in cancer risk assessment (ED_{01} study). Pathotox Publishers. Park Forest South, IL.
10. RAMAZZINI, B. 1713. *De morbis artificum* [Diseases of Workers]. The Latin text of 1713 revised with translation and notes by Wilmer Cave Wright. 1940. The University of Chicago Press. Chicago, IL.

Ramazzini Foundation Cancer Program

History and Major Projects, Life-Span Carcinogenicity Bioassay Design, Chemicals Studied, and Results

MORANDO SOFFRITTI, FIORELLA BELPOGGI, FRANCO MINARDI, AND CESARE MALTONI[†]

Cancer Research Center, European Ramazzini Foundation for Oncology and Environmental Sciences, Bologna, Italy

> ... I have dreamed of man's state, of his courteous
> and enlightened social state: beyond which,
> in the temple, the horrible blood-sacrifice
> was consummated.
> —THOMAS MANN
> *The Magic Mountain*

ABSTRACT: The Ramazzini Foundation research program was started over thirty years ago. The features of this program are: (1) systematic and integrated project design; (2) consistency over time; (3) homogeneity of approach: key members of the team remain unchanged; and (4) choice to work on new frontiers of scientific research. The program centers mainly on three projects: Project 1: experimental carcinogenicity bioassays; Project 2: experimental anticarcinogenesis assays to identify factors and active principles (compounds) capable of opposing the onset of tumors while being suitable for preventive/chemopreventive intervention; Project 3: epidemiological studies, both descriptive and analytical, on tumor incidence and mortality in persons professionally and environmentally exposed to industrial carcinogenic risks. The project involving experimental carcinogenicity bioassays for the identification of exogenous carcinogens (environmental and industrial above all) began in 1966. This project has included 398 experimental bioassays on 200 compounds/agents using some 148,000 animals monitored until their spontaneous death. Among the studies already concluded, 47 agents have shown "clear evidence" of carcinogenicity. The results have demonstrated for the first time that (1) vinyl chloride can cause liver angiosarcoma as well as other tumors; (2) benzene is carcinogenic in experimental animals for various tissues and organs; (3) formaldehyde may produce lymphomas and leukemias; and (4) methyl-*tert*-butyl ether (MTBE), the most common oxygenated additive used in gasolines, can cause lymphomas/leukemias. Many of the results achieved have led to the introduction of norms and measures of primary prevention.

KEYWORDS: Ramazzini Foundation; carcinogenicity; long-term bioassay; rat; benzene; vinyl chloride; MTBE; xylenes

---

Address for correspondence: Morando Soffritti, M.D., Cancer Research Center, European Ramazzini Foundation for Oncology and Environmental Sciences, Bentivoglio Castle, 40010 Bentivoglio (BO), Italy. Voice: +39-051-6640460; fax: +39-051-6640223.
crcfr@tin.it.
[†]Deceased.

INTRODUCTION

Cancer is one of the most important current health problems. One epidemiological datum is enough: in industrialized countries, cancer represents about 30% of all deaths. The problem will get worse because the disease is expected to increase: cancer is related to the aging of the population, to environmental pollution, and to the modified life styles which characterize our times, all of which will increase in coming years with the diffusion of the industrial model of development.

Cancer is an extremely complex disease, not easy to control, and one about which there is insufficient knowledge. Clinical, psychological, and economic aspects of the disease should all be taken into consideration: they cause heavy individual, familial, and social costs.

To face the problem, it is necessary to increase our knowledge to provide solid scientific bases for the prevention and clinical control of cancer. Basic as well as preventive and clinical research should be developed. In this research, experimental studies play a central role. Those who have known and followed the reality of oncological research are aware that most progress in oncology derives from experimental studies.

THE RESEARCH PROGRAM OF THE RAMAZZINI FOUNDATION

The Ramazzini Foundation (RF), founded in 1992, is a nonprofit private institution with governmental recognition. The aims of the Foundation include: (1) descriptive and analytical epidemiology; (2) experimental identification of toxic and carcinogenic risks; (3) experimental research on chemoprevention; (4) clinical monitoring of high-risk groups; and (5) terminal care of cancer patients.

The cancer research program of the Ramazzini Foundation was started 30 years ago as part of the research activity conducted by the "F. Addarii" Institute of Oncology in the town of Bologna.

The features of the program are: (1) systematic and integrated program design; (2) consistency over time; (3) homogeneity of approach, the key members of the team remaining as far as possible unchanged; (4) the decision to work on new frontiers of cancer research.

The program focuses mainly on three projects:

- Project 1: long-term carcinogenicity bioassays for the identification of xenobiotic carcinogenic agents—in particular, those of industrial origin;
- Project 2: experimental bioassays of anticarcinogenesis for the identification of factors, agents (also drugs) effective in contrasting the onset/progression of tumors (chemoprevention);
- Project 3: descriptive and analytical epidemiological studies on cancer incidence and mortality in people from various geographic areas of Italy environmentally and/or occupationally exposed to carcinogenic risks.

Project of Long-Term Carcinogenicity Bioassays

This project was started in 1966 and has been conducted with a systematic and integrated approach aimed at identifying exogenous carcinogens and quantifying their effects. This project is second only to that of the United States' National Toxicology Program, and has studied a greater number of agents than in any other single laboratory. Most agents have been selected on the basis the amount produced, their diffusion in the environment, and the number of people potentially exposed. Very few agents have been selected from those already proven to be carcinogenic in other laboratories.

Most CRC/RF long-term carcinogenicity bioassays are planned and conducted following a basic design protocol which is summarized as follows:

(1) *Aim*: to detect the chronic toxic and carcinogenic effects of chemical and physical agents.

(2) *Animals*: Sprague-Dawley rats from the colony of the CRC/RF, now used for more than 30 years. The basic tumor incidences of this strain are well known, and the cancer susceptibility is not very different from the human counterpart. Wistar rats and Swiss mice (and other strains), as well as golden hamsters, are also used on some occasions.

(3) *Experimental groups and group sizes*: two or more (depending on the importance of the agent for public health) experimental groups (comprehensive control groups) are employed in the various experiments; each group contains 50–60 or more animals for each sex.

(4) *Routes of exposure*: typically those mimicking human exposure. The most frequently used are: inhalation, injection, ingestion, and external exposure (radiation).

(5) *Concentration/dose/intensity of the agent studied*: for each agent studied, at least two, but generally three, dose levels are tested. The dose levels are the maximum tolerated level, the order of magnitude to which humans may be exposed, and one intermediate level. Data on maximum tolerated dose levels, when not available from the scientific literature, are determined by range-finding experiments.

(6) *Starting of treatment*: during embryonal life (12 days)/perinatal/6–8 weeks old/exceptionally at other ages.

(7) *Duration of treatment*: 104 weeks/life-span/other lengths infrequently.

(8) *Duration of experiment*: until spontaneous death.

(9) *Pathology*: on dying, animals undergo systematic necropsy. Histopathology is routinely performed on the following organs and tissues: skin and subcutaneous tissue, brain, pituitary gland, Zymbal glands, parotid glands, submaxillary glands, Harderian glands, cranium (with oral and nasal cavities and external and internal ear ducts) (5 sections of head), tongue, thyroid and parathyroid, pharynx, larynx, thymus, and mediastinal lymph nodes, trachea, lung and mainstem bronchi, heart, diaphragm, liver, spleen, pancreas, kidneys, adrenal glands, esophagus, stomach (fore and glandular), intestine (four levels), urinary bladder, prostate, gonads, interscapular fat pad, subcutaneous and mesenteric lymph nodes, and any other organs or tissues with pathological lesions.

In some cases long-term carcinogenicity bioassays are planned and conducted following a special design protocol. These experiments are called "mega-experiments" and are characterized as follows: (1) *aims*: to detect low/diffuse carcinogenic risks, defined as the exposure to single or multiple agents or mixtures that

TABLE 1. The Ramazzini Foundation Cancer Program

PROJECT OF LONG-TERM CARCINOGENICITY BIOASSAYS: AGENTS STUDIED

No.	Compounds/agents	No. of Bioassays	Animals Species	No.	Route of Exposure[a]
	Plastic monomers				
1.	Vinyl chloride	29	Rat, mouse, hamster	8,293	Ing,Inh,Ip,Sc,Tr
2.	Vinylidene chloride	7	Rat, mouse, hamster	3,164	Ing,Inh,Tr
3.	Vinylidene fluoride	1	Rat	190	Ing
4.	Acrylonitrile	4	Rat	1,361	Ing,Inh,Tr
5.	Vinyl acetate	3	Rat, mouse	1,672	Ing,Tr
6.	Ethylene	1	Rat	200	Ing
7.	Propylene	2	Rat	1,680	Inh
8.	Styrene	4	Rat	980	Ing,Inh,Ip,Sc
9.	Styrene oxide	1	Rat	240	Ing
10.	p-methylstyrene	4	Rat, mouse	1,422	Ing
	Polymers				
	Plastic disks				
11.	- Ivoclar	2	Rat	84	Imp
12.	- Acronite	2	Rat	84	Imp
13.	- Lucitone	2	Rat	84	Imp
14.	- Stellon	2	Rat	84	Imp
15.	- Teflon	4	Rat	179	Imp
	Polymeric fibres				
16.	- Kevlar	2	Rat	680	Ip,Ipl
17.	- Farlosa C-2-OM	1	Rat	240	Ip,Ipl,Sc
18.	PVC (granules)	1	Rat	230	Ing
19.	Water in PVC bottles	2	Rat	2,200	Ing
	Chlorinated organic intermediates				
20.	Dichloroethane	3	Rat, mouse	2,626	Inh
	Chlorinated solvents				
21.	Methylene chloride	3	Rat, mouse	1,346	Ing,Inh,Tr
22.	Carbon tetrachloride	1	Rat	160	Ing
23.	Methylchloroform	1	Rat	180	Ing
24.	Trichloroethylene	8	Rat, mouse	3,948	Ing,Inh
25.	Tetrachloroethylene	1	Rat	180	Ing
	Ethers				
26.	Phenylether-biphenyl ("Dowtherm")	1	Rat	360	Ing
	Aldehydes				
27.	Formaldehyde	4	Rat	1,447	Ing
28.	Acetaldehyde	4	Rat	870	Ing
	Propellants				
29.	Trichlorofluoromethane (FC11)	2	Rat, mouse	1,170	Inh
30.	Dichlorodifluoromethane (FC12)	2	Rat, mouse	1,080	Inh
31.	Chlorodifluoromethane (FC22)	2	Rat, mouse	720	Inh
	Detergents				
32.	Carbonate	3	Rat	400	Ing
33.	Tripolyphosphate	2	Rat	360	Ing
34.	Nitrilotriacetic acid (NTA)	3	Rat	400	Ing
35.	Zeolite MS 4A	3	Rat	580	Ing,Int,Ip
	Pesticides				
36.	N-(3,5-dichlorophenyl)-5-methyl-5-carbetoxy-1-3-oxazolydin-2,4-dione ("Serinal")	2	Rat, mouse	1,080	Ing
37.	Methyl 2-[n-phenylacetyl-N-(2,6-dimethylphenyl)amino]propanoate ("Galben")	3	Rat, mouse	960	Ing
38.	Phenthoate ("Cidial")	2	Rat, mouse	880	Ing
39.	Ethylene bisdithiocarbamate ("Mancozeb")	2	Rat	850	Ing
	Compounds used in leather industry				
	Chromium compounds				
40.	- Chromitan NA	1	Rat	80	Sc
41.	- Chromitan B	1	Rat	80	Sc
42.	- Chromitan MS	1	Rat	80	Sc
43.	- Chromorganik	1	Rat	80	Sc
44.	- Chromium alum	1	Rat	80	Sc
45.	- Baychrom	1	Rat	80	Sc
46.	- Chromopol	1	Rat	80	Sc
47.	- Coripol	1	Rat	80	Sc

—Continued

TABLE 1. *Continued*

PROJECT OF LONG-TERM CARCINOGENICITY BIOASSAYS: AGENTS STUDIED

No.	Compounds/agents	No. of bioassays	Animals Species	No.	Route of Exposure[a]
	Natural and man-made tannins				
48.	- Chestnut tannin	1	Rat	80	Sc
49.	- Quebraco	1	Rat	80	Sc
50.	- Mimosa	1	Rat	80	Sc
51.	- Tupasol	1	Rat	80	Sc
52.	- Tannigan BN	1	Rat	80	Sc
53.	- Tannigan P2 PLV	1	Rat	80	Sc
54.	- Tannigan 3LR	1	Rat	80	Sc
55.	- Tannigan CLS	1	Rat	80	Sc
56.	- Tannigan POL PAK	1	Rat	80	Sc
57.	- Tannesco HN	1	Rat	80	Sc
58.	- Blancoral	1	Rat	80	Sc
Drugs					
59.	Vitamin A	5	Rat	5,100	Ing
60.	Vitamin C	5	Rat	3,680	Ing
61.	Vitamin E	5	Rat	3,680	Ing
62.	N-(4-hydroxyphenyl)-retinamide	4	Rat	840	Ing
63.	Tamoxifen	17	Rat	6,008	Ing
64.	ICI 182780	1	Rat	450	Sc
65.	Toremifen	2	Rat	2,045	Ing
66.	Leuprolide	7	Rat	3,560	Sc
67.	Medroxyprogesterone acetate	4	Rat	1,710	Ing
68.	4-hydroxyandrostenedione	2	Rat	790	Sc
69.	Anastrozole	2	Rat	750	Ing
70.	Conjugated natural oestrogens	1	Rat	340	Ing
71.	Oestradiol	1	Rat	270	Sc
72.	Cyclophosphamide + methotrexate+ 5-fluorouracil (CMF)	1	Rat	300	Ip
73.	Alpha–interpheron	1	Rat	300	Sc
74.	Adriamycin	1	Rat	160	Sc
75.	Epirubicin	1	Rat	160	Sc
76.	Idarubicin	1	Rat	80	Sc
77.	MAK 4	3	Rat	1,000	Ing
78.	MAK 5	3	Rat	1,000	Ing
79.	MA631	3	Rat	1,000	Ing
Fuels: mixtures					
80.	Unleaded gasoline (1984)	1	Rat	300	Ing
81.	Unleaded gasoline (1993)	1	Rat	240	Ing
82.	Leaded gasoline	1	Rat	300	Ing
83.	Gasoline containing 3% methyl alcohol	1	Rat	240	Ing
84.	Gasoline containing 5% ethyl alcohol	1	Rat	240	Ing
85.	Gasoline containing 15% MTBE	1	Rat	240	Ing
86.	Gasoline containing 15% ETBE	1	Rat	240	Ing
87.	Kerosene	1	Rat	300	Ing
88.	Diesel fuel	1	Rat	300	Ing
89.	Naphtha	1	Rat	200	Ing
Fuels: aromatic hydrocarbons					
90.	Benzene	8	Rat, mouse	1,950	Ing,Inh,Tr
91.	Toluene	4	Rat	440	Ing
92.	Xylenes	2	Rat	380	Ing
93.	Ethylbenzene	2	Rat	380	Ing
94.	Trimethylbenzene	1	Rat	200	Ing
Fuels: isoparaffins					
95.	2,2,4-trimethylpentane	2	Rat	408	Ing
Fuels: oxygenated additives					
96.	Methyl alcohol	3	Rat	1,340	Ing
97.	Ethyl alcohol	4	Rat, mouse	1,458	Ing
98.	Methyl-tert-butyl ether (MTBE)	1	Rat	360	Ing
99.	Ethyl-tert-butyl ether (ETBE)	1	Rat	360	Ing
100.	Ter-amil-methyl ether (TAME)	1	Rat	600	Ing
101.	Di-isopropyl ether (DIPE)	1	Rat	600	Ing
Combustion Products (CP)					
102.	Automobile exhausts	3	Rat, mouse	400	Sc
103.	CP of domestic heating with oil	2	Rat, mouse	160	Sc
104.	CP of oil energy plant	2	Rat, mouse	160	Sc
105.	CP of carbon energy plant (including ashes)	2	Rat, mouse	240	Sc
106.	Welding fumes	1	Rat	880	Sc
107.	Ashes from waste incinerator	1	Rat	240	Int,Sc

—Continued

TABLE 1. Continued

No.	Compounds/agents	No. of Bioassays	Animals Species	No.	Route of Exposure[a]
Beverages and diet					
108.	"Coca-Cola"	4	Rat	1,999	Ing,Tr
109.	"Pepsi-Cola"	1	Rat	400	Ing
110.	Sucrose	1	Rat	400	Ing
111.	Caffeine	1	Rat	800	Ing
112.	Aspartame	1	Rat	1,800	Ing
Inorganic compounds					
113.	Chlorine (sodium hypochlorite)	2	Rat	480	Ing
114.	Arsenic (sodium arsenate)	2	Rat	823	Ing
115.	Arsenious anhydride	1	Rat	80	Sc
116.	Cadmium sulphide (cadmium yellow)	1	Rat	100	Sc
	Trivalent and hexavalent chromium compounds				
117.	- Lead chromate (chromium yellow)	1	Rat	100	Sc
118.	- Basic lead chromate (chromium orange)	1	Rat	100	Sc
119.	- Lead chromate sulphate and molybdate (molybdenum orange)	1	Rat	100	Sc
120.	- Chromite	1	Rat	100	Sc
121.	- Chromium alum	1	Rat	100	Sc
122.	- Basic chromium sulphate (Neochromium)	1	Rat	100	Sc
123.	- Silica-coated lead chromate	1	Rat	80	Sc
	Iron oxides				
124.	- Iron oxide hydrate (iron yellow)	1	Rat	100	Sc
125.	- Iron oxide (iron red)	1	Rat	100	Sc
126.	Magnesium oxide	1	Rat	120	Sc
127.	Silicone	2	Rat	600	Imp,Sc
	Titanium oxides				
128.	- Sample 1	1	Rat	80	Sc
129.	- Sample 2	1	Rat	80	Sc
130.	- Sample 3	1	Rat	80	Sc
	Vanadium				
131.	- Vanadium bioxide	1	Rat	120	Sc
132.	- Vanadium pentaoxide	1	Rat	120	Sc
133.	- Vanadium chloride	1	Rat	120	Sc
134.	- Ash A	1	Rat	120	Sc
135.	- Ash B	1	Rat	120	Sc
136.	Vitallium disks	3	Rat, mouse	252	Imp
137.	Vitallium disks (holed)	1	Rat	60	Imp
138.	Vitallium fragments	1	Rat	60	Imp
139.	Basic zinc chromate (yellow zinc)	1	Rat	120	Sc
Natural and man-made mineral particles					
	Asbestos				
140.	- Crocidolite	5	Rat	860	Ing,Int,Ip,Ipl,Sc
141.	- Chrysotile (Canada)	5	Rat, mouse	1,260	Int,Ip,Ipl,Sc
142.	- Chrysotile (Rhodesia)	1	Rat	80	Ip
143.	- Chrysotile (California)	1	Rat	80	Ip
144.	- Amosite	1	Rat	80	Ip
145.	- Anthophyllite	1	Rat	80	Ip
146.	- Asbestos-cement	1	Rat	240	Ip,Ipl,Sc
	Modified asbestos				
147.	- Compound 1	1	Rat	160	Ip, Ipl
148.	- Compound 2	1	Rat	160	Ip, Ipl
149.	- Compound 3	1	Rat	160	Ip, Ipl
150.	- Compound 4	1	Rat	160	Ip, Ipl
151.	- Compound 5	3	Rat, mouse	860	Ip, Ipl
152.	- Compound 6	3	Rat, mouse	860	Ip, Ipl
153.	- Compound 8	3	Rat, mouse	860	Ip, Ipl
154.	Wollastonite	1	Rat	240	Ip,Ipl,Sc
155.	Rock wool	3	Rat	640	Int,Ip,Ipl
156.	Ceramic fibres	1	Rat	440	Ip,Ipl
157.	Glass fibres	1	Rat	200	Int,Ip
158.	Crystalline silica	1	Rat	160	Ip,Sc
159.	Amorphous silica	1	Rat	160	Ip,Sc
160.	Alumina	1	Rat	160	Ip,Sc
161.	Talc (pure)	1	Rat	160	Ip,Sc
162.	Talc (industrial)	1	Rat	240	Ip,Ipl,Sc
163.	Kaolin	1	Rat	160	Ip,Sc
164.	Bentonite	1	Rat	240	Ip,Ipl
165.	Erionite	2	Rat, mouse	400	Ip,Ipl,Sc

—Continued

TABLE 1. *Continued*

PROJECT OF LONG-TERM CARCINOGENICITY BIOASSAYS: AGENTS STUDIED

No.	Compounds/agents	No. of Bioassays	Animals Species	No.	Route of Exposure[a]
	Other natural zeolites				
166.	- Mordenite (sed)	1	Rat	240	Ip,Ipl,Sc
167.	- Phillipsite	1	Rat	240	Ip,Ipl,Sc
168.	- Clinoptilolite	1	Rat	160	Ip,Sc
169.	- Cabasite	1	Rat	160	Ip,Sc
170.	- Ferrierite	1	Rat	160	Ip,Sc
171.	- Mordenite (crystalline)	1	Rat	240	Ip,Ipl,Sc
172.	- Heulandite	1	Rat	160	Ip,Sc
173.	- Mesolite	1	Rat	160	Ip,Sc
174.	- Natrolite	1	Rat	160	Ip,Sc
175.	- Solecite	1	Rat	160	Ip,Sc
176.	- Stilbite	1	Rat	160	Ip,Sc
177.	- Thomsonite	1	Rat	160	Ip,Sc
	Man-made zeolites and precursors				
178.	- TE-16460	1	Rat	360	Ip,Ipl,Sc
179.	- TE-16461	1	Rat	360	Ip,Ipl,Sc
180.	- TE-16462	1	Rat	360	Ip,Ipl,Sc
181.	- WGB 1189-2-1 (catalyst)	1	Rat	360	Ip,Ipl,Sc
182.	- WGB 1189-2-1 (elutriated)	1	Rat	360	Ip,Ipl,Sc
183.	- WGB 1190-2-1 (catalyst)	1	Rat	360	Ip,Ipl,Sc
184.	- WGB 1190-2-1 (elutriated)	1	Rat	360	Ip,Ipl,Sc
185.	- WGB 1191-2-1 (catalyst)	1	Rat	360	Ip,Ipl,Sc
186.	- WGB 1191-2-1 (elutriated)	1	Rat	360	Ip,Ipl,Sc
187.	- WGB 1192-1-1	1	Rat	360	Ip,Ipl,Sc
188.	- WGB 1193-1-1	1	Rat	360	Ip,Ipl,Sc
189.	- WGB 1194-1-1	1	Rat	360	Ip,Ipl,Sc
190.	- Paulsboro (catalyst)	1	Rat	360	Ip,Ipl,Sc
191.	- Paulsboro (catalyst fines)	1	Rat	360	Ip,Ipl,Sc
192.	- Kaiser alumina	1	Rat	360	Ip,Ipl,Sc
193.	- Mobil Joliet	1	Rat	360	Ip,Ipl,Sc
194.	- Joliet fresh	1	Rat	360	Ip,Ipl,Sc
195.	- MS 4A	1	Rat	160	Ip,Ipl,Sc
196.	- MS 5A	1	Rat	160	Ip,Ipl,Sc
197.	- MS 13X	1	Rat	160	Ip,Ipl,Sc
198.	Carbon fibers (disks)	1	Rat	140	Imp
Ionizing radiation					
199.	Gamma radiation	11	Rat	19,904	Total body irradiation by Co external source
Non-ionizing radiation					
200.	Extremely low frequency electromagnetic fields (50 Hz)	5	Rat	9,883	Total body irradiation

Abbreviations: Imp = subcutaneous implantation; Ing = ingestion; Inh = inhalation; Int = intratracheal instillation; Ip = intraperitoneal injection; Ipl = intrapleural injection; Sc = subcutaneous injection; Tr = transplacental route.

are expected to have a limited carcinogenic potential because of agent type (weak carcinogen) and/or low dose/concentration/intensity, yet involving large population groups and, in some cases, all of mankind; (2) *specific prerequisites*: they must reproduce as far as possible the various conditions of human exposure; they must include large groups of animals (at least 400 to 1,000) in order to express variations in effects; the study must be protracted for the life-span of the animals; animals should have as similar a basic tumorigram as possible to the human counterpart.

The results provided by "mega-experiments" serve to identify the risks and measure their levels; if negative, they do not necessarily mean absence of risk, but they then serve to determine the existence of a safeguard limit.

The number of compounds/agents and the long-term bioassays conducted per agent, the type of animal used, the route of exposure, and the numbers of animals used are described in TABLE 1.

In all, 398 long-term bioassays have been conducted on 200 compounds/agents using a total of 148,164 animals monitored until spontaneous death. Among the studies already concluded, the 47 agents showing "clear evidence" of carcinogenicity are detailed in TABLE 2.

The results obtained with this project have proved for the first time:
(1) the capacity of vinyl chloride to induce liver angiosarcomas, and other types of tumor;

TABLE 2. The Ramazzini Foundation Cancer Program

PROJECT OF LONG-TERM CARCINOGENICITY BIOASSAYS: RESULTS[a]

The 47 agents with "clear" evidence of carcinogenicity

- Plastic monomers
  - Vinyl chloride[1, 2]
  - Vinylidene chloride[3]
  - Acrylonitrile[4]
  - Vinyl acetate[5,6]
  - Styrene oxide[7]
- Plastic polymers
  - Ivoclar[8]
  - Acronite[8]
  - Lucitone[8]
  - Teflon[8]
- Chlorinated solvents
  - Trichloroethylene[9]
- Pesticides
  - Mancozeb[12]
- Inorganic compounds
  - Cadmium sulphide (cadmium yellow)[8]
  - Lead chromate (chromium yellow)[8]
  - Basic lead chromate (chromium orange)[8]
  - Lead chromate sulphate and molybdate (molybdenum orange)[8]
  - Chromium alum[8]
  - Basic chromium sulphate (Neochromium)[8]
  - Silica-coated lead chromate[8]
  - Vitallium disks[8]
  - Vitallium disks (holed)[8]
  - Basic zinc chromate (yellow zinc)[8]
- Drugs
  - Adriamycin[8]
  - Epirubicin[8]
  - Idarubicin[8]

- Fuels: aromatic hydrocarbons
  - Benzene[13, 14]
- Fuels: oxygenated additives
  - Methyl alcohol[15]
  - Ethyl alcohol[5]
  - Methyl tert-butyl ether (MTBE)[16]
  - Tertiary-amyl-methyl ether (TAME)[17]
  - Di-isopropyl ether (DIPE)[17]
- Aldehydes
  - Formaldehyde[10,11]
- Compounds used in leather industry
  - Chromitan B[8]
  - Baychrom[8]
  - Quebraco[8]
  - Tupasol[8]
- Natural & man-made mineral particles
  - Crocidolite[18]
  - Chrysotile (3 samples)[18]
  - Amosite[18]
  - Anthophyllite[18]
  - Asbestos-cement[18]
  - Modified asbestos (6 samples)[19]
  - Rock wool[20]
  - Ceramic fibers[21]
  - Talc (industrial)[22]
  - Erionite[18]
  - Carbon fibers (disks)[8]
- Ionizing radiation
  - Gamma radiation[23, 24]

[a] up to the year 2002

(2) the capacity of formaldehyde to cause lymphomas and leukemias;
(3) the carcinogenicity of benzene in experimental animals, producing various types of tumors in different tissues and organs;
(4) the capacity of methyl alcohol and ethyl alcohol to induce various types of tumor in different tissues and organs;
(5) the capacity of methyl tert-butyl ether (MTBE), the most utilized gasoline oxygenated additive, to cause lymphomas and leukemias;
(6) the carcinogenicity of Mancozeb, tertiary-amyl-methyl ether (TAME) and di-isopropyl ether (DIPE).

The results obtained have formed the basis of rules and regulations for primary prevention.

One distinctive characteristic of the CRC/RF long-term carcinogenicity bioassays is to keep experimental animals under observation until spontaneous death. The neoplastic response depends not only on the kind of agent, its physicochemical and toxicologic properties, the mode of exposure, and the type of animal, but also to a great extent, on the latency of the tumor, which varies and may be very long. Experimental findings agree that the latent neoplastic potential for causing a tumor increases with the length of the observation time or age. That is why we are convinced that experimental carcinogenicity trials should continue until spontaneous animal

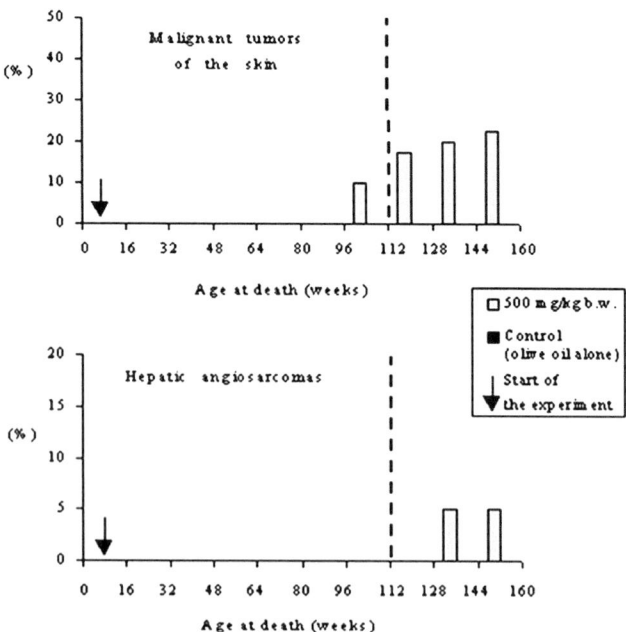

FIGURE 1. Cumulative prevalence of animals with malignant tumors of the skin and hepatic angiosarcomas, histopathologically observed, by age at death, in male Sprague-Dawley rats treated with benzene (BT 15).

death and not be cut short. Cutting short an experiment after two years may mask a possible carcinogenic response.

The following cases afford good examples.

Benzene

At the CRC/RF, benzene by ingestion and inhalation was studied. In experiment BT 902, benzene was administered by ingestion (stomach tube) in extra virgin olive oil. Benzene was administered at doses of 500 or 0 mg/kg b.w. once daily 4 or 5 days per week for 104 weeks to 40 male and 40 female Sprague-Dawley rats beginning at age 7 weeks. Animals were observed until spontaneous death. The control group received olive oil alone. An increase was observed in total malignant tumors; Zymbal gland, oral and nasal cavity, skin, forestomach carcinomas; and hepatic angiosarcomas. Skin carcinomas and hepatic angiosarcomas were observed after 112 weeks of age (FIG. 1).[25]

Xylenes

Xylenes (experiment BT 904) were administered by stomach tube at concentrations of 500 or 0 mg/kg b.w. using the protocol described above for benzene. The control group received olive oil alone. An increase was observed in total malignant tumors, mammary and oral cavity carcinomas, and hemolymphoreticular neoplasias. The increase in total malignant tumors, oral cavity carcinomas, and hemolymphoreticular neoplasias was only observed after 112 weeks of age (FIG. 2).[26] It should be

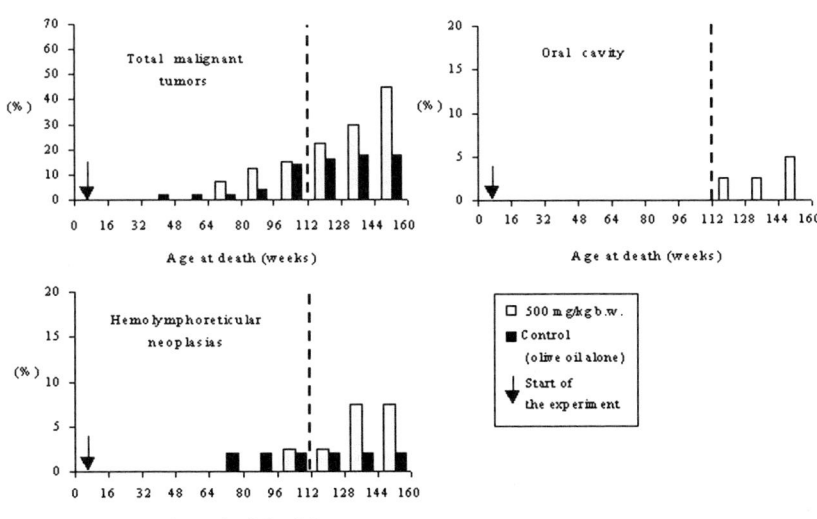

FIGURE 2. Cumulative prevalence of total malignant tumors, oral cavity carcinomas and hemolymphoreticular neoplasias, histopathologically observed, by age at death, in male Sprague-Dawley rats treated with xylenes (BT 904).

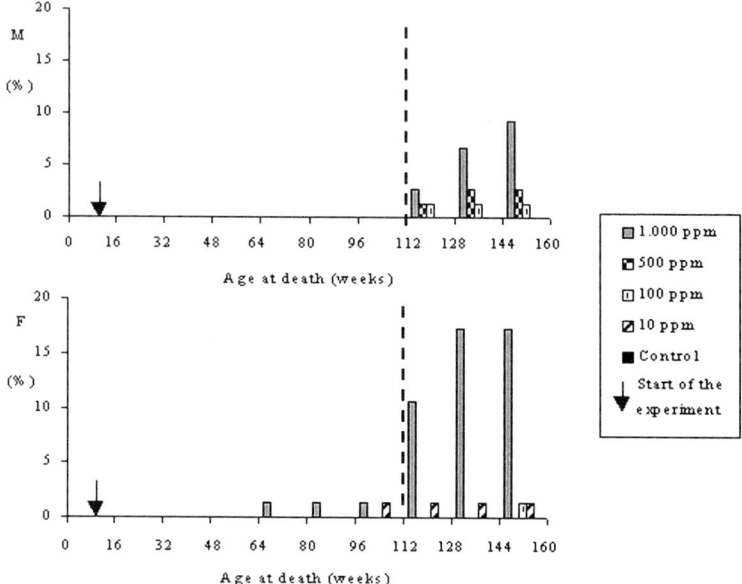

FIGURE 3. Cumulative prevalence of thyroid gland malignant tumors, histopathologically observed, by age at death, in male and female Sprague-Dawley rats treated with Mancozeb (BT 5007).

noted that the experiment with xylene performed by the NTP sacrificed rats after 104 weeks of treatment; no carcinogenic effect was found.[27]

Mancozeb

Mancozeb (experiment BT 5007) was administered to Sprague-Dawley rats with feed at concentrations of 1,000, 500, 100, 10, or 0 ppm supplied *ad libitum* for 104 weeks to 75 males and 75 females, 8 weeks old at start. All animals were observed until spontaneous death. Control animals received standard feed. Among other things, a strong increase in malignant tumors of the thyroid gland in males and females was observed after 112 weeks of age (FIG. 3).[12]

Vinyl Acetate Monomer

Vinyl acetate monomer (experiment BT 51) was administered in drinking water supplied *ad libitum*, at doses of 5,000, 1,000, and 0 ppm (v/v), to Sprague-Dawley rats, 17-week-old (breeders) or 12-day embryos (offspring) at the start of the experiment. Treatment lasted 104 weeks after which time the animals were kept under control conditions until spontaneous death. The increase in carcinomas of the oral cavity, tongue, esophagus and forestomach (considered altogether), in male and female offspring, was only clearly observed after 112 weeks of age (FIG. 4).[6]

In an other experiment (BT 52), vinyl acetate monomer was administered in drinking water, supplied *ad libitum*, at doses of 5,000, 1,000, and 0 ppm (v/v), to

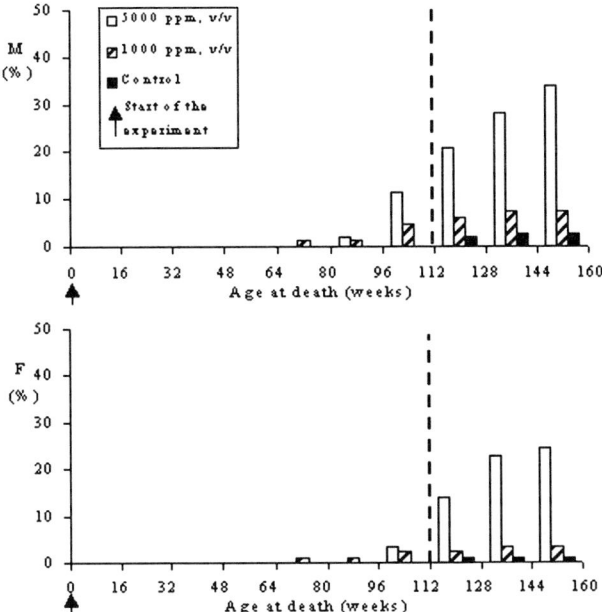

FIGURE 4. Cumulative prevalence of animals with tumors of the oral cavity, tongue, esophagus and forestomach, histopathologically observed, by age at death, in male and female offspring Sprague-Dawley rats treated with vinyl acetate monomer (BT 51).

Swiss mice, 17-week-old (breeders) or 12-day embryos (offspring) at the start of the experiment. The treatment lasted 78 weeks, the animals were kept under control conditions until spontaneous death. The increase in total malignant tumors in male and female breeders and offspring was again clearly observed after 112 weeks of age (FIGS. 5 and 6).[5]

Project of Experimental Anticarcinogenesis (Chemopreventive) Bioassays

By interventive prevention, we mean the use of factors and agents to contrast, block or inhibit the onset of tumors of different types and sites. When these factors or agents are chemicals, and in particular drugs, the interventive prevention is properly known as chemoprevention.

The project started in 1985. The aims of this project are: (1) to set up adequate animal models and experimental protocols; (2) to conduct experimental studies on compounds potentially eligible for chemoprevention; (3) to determine the minimal effective drug dose, the effect of the drug administration schedule, the effect of the interruption of drug administration, side effects (in particular carcinogenic), efficacy, and tolerability of the drug with or without other drugs.

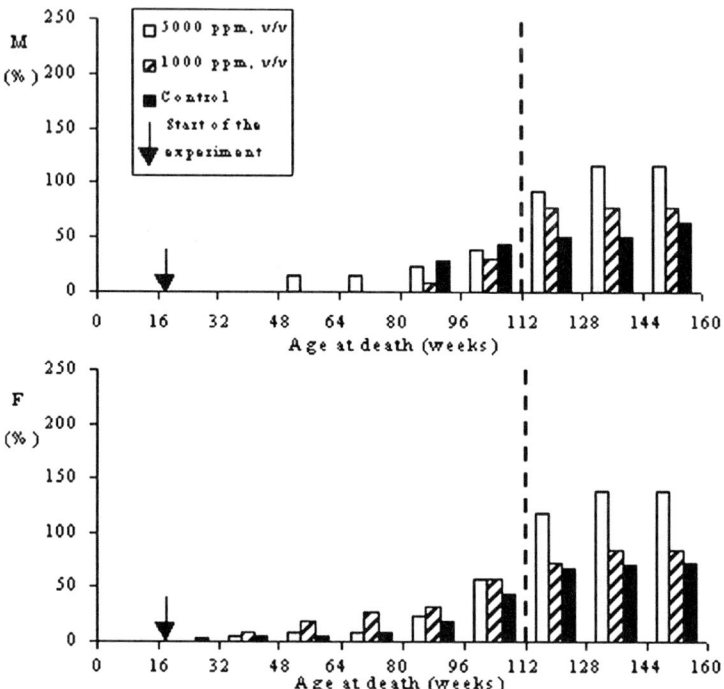

FIGURE 5. Cumulative prevalence of total malignant tumors *per* 100 animals, histopathologically observed, by age at death, in male and female breeder Swiss mice treated with vinyl acetate monomer (BT 52).

Chemopreventive bioassays are planned and conducted following two types of protocols:

- *Protocol I:* studies on limited groups of animals (90–150), at high risk (by virtue of age and other factors), with limited treatment and observation period (35–40 weeks), so as to reproduce clinical trials.
- *Protocol II:* studies on larger groups of animals (200–400), starting at a young age, with a varying duration of treatment (often life span), and with observation until spontaneous death, thus reproducing epidemiological studies on populations.

All studies are performed on Sprague-Dawley rats and conducted according to Good Laboratory Practices.

Twenty factors/compounds have been studied on Sprague-Dawley rats from the colony of the CRC/RF, which represents a human-equivalent model for various different tumors, and in particular mammary cancer. The factors/compounds studied are given in TABLE 3. Over 33,000 animals have been used for the study of these factors/compounds.

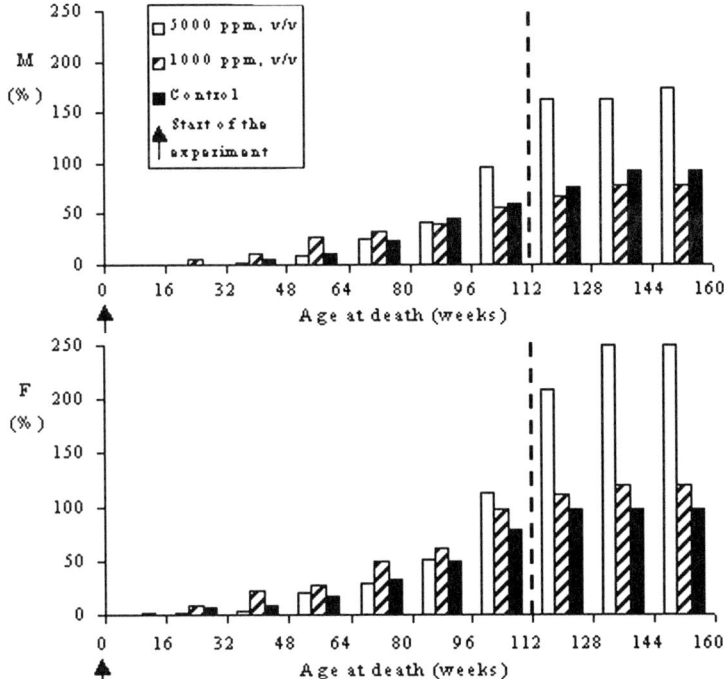

FIGURE 6. Cumulative prevalence of total malignant tumors per 100 animals, histopathologically observed, by age at death, in male and female offspring of Swiss mice treated with Vinyl acetate monomer (BT 52).

With these experiments the preventive effects of pregnancy, ovarectomy, leuprolide,[28] tamoxifen,[29,30] faslodex, medroxyprogesterone acetate,[31] and 4-hydroxyandrostenedione on mammary carcinogenesis have been shown.

In contrast, our experimental studies on Vitamin A (retinol palmitate and acetate) have shown that this vitamin increases the incidence of mammary malignant tumors.[32]

The results of this project not only indicate the chemopreventive effects of the factors/compounds studied, but may provide important information on their side effects. Furthermore they may be useful, as in the case of tamoxifen, to determine the lowest doses still effective for chemoprevention: this is an important datum in the case of trials on women, to avoid unnecessary overtreatment.

Project of Descriptive and Analytical Epidemiology

This project encompasses: (1) registries of nominative mortality, with particular regard to tumors, in the province of Bologna and other Italian geographic areas; and (2) analytical epidemiology studies on the tumor incidence and mortality in people professionally and environmentally exposed to carcinogens of environmental, industrial origin or both.

TABLE 3. The Ramazzini Foundation Cancer Program

PROJECT OF EXPERIMENTAL CHEMOPREVENTIVE BIOASSAYS: FACTORS / AGENTS STUDIED

1. Pregnancy	**Progestins**
	12. Medroxyprogesterone acetate (MPA)
2. Ovarectomy	
	Aromatase inhibitors
Vitamins	13. 4-hydroxyandrostenedione (4-OHA)
3. Vitamin A	14. Anastrozole
4. Vitamin C	
5. Vitamin E	**Estrogens**
	15. Conjugated estrogens (Premarin)
Retinoids	
6. N-(4-hydroxyphenil)retinamide (HPR)	**Chemotherapeutic drugs**
	16. Cyclophosphamide, methotrexate, 5-fluorouracil (CMF)
Antiestrogens	
7. Tamoxifen	**Immunomodulants**
8. Faslodex	17. Alpha-interpheron
9. Toremifene	
10. Raloxifene	**Drug Combinations**
	18. Tamoxifen-HPR
LHRH antagonists	19. Tamoxifen-MPA
11. Leuprolide (LEU)	20. Tamoxifen-LEU-MPA

The aims of the registry of nominative mortality are: (1) to produce information on the trends of tumor mortality and other causes of death; (2) to identify possible areas of higher mortality due to tumors and trace the causes; (3) to establish a database available for prevention and clinical practice; and (4) to furnish data for health programs.

All mortality data are acquired from photocopies of the original death certificate. Data are ordered, completed, recodified, elaborated, and published according to a controlled protocol, applied to the registries of all areas studied. The results of every geographic area are published in a consistent manner.

Data dealing with the project design and state-of-the-art are presented in TABLE 4. Mortality data for a 40-year period from Bologna, its province, and some other areas will be available in the next few years.

The nominative mortality registries, with particular regard to tumors, are precise instruments that define the current dimension of neoplastic disease and can be used to forecast future trends and to evaluate the social burden of tumors and the efficacy of current control strategies.

The studies on analytical epidemiology began in the early 1960s. The main areas of interest are tumors of the urinary tract due to aromatic amines,[33] hepatic angiosarcomas due to vinyl chloride,[34] mesotheliomas due to asbestos used in the railroads[35] and sugar refineries,[36] and to asbestos exposure in other circumstances.

Our data on pathologies related to exposure to asbestos are reported in TABLE 5.

TABLE 4. The Ramazzini Foundation Cancer Program

PROJECT OF DESCRIPTIVE AND ANALYTICAL EPIDEMIOLOGY: THE REGISTRIES OF NOMINATIVE MORTALITY WITH PARTICULAR REGARD TO TUMORS

	Geographical Area	Population	Years Considered
1.	Province of Bologna	908,631	1959-2000
2.	Province of Reggio Emilia	429,865	1959-1988
3.	LHU 32 of Portomaggiore (FE)	53,701	1986-1987
4.	Faenza (RA)	48,419	1960-1998
5.	LHU 75 of Acqui Terme (AL)	44,587	1959-1994
6.	Albignasego (PD)	18,452	1960-1996
7.	Cittadella (PD)	18,415	1960-1997
8.	LHU 5 of Urbino		1960-1994
	Province of Pesaro and Urbino	340,830	1995-2000
9.	Rocca di Papa (RM)	13,242	1960-1996
10.	S. Giovanni Rotondo (FG)	25,121	1960-2000
11.	Manfredonia (FG)	58,623	1960-2000
12.	Sannicandro Garganico (FG)	18,643	1960-2000
13.	Mountain Communities Appennino Dauno Settentrionale (FG)	24,296	1960-2000
14.	Foggia	155,237	1980-2000
15.	Bari	338,949	1980-2000
16.	S. Vito dei Normanni (BR)	20,877	1936-2000
17.	Brindisi	94,429	1990-2000
18.	Catanzaro	96,886	1980-1999
19.	Molise Region	327,987	1990-1999
	TOTAL	**3,037,190**	

Studies in these areas are ongoing. A mortality study has recently started on a cohort of over 1,400 workers in the vinyl chloride/polyvinyl chloride industries.

The results of these studies help to identify and quantify the risk of populations exposed to carcinogenic risks and can be used for preventive strategies and to establish the occupational origin in cases of compensation for neoplasias. They are an important area of study.

CONCLUSION

All the projects herein reported have, for the past 35 years, borne the strong leadership of Professor Cesare Maltoni, Director of the Institute of Oncology "F. Addarii" (1964–1997) and thereafter Scientific Director of the Ramazzini Foundation until his untimely death in January 2001. The commitment of his group of coworkers continues.

TABLE 5. The Ramazzini Foundation Cancer Program

PROJECT OF DESCRIPTIVE AND ANALYTICAL EPIDEMIOLOGY: REGISTRY OF THE PATHOLOGIES OF PEOPLE EXPOSED TO ASBESTOS

Category	Pathology			
	Mesothelioma	Lung cancer	Other tumors	Asbestosis[a]
Railway workers	207	146	157	173
- FS workers	92	26	27	37
- Workers of rolling-stock workshops not belonging to FS	97	119	128	132
- Workers not belonging to FS, exposed to work environment pollution	4	0	1	1
- Family members of FS workers	7	0	0	3
- Family members of workers in rolling-stock workshops not belonging to FS	3	1	1	0
- Environmental pollution in FS	4	0	0	0
Shipyard and port workers	17	14	15	2
- Workers	17	13	15	2
- Family members	0	1	0	0
Workers of asbestos-cement industry	18	7	3	2
- Workers	17	7	3	2
- Family members	1	0	0	0
Workers of sugar refineries	19	2	1	1
- Workers	18	2	0	1
- Family members	1	0	1	0
Seamen and sailors	18	1	0	1
- Workers	17	1	0	1
- Family members	1	0	0	0
Other categories[b]	238	45	28	114
TOTAL	517	215	204	293

[a] Pleural plaques, pleural thickenings, pleural asbestosis, pleuro-pulmonary asbestosis and pulmonary asbestosis are included.
[b] Over 60 work categories are included.

The project of primary experimental research into industrial carcinogenesis, anticarcinogenesis, and epidemiology performed by Ramazzini Foundation has been conducted in highly standardized control conditions by a team largely unchanged over time. For these reasons, it represents an unusual entity in oncological research, in our country and the world. Projects such as these involve considerable cost.

The meaning of this cost was very clearly expressed by Cesare Maltoni at the end of his presentation at the Conference on Vinyl Chloride in Bethesda in 1980, as reported in the proceedings published in the EHP.

> The cost of our project of long-term carcinogenicity bioassays on vinyl chloride includes the cost of (1) the planning and setting-up of experimental apparatus, including inhalation facilities, of a type uncommon in 1971, and the working out of a protocol for long-term bioassays; (2) the study of nearly 7,000 animals up to the point of their natural death; (3) ten years of work; (4) the routine examination of some 200,000 histological slides; (5) a financial commitment equivalent to more than $2,000,000 U.S. at present prices; (6) the availability of the same team of scientists throughout the entire 10 years of the project, a prerequisite which may be difficult or even impossible to ensure in many countries at the present time; (7) the highly motivated commitment of those scientists to a type of work which is long-lasting, onerous and often tedious; (8) the effort involved in maintaining the consistency of the methodology, which has as its reverse side the limits placed on the exercise of imagination (the most positive element in scientific life); (9) the effort involved in establishing and preserving objectivity and balance in the evaluation and interpretation of data; (10) and finally, the strength required to withstand the sense of loneliness arising from the lack of co-operation of many of those bodies which should properly be concerned with the progress of science in this field, not excluding part of the scientific community whose indifference sometimes degenerates into frank hostility.
>
> The high costs probably represent the reason why, in the field of experimental and environmental carcinogenesis, words overlap facts, opinions overlap data, and meetings and commissions reports submerge good laboratory work.

Twenty-two years after the paper appeared in EHP, our entire project, in number of animals, investments and commitment, is about 20 times greater than the research program on vinyl chloride; we can still only reconfirm those words. Indeed, we wish to state, clearly, that:

(1) primary research in the field of oncology continues to be disregarded, almost everywhere, including in Italy and the rest of Europe
(2) innovative research is understood and accepted with difficulty; and
(3) when primary and innovative research reveals that agents of fundamental importance for technological development, and therefore of great economic and political interest, may be hazardous for health and, specifically, may present a risk of cancer, obstacles will be put in its way, and the public is unlikely to know or understand.

Two clear perceptions give us cause for optimism however: (1) history teaches us that the search for truth, and the habit of bearing witness to it, in the end prevails; and (2) culture and science induce in their disciples a sort of mithridatization: in the words of our great poet Giacomo Leopardi, they come to learn an "aristocratic pleasure in displeasing," in the defense of truth.

REFERENCES

1. MALTONI, C., G. LEFEMINE, A. CILIBERTI, et al. 1984. Experimental research on vinyl chloride carcinogenesis. In Archives of Research on Industrial Carcinogenesis, Vol. 2. C. Maltoni & M.A. Mehlman, Eds. Princeton Scientific Publishing. Princeton, NJ.

2. MALTONI, C. 1974. Oncogenicity bioassays of vinyl chloride: plan, current results, and perspectives. *In* Proceedings of the 2nd International Symposium of Medichem on Chromosome Aberrations by Industrial Chemicals and Vinyl Chloride Toxicity.: 65–79. Milano, Italy, October 28–29, 1974.
3. MALTONI, C., G. LEFEMINE, G. COTTI, et al. 1985. Experimental research on vinylidene chloride carcinogenesis. *In* Archives of Research on Industrial Carcinogenesis, Vol. 3. C. Maltoni & M.A. Mehlman, Eds. Princeton Scientific Publishing. Princeton, NJ.
4. MALTONI, C., A. CILIBERTI, G. COTTI, et al. 1987. Experimental research on acrylonitrile carcinogenesis. *In* Archives of Research on Industrial Carcinogenesis, Vol. 6. C. Maltoni & M.A. Mehlman, Eds. Princeton Scientific Publishing. Princeton, NJ.
5. MALTONI, C., A. CILIBERTI, G. LEFEMINE, et al. 1997. Results of a long-term experimental study on the carcinogenicity of vinyl acetate monomer in mice. Ann. N.Y. Acad. Sci. **837:** 209–238.
6. MINARDI, F., F. BELPOGGI, M. SOFFRITTI, et al. 2002. Results of long-term carcinogenicity bioassay on vinyl acetate monomer in Sprague-Dawley rats. Ann. N.Y. Acad. Sci. **982:** this volume.
7. CONTI, B., C. MALTONI, G. PERINO, et al. 1988. Long-term carcinogenicity bioassays on styrene administered by inhalation, ingestion and injection and styrene oxide administered by ingestion, on Sprague-Dawley rats, and para-methylstyrene administered by ingestion in Sprague-Dawley rats and Swiss mice. Ann. N.Y. Acad. Sci. **534:** 203–234.
8. MINARDI, F., L. CORTESI & C. MALTONI. 1991. Subcutaneous sarcomatogenicity bioassays in the identification of environmental carcinogens: some results. Acta Oncol. **12:** 365–372.
9. MALTONI, C., G. LEFEMINE & G. COTTI. 1986. Experimental research on trichloroethylene carcinogenesis. *In* Archives of Research on Industrial Carcinogenesis, Vol. 5. C. Maltoni & M.A. Mehlman, Eds. Princeton Scientific Publishing. Princeton, NJ.
10. SOFFRITTI, M., C. MALTONI, F. MAFFEI & R. BIAGI. 1989. Formaldehyde: an experimental multipotential carcinogen. Toxicol. Ind. Health. **5:** 699–730.
11. SOFFRITTI, M., F. BELPOGGI, L. LAMBERTINI, et al. 2002. Results of long-term experimental studies on the carcinogenicity of formaldehyde and acetaldehyde in rats. Ann. N.Y. Acad. Sci. **982:** this volume.
12. BELPOGGI, F., M. SOFFRITTI, L. BUA, et al. 2002. Results of long-term experimental studies on the carcinogenicity on ethylene-bis-dithiocarbamate (Mancozeb) in rats. Ann. N.Y. Acad. Sci. **982:** this volume.
13. MALTONI, C. & C. SCARNATO. 1977. Le prime prove sperimentali dell'azione cancerogena del benzene. Gli Ospedali della Vita **4:** 111–113.
14. MALTONI, C., A. CILIBERTI, G. COTTI, et al. 1989. Benzene, an experimental multipotential carcinogen: results of the long-term bioassays performed at the Bologna Institute of Oncology. Environ. Health Perspect. **82:** 109–124.
15. SOFFRITTI, M., F. BELPOGGI, D. CEVOLANI, et al. 2002. Results of long-term experimental studies on the carcinogenicity of methyl alcohol and ethyl alcohol in rats. Ann. N.Y. Acad. Sci. **982:** this volume.
16. BELPOGGI F., M. SOFFRITTI & C. MALTONI. 1995. Methyl-tertiary-butyl ether (MTBE)—a gasoline additive—causes testicular and lymphohaematopoietic cancers in rats. Toxicol. Ind. Health. **11:** 119–149.
17. BELPOGGI F., M. SOFFRITTI, F. MINARDI, et al. 2002. Results of long-term carcinogenicity bioassays on *tert*-amyl-methyl-ether (TAME) and di-isopropyl-ether (DIPE) in rats. Ann. N.Y. Acad. Sci. **982:** this volume.
18. MALTONI, C. & F. MINARDI. 1989. Recent results of carcinogenicity bioassays of fibres and other particulate materials. *In* Non Occupational Exposure to Mineral Fibres. V. Bignon, V. Peto & R. Saracci, Eds. **90:** 46–53. IARC Scientific Publication. Lyon, France.
19. MINARDI, F. & C. MALTONI. 1988. Results of recent experimental research on the carcinogenicity of natural and modified asbestos. Ann. N.Y. Acad. Sci. **534:** 754–761.
20. MINARDI, F. & C. MALTONI. 1998. Results of long-term carcinogenicity bioassays of rockwool on Sprague-Dawley rats. Eur. J. Oncol. **3:** 251–260.
21. MINARDI, F. & C. MALTONI. 1998. Results of long-term carcinogenicity bioassays of ceramic fibres ("Fiberfrax") on Sprague-Dawley rats. Eur. J. Oncol. **3:** 241–249.

22. MINARDI, F., F. BELPOGGI, A. FRANCH, et al. 1990 La cancerogenesi da talco grezzo contaminato con amianto: primi risultati dei saggi sperimentali dell'Istituto di Oncologia di Bologna. *In* Recenti Progressi nelle Conoscenze e nel Controllo dei Tumori. E. Triggiani, G. Sammarco, G. Liguori, *et al.*, Eds.: 279–293. Monduzzi Editore. Bologna.
23. SOFFRITTI, M., F. BELPOGGI, F. MINARDI, *et al.* 1999. Mega-experiments to identify and assess diffuse carcinogenic risks. Uncertainty in the risk assessment of environmental and occupational hazards. An International Workshop (24–26 September 1998) Ann. N.Y. Acad. Sci. **895:** 34–55.
24. SOFFRITTI, M., F. BELPOGGI, F. MINARDI, *et al.* 1999. Mega-experiments on the carcinogenicity of γ-radiation on Sprague-Dawley rats at the Cancer Research Centre of the European Ramazzini Foundation of Oncology and Enviromental Sciences: plan and report of early results on mammary carcinogenesis. Eur. J. Oncol. **4:** 509–522.
25. MALTONI, C., B. CONTI, G. COTTI & F. BELPOGGI. 1985. Experimental studies on benzene carcinogenicity at the Bologna Institute of Oncology: current results and ongoing research. Am. J. Ind. Med. **7:** 415–446.
26. MALTONI, C., A. CILIBERTI, C. PINTO, *et al.* 1997. Results of long-term experimental carcinogenicity studies on the effects of gasoline, correlated fuels, and major gasoline aromatics on rats. Ann. N.Y. Acad. Sci. **837:** 15–52.
27. NATIONAL TOXICOLOGY PROGRAM. 1986. Toxicology and carcinogenesis studies of xylenes (mixed) (cas no. 1330-20-7) in F344/N rats and B6C3F$_1$ mice (gavage studies). Technical Report Series No. 327.
28. SOFFRITTI, M., F. MINARDI & C. MALTONI. 1996. Risultati degli studi sperimentali sugli effetti chemiopreventivi della vitamina A e di un LHRH agonista-antagonista (il leuprolide). *In* Recenti Progressi nel Controllo Farmacologico dei Tumori. G. Colucci, C. Maltoni & E. Maiello, Eds.: 39–42. Atti del XXII Congresso Nazionale di Oncologia —1° Forum di Oncologia dei Paesi Mediterranei (Bari, 27–30 Novembre 1996).
29. MALTONI, C., F. MINARDI & C. PINTO. 1996. Gli effetti chemiopreventivi del Tamoxifene sul cancro mammario: risultati sperimentali. *In* Recenti Progressi nel Controllo Farmacologico dei Tumori. G. Colucci, C. Maltoni & E. Maiello, Eds.: 31–37. Atti del XXII Congresso Nazionale di Oncologia—1° Forum di Oncologia dei Paesi Mediterranei (Bari, 27–30 Novembre 1996).
30. MALTONI, C., F. MINARDI, C. PINTO, *et al.* 1997. Results of three life-span experimental carcinogenicity and anticarcinogenicity studies on tamoxifen in rats. Ann. N.Y. Acad. Sci. **837:** 469–512.
31. MINARDI, F., L. BUA, A. SANTI, *et al.* 1996. Experimental studies on the chemopreventive effects of medroxyprogesterone acetate (MPA) on mammary cancer using a human-equivalent animal model. *In* The Scientific Bases of Cancer Chemoprevention. Proceedings of the International Forum.: 235–240. Bologna. 31 March–2 April 1996.
32. SOFFRITTI, M., A. MOBIGLIA, C. PINTO, *et al.* 1996. Results of experimental bioassays on the chemopreventive effects of vitamin A and N-(4-hydroxyphenyl)-retinamide (HPR) on mammary cancer. *In* The Scientific Bases of Cancer Chemoprevention. Proceedings of the International Forum.: 241–248. Bologna. 31 March–2 April 1996.
33. MALTONI, C. 1974. Occupational carcinogenesis. *In* International Symposium, Cancer Detection and Prevention. C. Maltoni, Y. Fassing, E.C. Hammond, *et al.*, Eds.: 19–26. Excerpta Medica. Amsterdam.
34. CILIBERTI, A., G. LEFEMINE, F. BELPOGGI, *et al.* 1994. Incidenza spontanea dei tumori epatici (epatocarcinomi ed angiosarcomi) e dei loro precursori in ratti Sprague-Dawley tenuti sotto osservazione fino a morte naturale. *In* I Tumori Primitivi e Secondari del Fegato. C. Natale, R. Laricchiuta, D. Perfetto, *et al.*, Eds.: **1:** 31–37. XIX Congresso Nazionale di Oncologia (Foggia, 26–28 Maggio 1994). Monduzzi Editore. Bologna.
35. MALTONI, C., L. LAMBERTINI, D. CEVOLANI, *et al.* 2002. I mesoteliomi da amianto usato nelle ferrovie: resoconto di 199 casi. Eur. J. Oncol. **7:** 51–55.
36. MALTONI, C., C. PINTO, D. VALENTI, *et al.* 1998. Mesotheliomas following exposure to asbestos used in sugar refineries: report of 17 Italian cases. *In* 25° Anniversario degli Istituti di Chirurgia dell'Università di Chieti Diretti dal Professor Vanni Beltrami. Chieti 1973–1998. Vecchio Faggio, Ed. Scritti degli allievi e degli amici.

Results of Long-Term Experimental Studies on the Carcinogenicity of Methyl Alcohol and Ethyl Alcohol in Rats

MORANDO SOFFRITTI, FIORELLA BELPOGGI, DANIELA CEVOLANI, MARINA GUARINO, MICHELA PADOVANI, AND CESARE MALTONI[†]

Cancer Research Center, European Ramazzini Foundation for Oncology and Environmental Sciences, Bologna, Italy

ABSTRACT: Methyl alcohol was administered in drinking water supplied *ad libitum* at doses of 20,000, 5,000, 500, or 0 ppm to groups of male and female Sprague-Dawley rats 8 weeks old at the start of the experiment. Animals were kept under observation until spontaneous death. Ethyl alcohol was administered by ingestion in drinking water at a concentration of 10% or 0% supplied *ad libitum* to groups of male and female Sprague-Dawley rats; breeders and offspring were included in the experiment. Treatment started at 39 weeks of age (breeders), 7 days before mating, or from embryo life (offspring) and lasted until their spontaneous death. Under tested experimental conditions, methyl alcohol and ethyl alcohol were demonstrated to be carcinogenic for various organs and tissues. They must also be considered multipotential carcinogenic agents. In addition to causing other tumors, ethyl alcohol induced malignant tumors of the oral cavity, tongue, and lips. These sites have been shown to be target organs in man by epidemiologic studies.

KEYWORDS: methyl alcohol; ethyl alcohol; carcinogenicity; long-term bioassay; rat

INTRODUCTION

Automobiles and gasoline are two of the main consumer products characterizing the modern age. The annual consumption of gasoline has been estimated at over 600,000,000 tons throughout the world. With its vapors and combustion products, gasoline contributes in a decisive way to polluting the biosphere. The gasoline-car combination has turned progressively into an ecological and health problem, affecting the physical, chemical, and biological equilibria of the earth's biosphere, likely with a parallel effect on the health of man. Our awareness of the size and urgency of the ecological and health hazard calls for the promotion of strategies to deal with the problem.

Address for correspondence: Morando Soffritti, M.D., Cancer Research Center, European Ramazzini Foundation for Oncology and Environmental Sciences, Bentivoglio Castle, 40010 Bentivoglio (BO), Italy. Voice: +39-051-6640460; fax: +39-051-6640223.
crcfr@tin.it
[†]Deceased.

Technological improvements may include: (1) producing less polluting gasolines in terms of both vapors and compounds generated by combustion; (2) designing engines geared to a lower emission of combustion products; and (3) producing systems for containing combustion products (such as catalytic converters). Producing less polluting gasolines calls for biomedical research aimed at identifying the potential toxic and carcinogenic effects of various kinds of gasoline, their components and combustion products, gauging what point such effects have currently reached, and producing a comparative evaluation of the risks. While the introduction of new fuels and gasoline additives may decrease certain components that are toxic to health and to the environment, it may generate new ones. No new fuel should be introduced as an improvement until research has proven it safe.

In 1975, the laboratory of the Cancer Research Center of the Ramazzini Foundation (CRC/RF) started a systematic project of carcinogenicity bioassays on: (1) various types of gasolines and other petroleum-derived fuels, namely, European unleaded gasoline (with a high content of aromatics), unleaded reformulated gasoline, leaded gasoline, gasoline containing 3% methyl alcohol, gasoline containing 5% ethyl alcohol, gasoline containing 15% methyl *tert*-butyl ether (MTBE), gasoline containing 15% ethyl *tert*-butyl ether (ETBE), kerosene, diesel fuel, and naphtha; (2) some of the major aromatic components such as benzene, toluene, xylenes, ethylbenzene, and 1,2,4-trimethylbenzene; (3) various types of octane enhancers such as the oxygenated additives methyl alcohol, ethyl alcohol, MTBE, ETBE, tertiary-amyl-methyl ether (TAME), di-isopropyl ether (DIPE), and the isoparaffin 2,2,4-trimethylpentane (TMP); (4) automobile exhaust and combustion products of note in oxygenated additive-containing gasolines such as formaldehyde and acetaldehyde; and (5) catalysts such as artificial zeolites and precursors used in petroleum refining.

For the CRC/RF project on fuels, 51 carcinogenicity bioassays have been performed, 42 industrial products studied, and more than 20,000 experimental animals used. This project is, to our knowledge, unique. The results of the experiments on European unleaded gasoline with a high content of aromatics,[1] leaded gasoline,[1] kerosene,[1] diesel fuel,[1] benzene,[2] toluene,[1] xylenes,[1] ethylbenzene,[1] 1,2,4-trimethyl benzene,[1] MTBE,[3] ETBE,[4] and preliminary results on formaldehyde[5] have been published. This report outlines the final results of the carcinogenicity experiments performed at the CRC/RF on methyl alcohol and ethyl alcohol, two compounds proposed as oxygenated additives/alternative fuels.

Methyl alcohol is a clear, colorless, volatile, flammable liquid with a mild alcoholic odor. It is miscible with water and many organic solvents and forms many binary and zeotropic mixtures.[6] Methyl alcohol (CH_3OH) has a molecular weight of 32.04. Most methyl alcohol is produced by catalytic conversion of pressurized synthesis gas (hydrogen, carbon monoxide, and carbon dioxide) in the presence of metallic heterogeneous catalysts.[7,8] Since 1979, the world production of methyl alcohol has steadily increased and is now greater than 30 million tons per year.[9]

Approximately 70% of the methyl alcohol produced worldwide is used as feedstock and for the chemical synthesis of formaldehyde, MTBE, acetic acid, methyl metacrylate, and dimethyl terphtalate.[6] Methyl alcohol is a potential substitute for petroleum. It can be directly used in fuels as a replacement for gasoline and as an additive in gasoline and diesel fuel. Methyl alcohol is favored over conventional fuels because of its lower ozone-forming potential, lower emission of some pollutants, particularly benzene, polycyclic aromatic hydrocarbons, and sulfur compounds, and its low evaporative emissions. On the other hand, the possibility of higher formaldehyde emissions, its higher acute toxicity, and, at present, lower cost-efficiency favor conventional fuels.[10] Methyl alcohol is not usually used alone but is included in solvent mixtures.[11]

Methyl alcohol occurs naturally in humans, animals, and plants.[12–20] Natural emission sources of methyl alcohol include volcanic gasses, vegetation, microbes, and insects.[21–23] In 1994, the U.S. EPA reported that methyl alcohol was the most released chemical to the environment.[24]

Urban air levels of methyl alcohol of 10.5–131 $\mu g/m^3$ (8–100 ppb) have been reported.[23] If methyl alcohol, either 100% or in gasoline blends, becomes a major automotive fuel, emissions of methyl alcohol may arise as uncombusted fuel in exhaust or from evaporation during refueling.[25] Some methyl alcohol exposure concentrations have been postulated for various scenarios. For instance, in a public garage, if 100% of vehicles were fueled with methyl alcohol, air concentrations were projected to be 150 ppm. In most cases, exposure of the general public would be brief but repeated over time.[26]

Data on the occurrence of methyl alcohol in water, particularly drinking water, are limited. Methyl alcohol was identified in water in 24 locations in the United States during the period 1974–1976.[27] The frequency of occurrence was as follows: finished drinking water, 12; effluents from chemical plants, 6; effluent from sewage treatment, 4; effluent from paper and latex production, 1.

Dietary methyl alcohol can arise in large part from fresh fruit and vegetables. The methyl alcohol content of fresh and canned fruit juices, principally orange and grapefruit juices, varies considerably and may range from 1–43 mg/L to 12–640 mg/L, with an average of 140 mg/L.[28–30] Fermented and distilled beverages can contain high levels of methyl alcohol.[31] The sweetening agent aspartame hydrolyzes in the gastrointestinal tract to become free methyl alcohol.[25]

The primary routes of methyl alcohol exposure are inhalation and ingestion. Methyl alcohol distributes readily and uniformly in tissues in direct relation to their water content. In all mammalian species studied, methyl alcohol is metabolized in the liver by sequential oxidative steps to formaldehyde, formic acid, and CO_2.[6] However, there are wide differences in the route of formate oxidation among different species that determine the sensitivity to methyl alcohol.[6] Oral administration to rats showed an LD_{50} in the range of 7.4–13 g/kg bw.[32]

In vitro and *in vivo* mutagenicity studies on methyl alcohol, such as the Ames test, somatic mutation assay in CH-V79 cells, chromosome aberrations, sister chromatide exchanges, and the micronucleus test in mice, were all reported to be negative.[33,34] There are no reports of genotoxic, reproductive, or developmental effects in humans from methyl alcohol exposure.[6]

Although more than 30 million tons per year of methyl alcohol are produced, carcinogenicity studies are less than adequate. In two carcinogenicity studies, performed by the New Energy Development Organization (NEDO) in Japan, in which $B6C3F_1$ mice and Fisher 344 rats of both sexes were exposed by inhalation to 10, 100, and 1,000 ppm methyl alcohol for 20 hours/day for 18 and 24 months, respectively, no evidence of carcinogenicity was found in either species.[33, 34]

In a pilot study performed at the CRC/RF, Sprague-Dawley rats were exposed to 15 or 0 ppm methyl alcohol administered in drinking water for 104 weeks and then observed until spontaneous death. An increase in the incidence of total malignant tumors and leukemia (mostly in males) was observed.[5] The final results of this study are reported in this volume in the report on formaldehyde.

Ethyl alcohol (CH_3CH_2OH) has a molecular weight of 46.07. Ethyl alcohol may be produced from fermentation and petroleum ethylene synthesis.[35] Ethyl alcohol is one of the most widely produced, used, and diffused compounds at the global level.

The ethyl alcohol contained in alcoholic beverages (wine, beer, spirits, etc.), amounted to an annual world production in the mid-1980s of over 110 million hectoliters.[36] The total fuel ethyl alcohol production worldwide was around 200 million hectoliters in 2001. If all recently announced ethyl alcohol projects are implemented, the total worldwide ethyl alcohol fuel production could grow to 310 million hectoliters by 2006.[37]

Ethyl alcohol is used in many industrial processes, namely, the pharmaceutical, cosmetic, and synthetic rubber industries, as an antifreeze and as a solvent or processing agent for various purposes.[38] Ethyl alcohol is naturally present in alcoholic beverages as a consequence of fermentation of carbohydrates with yeast.[39]

Ethyl alcohol is proposed as a fuel, as an oxygenated additive of gasoline, and as a precursor of ETBE, a synthetic additive of gasoline, an alternative to/competing with the oxygenated additive most commonly used today, MTBE. The possibility that ethyl alcohol may be widely used as a fuel, as an oxygenated additive of gasoline, or in the production of ETBE increases the risk of its diffusion in surface and groundwater during production, storage, transportation, and use, as from the emission of exhaust from vehicles as an unburned product of gasoline combustion. The combustion of ethyl alcohol and its metabolic transformation in the body produce acetaldehyde which, according to data in the literature and the results of experiments conducted at CRC/RF reported in this volume, has a carcinogenic potential in laboratory animals.

Data on the air concentration of ethyl alcohol are few. The average ambient level in air in the city of Porto Alegre, Brazil, where vehicles run entirely on ethyl alcohol, is 0.023 mg/m^3 (12 ppb).[40] Atmospheric degradation is predicted to be rapid.[41] Ethyl alcohol rapidly degrades in groundwater and is not expected to persist beyond source areas. Ethyl alcohol in surface water is expected to undergo rapid biodegradation as long as it is not present in concentrations directly toxic to microorganisms.[41,42] Ethyl alcohol is not likely to accumulate or persist long in the environment.[39]

Although a great deal of information on the toxicological and health effects of ingested ethyl alcohol as a beverage is available, relatively little is known about its effects by inhalation exposure, which is relevant to its use as a fuel. About 60% of inhaled ethyl alcohol is retained by the body; the gastrointestinal tract completely absorbs ethyl alcohol in 2 to 6 hours; dermal absorption is insignificant.[43,44] From the portals of entry, ethyl alcohol distributes fairly uniformly throughout all tissues and organs, including the cerebrospinal fluid, brain, and, in pregnant human and laboratory animals, placenta and fetal tissues.[43,44] Ethyl alcohol is metabolized to acetaldehyde, which is then metabolized to CO_2 and water.[45]

Many epidemiologic studies have shown a positive relation between alcohol intake and excess tumors of the oral cavity, pharynx, larynx, esophagus, and liver.[46] These studies, however, do not show the total carcinogenic potential of ethyl alcohol. Ethyl alcohol has been the subject of 18 experimental studies on rodents, all of them considered less than adequate for the evaluation of carcinogenic potential.[46] Ethyl alcohol, given in association with nitrosamine or vinyl chloride, increases the carcinogenic potential of these compounds.[46] Because of the expansion in the use and diffusion of ethyl alcohol in the workplace and environment and the lack of adequate experimental data to evaluate its carcinogenicity, experiments on ethyl alcohol described herein were performed.

MATERIALS AND METHODS

Methyl alcohol is produced by J.T. Baker, Deventer, Holland, and has a purity grade of 99.8%. Ethyl alcohol was supplied by "CARLO ERBA" pharmaceutical products, Milan, Italy, in 1-liter glass bottles every 3 months; its purity was higher than 99.8%. The impurities were the following: acidity (acetic acid) $\leq 0.001\%$; alkalinity (NH_3) $\leq 0.0001\%$; carbonyl compound (CO) $\leq 0.0005\%$; isopropyl alcohol $\leq 0.003\%$; methanol (CH_3OH) $\leq 0.01\%$; residue on evaporation $\leq 0.001\%$; H_2O $\leq 0.2\%$.

During the experiment, both compounds were stored at a temperature of 4°C. Methyl alcohol was administered in drinking water at concentrations of 20,000, 5000, 500, or 0 ppm supplied *ad libitum* for 104 weeks to groups of male and female Sprague-Dawley rats beginning at 8 weeks of age. Control animals received tap water. The experiment on methyl alcohol started in April 1990 and ended after 153 weeks with the death of the last animal at 161 weeks of age.

Ethyl alcohol was administered in drinking water at concentrations of 10% or 0% supplied *ad libitum* to groups of male and female Sprague-Dawley rats; breeders and offspring were included in the experiment. Treatment started at 39 weeks of age (breeders), 7 days before mating, or from embryo life (offspring) and lasted until their spontaneous death. Control animals received tap water. The experiment on ethyl alcohol started in January 1986 and ended with the death of the last offspring at 179 weeks of age. Experiments were performed according to Good Laboratory Practices (GLP) and Standard Operating Procedure (SOP) of the CRC/RF.

Animals were identified by ear punch and housed in groups of five in makrolon cages ($41 \times 25 \times 15$ cm) with a stainless steel wire top; a shallow layer of white wood shavings served as bedding. The animals were kept in a single room at $23 \pm 2°C$ and 50–60% relative humidity.

Each morning, residual liquids from the previous day were removed, and the glass drinking bottles were washed and filled with fresh solution. Mean daily drinking water and feed consumption and weight were determined once weekly for the first 13 weeks and then every 2 weeks for 104 weeks. Thereafter, animals were weighed every 8 weeks until the end of the experiment. Status and behavior of animals were examined 3 times daily, and they were submitted to clinical examination for gross changes every 2 weeks.

Upon death, animals underwent systematic necropsy. Histopathology was routinely performed on the following organs and tissues: skin and subcutaneous tissue, brain, pituitary gland, Zymbal glands, parotid glands, submaxillary glands, Harderian glands, cranium (with oral and nasal cavities and external and internal ear ducts) (5 sections of head), tongue, thyroid and parathyroid, pharynx, larynx, thymus and mediastinal lymph nodes, trachea, lung and mainstem bronchi, heart, diaphragm, liver, spleen, pancreas, kidneys, adrenal glands, esophagus, stomach (fore and glandular), intestine (four levels), urinary bladder, prostate, gonads, interscapular fat pad,

TABLE 1. Long-term carcinogenicity bioassays on methyl alcohol administered with drinking water supplied *ad libitum* to male (M) and female (F) Sprague-Dawlwy rats

NUMBER AND PERCENTAGE OF MALE AND FEMALE SPRAGUE-DAWLEY RATS BEARING VARIOUS TYPES OF BENIGN AND MALIGNANT TUMORS[a]

Site		I: 20,000 ppm, v/v				II: 5,000 ppm, v/v				III: 500 ppm, v/v				IV: 0 (control)			
		Male		Female		Male		Female		Male		Female		Male		Female	
Histotype		No.	%	No.	%	No.	%	No.	%	No.	%	No.	%	No.	%	No.	%
Skin																	
Acanthoma		0	-	0	-	1	1.0	0	-	0	-	0	-	1	1.0	0	-
Dermatofibroma		0	-	0	-	1	1.0	0	-	1	1.0	0	-	1	1.0	0	-
Squamous cell carcinoma		0	-	0	-	0	-	0	-	1	1.0	0	-	0	-	0	-
Sebaceous adenocarcinoma		0	-	1	1.0	0	-	0	-	0	-	0	-	0	-	0	-
Subcutaneous tissue																	
Fibroma		0	-	0	-	0	-	0	-	0	-	0	-	1	1.0	0	-
Fibrolipoma		0	-	0	-	0	-	0	-	1	1.0	0	-	0	-	0	-
Fibroangioma		0	-	0	-	0	-	0	-	0	-	0	-	1	1.0	0	-
Fibrosarcoma		1	1.0	0	-	0	-	0	-	0	-	0	-	0	-	0	-
Liposarcoma		0	-	0	-	2	2.0	0	-	0	-	0	-	0	-	0	-
Rhabdomyosarcoma		0	-	0	-	0	-	0	-	0	-	0	-	1	1.0	0	-
Mammary glands																	
Fibroma and fibroadenoma		6(7)	6.0	40(72)	40.0	6	6.0	42(53)	42.0	6	6.0	47(63)	47.0	3	3.0	48(82)	48.0
Fibrolipoma		0	-	0	-	0	-	1	1.0	0	-	0	-	0	-	1	1.0
Lipoma		1	1.0	2	2.0	2	2.0	0	-	1	1.0	0	-	1	1.0	0	-
Adenocarcinoma		2	2.0	7(11)	7.0	2	2.0	5	5.0	0	-	8(10)	8.0	0	-	8(9)	8.0
Carcinosarcoma		0	-	2	2.0	2	2.0	0	-	0	-	0	-	0	-	0	-
Fibrosarcoma		0	-	1	1.0	1	1.0	0	-	0	-	0	-	1	1.0	0	-
Liposarcoma		0	-	0	-	2	2.0	1	1.0	3	3.0	1	1.0	2(3)	2.0	3	3.0
Angiosarcoma		0	-	0	-	0	-	1	1.0	1	1.0	0	-	0	-	0	-
Zymbal glands[b]																	
Sebaceous adenoma		0	-	2	2.0	0	-	0	-	0	-	0	-	0	-	0	-
Carcinoma		3	3.0	4	4.0	2	2.0	3	3.0	2	2.0	3	3.0	1	1.0	1	1.0
Ear ducts[b]																	
Carcinoma		24(29)	24.0	19(21)	19.0	17(20)	17.0	16(20)	16.0	13(16)	13.0	8(10)	8.0	9(10)	9.0	9(10)	9.0
Nasal cavities[b]																	
Carcinoma		0	-	1	1.0	0	-	0	-	1	1.0	1	1.0	1	1.0	0	-
Neuroblastoma		0	-	0	-	1	1.0	0	-	0	-	0	-	0	-	0	-

—continued

TABLE 1. *Continued*

NUMBER AND PERCENTAGE OF MALE AND FEMALE SPRAGUE-DAWLEY RATS BEARING VARIOUS TYPES OF BENIGN AND MALIGNANT TUMORS[a]

Site / Histotype	I: 20,000 ppm, v/v				II: 5,000 ppm, v/v				III: 500 ppm, v/v				IV: 0 (control)			
	Male		Female		Male		Female		Male		Female		Male		Female	
	No.	%	No.	%	No.	%	No.	%	No.	%	No.	%	No.	%	No.	%
Oral cavity, tongue and lips[b]																
Acanthoma	0	-	0	-	0	-	0	-	1	1.0	0	-	0	-	0	-
Carcinoma	1	1.0	1	1.0	2	2.0	0	-	0	-	0	-	1	1.0	3	3.0
Pharynx[b]																
Carcinoma	0	-	0	-	1	1.0	0	-	0	-	1	1.0	0	-	0	-
Larynx[b]																
Carcinoma	1	1.0	0	-	0	-	0	-	0	-	0	-	0	-	0	-
Trachea																
Polyp	0	-	1	1.0	0	-	0	-	0	-	0	-	0	-	0	-
Lung																
Adenoma	0	-	1	1.0	1	1.0	0	-	0	-	0	-	1	1.0	1	1.0
Adenocarcinoma	1	1.0	0	-	0	-	0	-	0	-	0	-	0	-	0	-
Stomach																
- Forestomach																
Acanthoma	1	1.0	2	2.0	2	2.0	2	2.0	2	2.0	1	1.0	0	-	1	1.0
Squamous cell carcinoma	0	-	1	1.0	0	-	0	-	0	-	1	1.0	0	-	0	-
Intestine																
Adenocarcinoma	0	-	0	-	0	-	1	1.0	0	-	0	-	0	-	0	-
Leiomyosarcoma	0	-	1	1.0	0	-	0	-	0	-	0	-	0	-	0	-
Liver																
Hepatocarcinoma	3	3.0	0	-	2	2.0	0	-	2	2.0	1	1.0	0	-	0	-
Angiosarcoma	0	-	0	-	0	-	0	-	0	-	1	1.0	0	-	0	-
Pancreas																
Exocrine adenoma	5	5.0	1	1.0	0	-	0	-	0	-	1	1.0	1	1.0	1	1.0
Islet cell adenoma	9	9.0	5	5.0	13	13.0	4	4.0	16	16.0	8	8.0	7	7.0	7	7.0
Islet cell adenocarcinoma	0	-	0	-	2	2.0	1	1.0	0	-	0	-	0	-	0	-
Kidneys																
Adenoma	0	-	0	-	0	-	0	-	1	1.0	0	-	0	-	1	1.0
Lipoma	0	-	1	1.0	0	-	1	1.0	0	-	0	-	0	-	0	-
Liposarcoma	0	-	0	-	1	1.0	0	-	0	-	0	-	0	-	0	-
Pelvis																
Transitional cell carcinoma	0	-	0	-	1	1.0	0	-	0	-	0	-	0	-	0	-
Bladder																
Leiomyoma	0	-	0	-	0	-	0	-	0	-	1	1.0	0	-	0	-
Prostate																
Carcinoma	1	1.0			0	-			0	-			1	1.0		

—continued

TABLE 1. *Continued*

NUMBER AND PERCENTAGE OF MALE AND FEMALE SPRAGUE-DAWLEY RATS BEARING VARIOUS TYPES OF BENIGN AND MALIGNANT TUMORS[a]

Site Histotype	I: 20,000 ppm, v/v				II: 5,000 ppm, v/v				III: 500 ppm, v/v				IV: 0 (control)			
	Male		Female		Male		Female		Male		Female		Male		Female	
	No.	%	No.	%	No.	%	No.	%	No.	%	No.	%	No.	%	No.	%
Seminal vesicles																
Adenoma	1	1.0			0	-			0	-			0	-		
Adenocarcinoma	0	-			0	-			1	1.0			0	-		
Testes																
Interstitial cell adenoma	17(24)	17.0			13(17)	13.0			9	9.0			12	12.0		
Ovaries																
Theca cell tumor			1	1.0			0	-			0	-			0	-
Granulosa cell tumor			0	-			0	-			1	1.0			0	-
Sertoli cell tumor			1	1.0			2	2.0			3	3.0			0	-
Malignant granulosa cell tumor			0	-			1	1.0			0	-			0	-
Uterus																
Polyp			5	5.0			14	14.0			20	20.0			12	12.0
Granular cell tumor (Abrikosoff's tumor)			0	-			0	-			1	1.0			0	-
Leiomyoma			0	-			0	-			1	1.0			4	4.0
Fibroangioma			0	-			0	-			0	-			1	1.0
Fibromyxoma			1	1.0			0	-			0	-			0	-
Adenocarcinoma			2	2.0			3	3.0			3	3.0			5	5.0
Leiomyosarcoma			0	-			1	1.0			0	-			0	-
Malignant Schwannoma			1	1.0			0	-			1	1.0			0	-
Uterus & Vagina																
Adenocarcinoma			0	-			2	2.0			0	-			0	-
Malignant Schwannoma			5	5.0			3	3.0			3	3.0			3	3.0
Vagina																
Leiomyoma			0	-			0	-			0	-			1	1.0
Malignant Schwannoma			1	1.0			0	-			0	-			0	-
Peritoneum																
Lipoma	0	-	0	-	0	-	0	-	0	-	1	1.0	0	-	1	1.0
Liposarcoma	0	-	0	-	1	1.0	0	-	1	1.0	0	-	1	1.0	0	-
Mesothelioma	1	1.0	1	1.0	1	1.0	0	-	0	-	0	-	0	-	0	-

—continued

TABLE 1. *Continued*

NUMBER AND PERCENTAGE OF MALE AND FEMALE SPRAGUE-DAWLEY RATS BEARING VARIOUS TYPES OF BENIGN AND MALIGNANT TUMORS[a]

Site		Groups														
		I: 20,000 ppm, v/v				II: 5,000 ppm, v/v				III: 500 ppm, v/v				IV: 0 (control)			
		Male		Female		Male		Female		Male		Female		Male		Female	
Histotype		No.	%	No.	%	No.	%	No.	%	No.	%	No.	%	No.	%	No.	%
Pituitary gland																	
Adenoma		34	34.0	54	54.0	36	36.0	49	49.0	30	30.0	46	46.0	34	34.0	59	59.0
Adenocarcinoma		0	-	1	1.0	0	-	0	-	0	-	1	1.0	0	-	0	-
Thyroid gland																	
Follicular adenoma		1	1.0	1	1.0	0	-	0	-	0	-	0	-	1	1.0	1	1.0
C-cell adenoma		0	-	2	2.0	1	1.0	1	1.0	0	-	2	2.0	1	1.0	3	3.0
Follicular carcinoma		1	1.0	0	-	0	-	1	1.0	1	1.0	0	-	0	-	0	-
C-cell carcinoma		0	-	0	-	0	-	0	-	1	1.0	0	-	0	-	0	-
Parathyroid glands																	
Adenoma		1	1.0	0	-	0	-	0	-	1	1.0	0	-	0	-	0	-
Adrenal glands																	
Cortical adenoma		1	1.0	3	3.0	1	1.0	3	3.0	1	1.0	3	3.0	0	-	7	7.0
Pheochromocytoma		19(26)	19.0	10(13)	10.0	32(44)	32.0	8(10)	8.0	38(51)	38.0	15(20)	15.0	26(38)	26.0	21(26)	21.0
Cortical adenocarcinoma		0	-	1	1.0	0	-	0	-	0	-	2	2.0	0	-	2	2.0
Pheochromoblastoma		3	3.0	0	-	3	3.0	1	1.0	1	1.0	1(2)	1.0	6	6.0	4(5)	4.0
Central nervous system																	
- Brain																	
Oligodendroglioma		0	-	0	-	1	1.0	0	-	2	2.0	1	1.0	0	-	3	3.0
Ependymoma		0	-	0	-	0	-	0	-	0	-	0	-	1	1.0	0	-
- Meninges																	
Benign meningioma		1	1.0	0	-	0	-	1	1.0	2	2.0	0	-	0	-	0	-
Malignant meningioma		0	-	0	-	0	-	0	-	1	1.0	0	-	0	-	0	-
Peripheral nervous system																	
- Major peripheral nerves																	
Malignant Schwannoma		1	1.0	1	1.0	0	-	0	-	1	1.0	0	-	2	2.0	0	-
- Ganglia																	
Benign Schwannoma		0	-	0	-	0	-	0	-	0	-	1	1.0	0	-	0	-
Pheochromocytoma		0	-	0	-	0	-	0	-	0	-	1	1.0	0	-	0	-
Bones[c]																	
- Head																	
Osteoma		0	-	0	-	1	1.0	0	-	0	-	0	-	0	-	0	-
Osteosarcoma		11	11.0	6	6.0	13	13.0	3	3.0	6	6.0	4	4.0	6	6.0	1	1.0
- Other sites																	
Osteosarcoma		1	1.0	0	-	0	-	0	-	1	1.0	1	1.0	2	2.0	0	-

—continued

TABLE 1. *Continued*

NUMBER AND PERCENTAGE OF MALE AND FEMALE SPRAGUE-DAWLEY RATS BEARING VARIOUS TYPES OF BENIGN AND MALIGNANT TUMORS[a]

Site Histotype	I: 20,000 ppm, v/v				II: 5,000 ppm, v/v				III: 500 ppm, v/v				IV: 0 (control)			
	Male		Female		Male		Female		Male		Female		Male		Female	
	No.	%	No.	%	No.	%	No.	%	No.	%	No.	%	No.	%	No.	%
Soft tissues																
Lipoma	1	1.0	0	-	0	-	0	-	0	-	0	-	0	-	0	-
Fibrosarcoma	0	-	0	-	0	-	0	-	0	-	0	-	1	1.0	0	-
Liposarcoma	0	-	1	1.0	0	-	0	-	1	1.0	0	-	0	-	0	-
Pericytosarcoma	0	-	1	1.0	0	-	0	-	0	-	0	-	0	-	0	-
Heart																
Myxoma	0	-	0	-	0	-	0	-	1	1.0	1	1.0	0	-	0	-
Malignant Schwannoma	2	2.0	0	-	1	1.0	0	-	0	-	0	-	1	1.0	1	1.0
Thymus																
Benign thymoma[d]	1	1.0	1	1.0	0	-	0	-	0	-	1	1.0	0	-	2	2.0
Fibroangioma	0	-	0	-	0	-	0	-	0	-	0	-	1	1.0	0	-
Malignant thymoma[d]	0	-	1	1.0	0	-	0	-	0	-	0	-	0	-	1	1.0
Spleen																
Osteoma	0	-	0	-	1	1.0	0	-	0	-	0	-	0	-	0	-
Fibroangioma	0	-	1	1.0	0	-	0	-	0	-	0	-	0	-	1	1.0
Leiomyosarcoma	0	-	0	-	1	1.0	0	-	0	-	0	-	0	-	0	-
Angiosarcoma	2	2.0	0	-	1	1.0	1	1.0	0	-	1	1.0	0	-	0	-
Mediastinal and mesenteric lymph nodes																
Fibroangioma	0	-	0	-	1	1.0	0	-	1	1.0	0	-	1	1.0	0	-
Hemolymphoreticular tissues[e, f]																
Lymphomas and leukemias	40	40.0	28	28.0	36	36.0	24(25)	24.0	35	35.0	24	24.0	28	28.0	13	13.0
Unknown																
-Abdominal lymph nodes metastases																
Adenocarcinoma	0	-	1	1.0	0	-	0	-	0	-	0	-	0	-	0	-

[a] Between brackets the number of tumors (one animal can bear more than one tumor)
[b] See table 3
[c] See table 4
[d] In 96% of cases the tumor itself is composed of a mixture in varying proportions of epithelial cells and lymphocytes. In the remaining 4%, only epithelial cells are present. We consider that a tumor composed exclusively of lymphocytes should not be classified as a thymoma but as a lymphoma involving the thymus.
[e] Including thymus, spleen, mediastinal and mesenteric lymph nodes
[f] See table 5

TABLE 2. Long-term carcinogenicity bioassays on methyl alcohol administered with drinking water supplied *ad libitum* to male (M) and female (F) Sprague-Dawley rats

TOTAL MALIGNANT TUMORS

Group No.	Concentration (ppm, v/v)	Animals		Malignant tumors			
				Tumor-bearing animals		Tumors	
		Sex	No.	No.	%	No.	Per 100 animals
I	20,000	M	100	70	70.0 ***♦♦	104	104.0 ***
		F	100	63	63.0 ***♦♦	95	95.0 ***
		M+F	200	133	66.5	199	99.5
II	5,000	M	100	64	64.0 ♦♦	97	97.0 ***
		F	100	48	48.0 ♦♦	73	73.0
		M+F	200	112	56.0	170	85.0
III	500	M	100	55	55.0 ♦♦	78	78.0
		F	100	48	48.0 ♦♦	72	72.0
		M+F	200	103	51.5	150	75.0
IV	0	M	100	50	50.0	66	66.0
		F	100	43	43.0	60	60.0
		M+F	200	93	46.5	126	63.0

*** $p<0.01$ using χ^2 test
♦♦ $p<0.01$ using Cochrane-Armitage test for dose-response relationship

subcutaneous and mesenteric lymph nodes, and any other organs or tissues with pathologic lesions. All slides were examined microscopically by the same group of pathologists; a senior pathologist reviewed all tumors and any other lesion of oncologic interest. All pathologists followed the same criteria of histopathological evaluation and classification. Multiple tumors of different type and site, of different type in the same site, of the same type in bilateral organs, of the same type in the skin, in the subcutaneous tissue, and in mammary glands, or at distant sites of diffuse tissue (i.e., bones and skeletal muscle) were plotted as single/independent tumors. Multiple tumors of the same type in the same tissue and organ (including those of the bilateral organs) were plotted only once.

Statistical analysis was performed using the χ^2 test to evaluate differences in tumor incidence between treated and control groups. The Cochrane Armitage test was used to evaluate dose-response relations.

RESULTS

Methyl Alcohol

There were no noteworthy changes in beverage or feed consumption apart from a decrease in water consumption in females treated with the highest dose between 8 and 56 weeks of age. A slight increase was observed in the body weight of males

TABLE 3. Long-term carcinogenicity bioassays on methyl alcohol administered with drinking water supplied *ad libitum* to male (M) and female (F) Sprague-Dawley rats

CARCINOMAS OF THE HEAD AND NECK

Group No.	Concentration (ppm, v/v)	Animals Sex	Animals No.	Zymbal glands No.	Zymbal glands %	Ear ducts[a] No.	Ear ducts[a] %	Animals with carcinomas Nasal cavities No.	Nasal cavities %	Oral cavity, tongue and lips No.	Oral cavity, tongue and lips %	Pharynx No.	Pharynx %	Larynx No.	Larynx %	Total No.	Total %
I	20,000	M	100	3	3.0	24(5) **24.0** ***◆◆		0	-	1	1.0	0	-	1	1.0	29 **29.0** ***◆◆	
		F	100	4	4.0	19(2) **19.0** ◆◆		1	1.0	1	1.0	0	-	0	-	25 **25.0** ***◆◆	
		M+F	200	7	3.5	43	21.5	1	0.5	2	1.0	0	-	1	0.5	54	27.0
II	5,000	M	100	2	2.0	17(3) **17.0** ◆◆		0	-	2	2.0	1	1.0	0	-	22 **22.0** ◆◆	
		F	100	3	3.0	16(4) **16.0** ◆◆		0	-	0	-	0	-	0	-	19 **19.0** ◆◆	
		M+F	200	5	2.5	33	16.5	0	-	2	1.0	1	0.5	0	-	41	20.5
III	500	M	100	2	2.0	13(3) **13.0** ◆◆		1	1.0	0	-	0	-	0	-	16 **16.0** ◆◆	
		F	100	3	3.0	8(2) 8.0 ◆◆		1	1.0	0	-	1	1.0	0	-	13 **13.0** ◆◆	
		M+F	200	5	2.5	21	10.5	2	1.0	0	-	1	0.5	0	-	29	14.5
IV	0	M	100	1	1.0	9(1) 9.0		1	1.0	1	1.0	0	-	0	-	12	12.0
		F	100	1	1.0	9(1) 9.0		0	-	3	3.0	0	-	0	-	13	13.0
		M+F	200	2	1.0	18	9.0	1	0.5	4	2.0	0	-	0	-	25	12.5

[a] Between brackets the number of animals with bilateral tumors

* $p<0.05$ using χ^2 test
*** $p<0.01$ using χ^2 test
◆ $p<0.05$ using Cochrane-Armitage test for dose-response relationship
◆◆ $p<0.01$ using Cochrane-Armitage test for dose-response relationship

TABLE 4. Long-term carcinogenicity bioassays on methyl alcohol administered with drinking water supplied *ad libitum* to male (M) and female (F) Sprague-Dawley rats

OSTEOSARCOMAS OF THE HEAD AND OTHER SITES

Group No.	Concentration (ppm, v/v)	Animals Sex	Animals No.	Animals with osteosarcomas Head No.	Animals with osteosarcomas Head %	Animals with osteosarcomas Other sites No.	Animals with osteosarcomas Other sites %	Animals with osteosarcomas Total No.	Animals with osteosarcomas Total %
I	20,000	M	100	11	11.0	1	1.0	12	12.0
		F	100	6	6.0	0	-	6	6.0
		M+F	200	17	8.5	1	0.5	18	9.0
II	5,000	M	100	13	13.0	0	-	13	13.0
		F	100	3	3.0	0	-	3	3.0
		M+F	200	16	8.0	0	-	16	8.0
III	500	M	100	6	6.0	1	1.0	7	7.0
		F	100	4	4.0	1	1.0	5	5.0
		M+F	200	10	5.0	2	1.0	12	6.0
IV	0	M	100	6	6.0	2	2.0	8	8.0
		F	100	1	1.0	0	-	1	1.0
		M+F	200	7	3.5	2	1.0	9	4.5

and, to a lesser extent, of females treated with the highest dose. No substantial changes in survival or behavioral changes were observed among the groups. No treatment-related nononcologic pathological changes were detected by gross inspection or histopathological examination.

The occurrence of benign and malignant tumors is shown in TABLE 1. Differences observed between treated and control animals were: (1) a dose-related increase of total malignant tumors in males and females of treated groups (TABLE 2); (2) a dose-related increase of carcinomas of the head and neck, mainly in the ear ducts, in males of treated groups and in females treated with 20,000 and 5,000 ppm (TABLE 3); (3) a statistically significant increase ($P < 0.01$) of testicular interstitial cell hyperplasias and adenomas in the group treated with the highest dose; (4) an increase in sarcomas of the uterus at the highest dose; (5) a dose-related increase in osteosarcomas of the head in males and females of the treated groups (TABLE 4); and (6) a dose-related increase in hemolymphoreticular neoplasias in males and females of the treated groups (TABLE 5).

Ethyl Alcohol

The intake of beverages and feed was lower in treated than control animals. No significant differences in body weight or behavior were observed between treated and control animals. No significant differences occurred in survival between treated and control animals, with the exception of a lower survival of treated female offspring in the period from 104 to 152 weeks of age. No treatment-related nononco-

TABLE 5. Long-term carcinogenicity bioassays on methyl alcohol administered with drinking water supplied *ad libitum* to male (M) and female (F) Sprague-Dawley rats

HEMOLYMPHORETICULAR NEOPLASIAS AND THEIR DISTRIBUTION BY HISTOCYTOTYPE

Group No.	Concentration (ppm, v/v)	Animals Sex	Animals No.	Total[a] No.	Total[a] %	Lymphoblastic lymphoma[b] No.	Lymphoblastic lymphoma[b] %	Lymphoblastic leukemia No.	Lymphoblastic leukemia %	Lymphocytic lymphoma[b] No.	Lymphocytic lymphoma[b] %	Lymphoimmuno-blastic lymphoma[b] No.	Lymphoimmuno-blastic lymphoma[b] %	Histiocytic sarcoma monocytic leukemia[b] No.	Histiocytic sarcoma monocytic leukemia[b] %	Myeloid leukemia[b] No.	Myeloid leukemia[b] %
I	20,000	M	100	40	**40.0**	1	2.5	0	-	0	-	37	92.5	1	2.5	1	2.5
		F	100	28	**28.0** ♦	1	3.6	0	-	0	-	21	75.0	3	10.7	3	10.7
		M+F	200	68	**34.0**	2	2.9	0	-	0	-	58	85.3	4	5.9	4	5.9
II	5,000	M	100	36	**36.0**	1	2.8	0	-	0	-	28	77.8	1	2.8	6	16.7
		F	100	24[c]	**24.0** ♦	1	4.2	0	-	0	-	19	79.2	2	8.3	3	12.5
		M+F	200	60[c]	**30.0**	2	3.3	0	-	0	-	47	78.3	3	5.0	9	15.0
III	500	M	100	35	**35.0**	3	8.6	0	-	0	-	24	68.6	4	11.4	4	11.4
		F	100	24	**24.0** ♦	1	4.2	0	-	1	4.2	17	70.8	2	8.3	3	12.5
		M+F	200	59	**29.5**	4	6.8	0	-	1	1.7	41	69.5	6	10.2	7	11.9
IV	0	M	100	28	28.0	1	3.6	0	-	0	-	16	57.1	3	10.7	8	28.6
		F	100	13	13.0	0	-	0	-	0	-	9	69.2	1	7.7	3	23.1
		M+F	200	41	20.5	1	2.4	0	-	0	-	25	61.0	4	9.8	11	26.8

[a] Percentages refer to the number of animals at start
[b] Percentages refer to the number of animals bearing hemolymphoreticular neoplasias
[c] One animal bore a lymphoimmunoblastic lymphoma and myeloid leukemia
* $p < 0.05$ using χ^2 test
♦ $p < 0.05$ using Cochrane-Armitage test for dose-response relationship

TABLE 6. Long-term carcinogenicity bioassays on ethyl alcohol administered with drinking water supplied *ad libitum* to male (M) and female (F) Sprague-Dawley rats

NUMBER AND PERCENTAGE OF MALE AND FEMALE SPRAGUE-DAWLEY RATS BEARING VARIOUS TYPES OF BENIGN AND MALIGNANT TUMORS[a]

Site Histotype	I: Ethyl alcohol 10% (Breeders)				II: Drinking water (Breeders)				III: Ethyl alcohol 10% (Offspring)				IV: Drinking water (Offspring)			
	Male		Female		Male		Female		Male		Female		Male		Female	
	No.	%	No.	%	No.	%	No.	%	No.	%	No.	%	No.	%	No.	%
Skin																
Acanthoma	0	-	0	-	1	0.9	0	-	1	3.3	0	-	0	-	0	-
Squamous cell carcinoma	0	-	0	-	1	0.9	0	-	0	-	0	-	0	-	0	-
Basocellular carcinoma	0	-	0	-	0	-	1	0.9	0	-	0	-	0	-	0	-
Subcutaneous tissue																
Fibroma	0	-	0	-	2	1.8	0	-	0	-	0	-	0	-	0	-
Rhabdomyosarcoma	0	-	0	-	0	-	0	-	0	-	0	-	0	-	1(2)	1.8
Pericytosarcoma	0	-	0	-	1	0.9	0	-	0	-	0	-	0	-	0	-
Mammary glands																
Fibroma and fibroadenoma	7(9)	6.4	30(43)	27.3	3	2.7	42(63)	38.2	0	-	21(33)	53.8	3	6.1	23(35)	41.8
Fibrolipoma	0	-	0	-	0	-	0	-	1	3.3	0	-	0	-	0	-
Lipoma	0	-	0	-	0	-	1	0.9	0	-	0	-	1	2.0	0	-
Adenocarcinoma	1	0.9	11(14)	10.0	0	-	8	7.3	0	-	7(10)	17.9	0	-	5(6)	9.1
Fibrosarcoma	0	-	1	0.9	0	-	0	-	0	-	0	-	0	-	1	1.8
Leiomyosarcoma	0	-	0	-	0	-	0	-	0	-	0	-	0	-	1	1.8
Liposarcoma	0	-	0	-	0	-	0	-	0	-	0	-	0	-	1	1.8
Angiosarcoma	0	-	0	-	1	0.9	0	-	0	-	0	-	0	-	0	-
Harderian glands																
Adenoma	0	-	1	0.9	0	-	0	-	0	-	0	-	0	-	0	-
Zymbal glands[b]																
Sebaceous adenoma	0	-	0	-	0	-	0	-	0	-	0	-	0	-	1	1.8
Carcinoma	3	2.7	6(7)	5.5	2	1.8	2(3)	1.8	0	-	1	2.6	0	-	3	5.5
Ear ducts[b]																
Carcinoma	2	1.8	7(9)	6.4	6	5.5	9	8.2	5	16.7	6	15.4	3(4)	6.1	5(6)	9.1
Nasal cavities[b]																
Carcinoma	3	2.7	1	0.9	2	1.8	1	0.9	0	-	0	-	0	-	1	1.8
Olfactory neuroblastoma	1	0.9	2	1.8	2	1.8	2	1.8	1	3.3	0	-	0	-	0	-

—continued

TABLE 6. Continued

NUMBER AND PERCENTAGE OF MALE AND FEMALE SPRAGUE-DAWLEY RATS BEARING VARIOUS TYPES OF BENIGN AND MALIGNANT TUMORS[a]

Site Histotype	I: Ethyl alcohol 10% (Breeders)				II: Drinking water (Breeders)				III: Ethyl alcohol 10% (Offspring)				IV: Drinking water (Offspring)			
	Male		Female		Male		Female		Male		Female		Male		Female	
	No.	%	No.	%	No.	%	No.	%	No.	%	No.	%	No.	%	No.	%
Oral cavity and lips[b]																
Acanthoma	0	-	0	-	0	-	1	0.9	0	-	1	2.6	0	-	0	-
Carcinoma	11	10.0	12	10.9	3	2.7	2	1.8	9	30.0	15	38.5	2	4.1	3	5.5
Tongue[b]																
Carcinoma	4	3.6	0	-	0	-	0	-	1	3.3	1	2.6	0	-	0	-
Pharynx[b]																
Carcinoma	0	-	0	-	1	0.9	1	0.9	0	-	0	-	0	-	0	-
Larynx[b]																
Carcinoma	0	-	1	0.9	0	-	0	-	0	-	0	-	0	-	1	1.8
Trachea																
Carcinoma	0	-	1	0.9	0	-	0	-	0	-	0	-	0	-	0	-
Lung																
Fibroma	0	-	1	0.9	0	-	0	-	0	-	0	-	0	-	0	-
Fibroangioma	0	-	0	-	0	-	0	-	1	3.3	0	-	0	-	0	-
Squamous cell carcinoma	0	-	1	0.9	0	-	0	-	0	-	0	-	0	-	0	-
Adenocarcinoma	0	-	0	-	0	-	1	0.9	0	-	0	-	0	-	0	-
Leiomyosarcoma	0	-	0	-	0	-	0	-	0	-	0	-	0	-	1	1.8
Esophagus																
Carcinoma	0	-	1	0.9	0	-	0	-	0	-	0	-	0	-	0	-
Stomach																
- Forestomach																
Acanthoma	8	7.3	3	2.7	1	0.9	1	0.9	0	-	2	5.1	1	2.0	2	3.6
Carcinoma	2	1.8	3	2.7	0	-	0	-	1	3.3	1	2.6	0	-	0	-
- Glandular stomach																
Adenocarcinoma	0	-	0	-	1	0.9	1	0.9	0	-	0	-	0	-	0	-
Liver																
Cholangioma	0	-	2	1.8	0	-	0	-	0	-	0	-	0	-	1	1.8
Hepatocarcinoma	1	0.9	2	1.8	0	-	0	-	1	3.3	2	5.1	5	10.2	1	1.8
Angiosarcoma	0	-	2	1.8	0	-	0	-	0	-	0	-	0	-	1	1.8
Pancreas																
Exocrine adenoma	2	1.8	2	1.8	1	0.9	1	0.9	0	-	0	-	0	-	0	-
Islet cell adenoma	6	5.5	1	0.9	4	3.6	2	1.8	1	3.3	2	5.1	5	10.2	1	1.8
Islet cell carcinoma	0	-	1	0.9	1	0.9	0	-	0	-	1	2.6	0	-	0	-
Kidneys																
Adenocarcinoma	1	0.9	1	0.9	0	-	1	0.9	0	-	0	-	0	-	1	1.8
Nephroblastoma	0	-	1	0.9	0	-	0	-	0	-	0	-	0	-	0	-
Liposarcoma	0	-	0	-	0	-	0	-	0	-	1	2.6	0	-	0	-
Angiosarcoma	1	0.9	0	-	0	-	0	-	0	-	0	-	0	-	0	-

—continued

TABLE 6. Continued

NUMBER AND PERCENTAGE OF MALE AND FEMALE SPRAGUE-DAWLEY RATS BEARING VARIOUS TYPES OF BENIGN AND MALIGNANT TUMORS[a]

Site / Histotype	I: Ethyl alcohol 10% (Breeders)				II: Drinking water (Breeders)				III: Ethyl alcohol 10% (Offspring)				IV: Drinking water (Offspring)			
	Male		Female		Male		Female		Male		Female		Male		Female	
	No.	%	No.	%	No.	%	No.	%	No.	%	No.	%	No.	%	No.	%
Seminal vesicles																
Adenocarcinoma	1	0.9			0	-			1	3.3			0	-		
Prostate																
Adenocarcinoma	0	-			1	0.9			0	-			0	-		
Testes																
Interstitial cell adenoma	23(34)	20.9			9(12)	8.2			4(6)	13.3			4	8.2		
Ovaries																
Cystadenoma			1	0.9			4	3.6			0	-			0	-
Granulosa cell tumor			0	-			1	0.9			0	-			0	-
Theca cell tumor			1	0.9			0	-			0	-			0	-
Sertoli cell tumor			2(3)	1.8			1(2)	0.9			3(6)	7.7			0	-
Fibroangioma			1	0.9			0	-			0	-			0	-
Adenocarcinoma			2(4)	1.8			0	-			0	-			0	-
Uterus																
Polyp			10	9.1			8	7.3			6	15.4			7	12.7
Granular cell tumor (Abrikosoff's tumor)			0	-			0	-			1	2.6			0	-
Leiomyoma			0	-			2	1.8			0	-			0	-
Fibroangioma			1	0.9			0	-			1	2.6			1	1.8
Squamous cell carcinoma			0	-			4	3.6			0	-			0	-
Adenocarcinoma			9	8.2			2	1.8			8	20.5			6	10.9
Chorionepithelioma			1	0.9			0	-			0	-			0	-
Fibrosarcoma			1	0.9			0	-			1	2.6			0	-
Leiomyosarcoma			0	-			0	-			0	-			0	-
Angiosarcoma			1	0.9			0	-			0	-			0	-
Uterus & Vagina																
Malignant Schwannoma			1	0.9			1	0.9			0	-			1	1.8
Peritoneum																
Mesothelioma	1	0.9	2	1.8	0	-	0	-	0	-	0	-	1	2.0	0	-

—continued

TABLE 6. Continued

NUMBER AND PERCENTAGE OF MALE AND FEMALE SPRAGUE-DAWLEY RATS BEARING VARIOUS TYPES OF BENIGN AND MALIGNANT TUMORS[a]

Site / Histotype	I: Ethyl alcohol 10% (Breeders)				II: Drinking water (Breeders)				III: Ethyl alcohol 10% (Offspring)				IV: Drinking water (Offspring)			
	Male		Female		Male		Female		Male		Female		Male		Female	
	No.	%	No.	%	No.	%	No.	%	No.	%	No.	%	No.	%	No.	%
Pituitary gland																
Adenoma	20	18.2	40	36.4	15	13.6	38	34.5	5	16.7	23	59.0	8	16.3	29	52.7
Thyroid gland																
C-cell adenoma	3	2.7	2	1.8	2	1.8	3	2.7	1	3.3	1	2.6	2	4.1	5	9.1
Follicular carcinoma	0	-	1	0.9	0	-	1	0.9	0	-	1	2.6	0	-	1	1.8
C-cell carcinoma	0	-	1	0.9	0	-	0	-	0	-	1	2.6	0	-	0	-
Adrenal glands																
Cortical adenoma	1(2)	0.9	5	4.5	0	-	2	1.8	0	-	4(6)	10.3	0	-	4	7.3
Pheochromocytoma	49(75)	44.5	45(60)	40.9	44(67)	40.0	29(43)	26.4	17(25)	56.7	18(25)	46.2	26(43)	53.1	15(16)	27.3
Cortical adenocarcinoma	0	-	1	0.9	0	-	1	2.7	0	-	2	5.1	0	-	1	1.8
Pheochromoblastoma	9(10)	8.2	4(5)	3.6	3(4)	2.7	3(4)	2.7	4(5)	13.3	0	-	1(2)	2.0	1	1.8
Sympatoblastoma	1	0.9	0	-	0	-	0	-	0	-	0	-	0	-	0	-
Central nervous system																
-Brain																
Oligodendroglioma	4	3.6	1	0.9	5	4.5	1	0.9	0	-	0	-	2	4.1	0	-
- Meninges																
Benign meningioma	0	-	0	-	0	-	0	-	0	-	1	2.6	0	-	0	-
Peripheral nervous system nerves																
- Major peripheral nerves																
Benign Schwannoma	0	-	0	-	1	0.9	0	-	0	-	0	-	1	2.0	0	-
- Ganglia																
Ganglioneuroma	0	-	0	-	0	-	0	-	0	-	0	-	1	2.0	0	-
- Paraganglia																
Pheochromocytoma	0	-	0	-	0	-	0	-	0	-	0	-	1	2.0	0	-
Skeletal muscle																
Rhabdomyosarcoma	0	-	1	0.9	0	-	0	-	0	-	0	-	0	-	0	-

—*continued*

TABLE 6. Continued

NUMBER AND PERCENTAGE OF MALE AND FEMALE SPRAGUE-DAWLEY RATS BEARING VARIOUS TYPES OF BENIGN AND MALIGNANT TUMORS[a]

Site								Groups								
	I: Ethyl alcohol 10% (Breeders)				II: Drinking water (Breeders)				III: Ethyl alcohol 10% (Offspring)				IV: Drinking water (Offspring)			
	Male		Female		Male		Female		Male		Female		Male		Female	
Histotype	No.	%	No.	%	No.	%	No.	%	No.	%	No.	%	No.	%	No.	%
Bones																
- Head																
Osteosarcoma	8	7.3	6	5.5	0	-	4	3.6	6	20.0	4	10.3	4	8.2	3	5.5
- Other sites																
Osteosarcoma	4	3.6	0	-	1	0.9	0	-	0	-	1	2.6	0	-	0	-
Chondrosarcoma	0	-	0	-	0	-	0	-	1	3.3	0	-	0	-	0	-
Soft tissues																
Liposarcoma	0	-	0	-	0	-	0	-	0	-	0	-	0	-	0	-
Heart																
Malignant Schwannoma	0	-	2	1.8	1	0.9	1	0.9	0	-	0	-	0	-	0	-
Thymus																
Malignant thymoma[c]	0	-	0	-	0	-	1	0.9	0	-	0	-	0	-	0	-
Spleen																
Fibroma	0	-	0	-	1	0.9	0	-	0	-	0	-	0	-	0	-
Fibroangioma	2	1.8	0	-	1	0.9	0	-	0	-	1	2.6	1	2.0	0	-
Mesenteric lymph nodes																
Fibroangioma	1	0.9	0	-	1	0.9	0	-	2	6.7	1	2.6	1	2.0	1	1.8
Hemolymphoreticular tissue[d]																
Lymphomas and leukemias	39	35.5	46	41.8	35	31.8	17	15.5	9	30.0	8	20.5	8(10)	20.4	11	20.0

[a] Between brackets the number of tumors (one animal can bear more than one tumor)
[b] See table 8
[c] In 96% of cases the tumor itself is composed of a mixture in varying proportions of epithelial cells and lymphocytes. In the remaining 4%, only epithelial cells are present. We consider that a tumor composed exclusively of lymphocytes should not be classified as a thymoma but as a lymphoma involving the thymus.
[d] Including thymus, spleen and mesenteric lymph nodes

TABLE 7. Long-term carcinogenicity bioassays on ethyl alcohol administered with drinking water supplied *ad libitum* to male (M) and female (F) Sprague-Dawley rats

TOTAL MALIGNANT TUMORS

Group No.	Concentration (%, v/v)	Animals			Malignant tumors				
		Age	Sex	No.	Tumor-bearing animals		Tumors		
					No.	%	No.	Per 100 animals	
I	10	39 weeks (breeders)	M	110	66	60.0	98	89.1 **	
			F	110	79	71.8 **	143	130.0 **	
			M+F	220	145	65.9	241	109.5	
II	0	39 weeks (breeders)	M	110	51	46.4	68	61.8	
			F	110	48	43.6	67	60.9	
			M+F	220	99	45.0	135	61.4	
III	10	Embryos (offspring)	M	30	23	76.7 *	41	136.7 **	
			F	39	26	66.7	64	164.1 **	
			M+F	69	49	71.0	105	152.2	
IV	0	Embryos (offspring)	M	49	23	46.9	30	61.2	
			F	55	31	56.4	53	96.4	
			M+F	104	54	51.9	83	79.8	

* $p<0.05$ using χ^2 test
** $p<0.01$ using χ^2 test

TABLE 8. Long-term carcinogenicity bioassays on ethyl alcohol administered with drinking water supplied *ad libitum* to male (M) and female (F) Sprague-Dawley rats

CARCINOMAS OF THE HEAD AND NECK

Group No.	Concentration (%, v/v)	Animals Age	Sex	No.	Zymbal glands[a] No.	Zymbal glands[a] %	Ear ducts[a] No.	Ear ducts[a] %	Animals with carcinomas — Nasal cavities No.	Nasal cavities %	Oral cavity, tongue and lips No.	Oral cavity, tongue and lips %	Pharynx No.	Pharynx %	Larynx No.	Larynx %	Total No.	Total %
I	10	39 weeks (breeders)	M	110	3	2.7	2	1.8	3	2.7	15	13.6**	0	-	0	-	23	20.9
			F	110	6 (1)	5.5	7 (2)	6.4	1	0.9	12	10.9*	0	-	1	0.9	27	24.5
			M+F	220	9	4.1	9	4.1	4	1.8	27	12.3	0	-	1	0.5	50	22.7
II	0	39 weeks (breeders)	M	110	2	1.8	6	5.5	2	1.8	3	2.7	1	0.9	0	-	14	12.7
			F	110	2 (1)	1.8	9	8.2	1	0.9	2	1.8	1	0.9	0	-	15	13.6
			M+F	220	4	1.8	15	6.8	3	1.4	5	2.3	2	0.9	0	-	29	13.2
III	10	Embryos (offspring)	M	30	0	-	5	16.7	0	-	10	33.3**	0	-	0	-	15	50.0**
			F	39	1	2.6	6	15.4	0	-	16	41.0**	0	-	0	-	23	59.0**
			M+F	69	1	1.4	11	15.9	0	-	26	37.7	0	-	0	-	38	55.1
IV	0	Embryos (offspring)	M	49	0	-	3 (1)	6.1	0	-	2	4.1	0	-	0	-	5	10.2
			F	55	3	5.5	5 (1)	9.1	1	1.8	3	5.5	0	-	1	1.8	13	23.6
			M+F	104	3	2.9	8	7.7	1	1.0	5	4.8	0	-	1	1.0	18	17.3

[a] Between brackets the number of animals with bilateral tumors

* $p<0.05$ using χ^2 test
** $p<0.01$ using χ^2 test

logical pathological changes were detected by gross inspection or histopathological examination.

The occurrence of benign and malignant tumors is shown in TABLE 6. Differences observed between treated and control animals were: (1) an increase in total malignant tumors in males and females, breeders, and offspring (TABLE 7); (2) an increase in total malignant mammary tumors per 100 animals in females, breeders, and offspring; (3) an increase in head and neck carcinomas, especially of the oral cavity, lips, and tongue, in males and females, breeders, and offspring (TABLE 8); (4) an increase in squamous cell carcinomas of the forestomach in males and females, breeders, and offspring; (5) an increase in interstitial cell adenomas of the testis in male breeders ($P < 0.05$) and offspring; (6) an increase in Sertoli cell tumors (ovary) in female offspring; (7) an increase in adenocarcinomas of the uterus in breeders and offspring; (8) an increase in pheochromoblastoma in male and female breeders and male offspring; and (9) an increase in osteosarcomas of the head and other sites in male breeders ($P < 0.01$) and offspring and in female breeders and offspring.

CONCLUSIONS

Methyl alcohol and ethyl alcohol were found to be carcinogenic for various tissues and organs. Based on these findings, methyl alcohol and ethyl alcohol must be considered multipotential carcinogenic agents.

Whether and to what extent methyl alcohol and ethyl alcohol exert their carcinogenic effects directly or through their metabolic products, formaldehyde and acetaldehyde, respectively, or by enhancing the effects of endogenous and exogenous carcinogenic factors are not known. Based on our data, the use and diffusion of methyl alcohol and ethyl alcohol must take into account these pathological effects for the protection of public health.

It is noteworthy that in the tested experimental conditions, ethyl alcohol was shown, for the first time, to be carcinogenic to the oral cavity, tongue, and lips. These sites have been shown to be target organs in man by epidemiologic studies.

ACKNOWLEDGMENTS

This research was partially supported by the Regional Agency for Prevention and Environment (Agenzia Regionale Prevenzione e Ambiente, ARPA) of the Emilia-Romagna Region, Italy.

REFERENCES

1. MALTONI, C., A. CILIBERTI, C. PINTO, et al. 1997. Results of long-term experimental carcinogenicity studies of the effects of gasoline, correlated fuels, and major gasoline aromatics on rats. Ann. N.Y. Acad. Sci. **837:** 15–52.
2. MALTONI, C., A. CILIBERTI, G. COTTI, et al. 1989. Benzene, an experimental multipotential carcinogen: results of the long-term bioassays performed at the Bologna Institute of Oncology. Environ. Health Perspect. **82:** 109–124.
3. BELPOGGI, F., M. SOFFRITTI & C. MALTONI. 1995. Methyl-tertiary-butyl ether (MTBE), a gasoline additive, causes testicular and lymphohaematopoietic cancers in rats. Toxicol. Ind. Health. **11:** 119–149.

4. MALTONI, C., F. BELPOGGI, M. SOFFRITTI, et al. 1999. Comprehensive long-term experimental project of carcinogenicity bioassays on gasoline oxygenated additives: plan and first report of results from the study on ethyl-tertiary butyl ether (ETBE). Eur. J. Oncol. **4:** 493–508.
5. SOFFRITTI, M., C. MALTONI, F. MAFFEI, et al. 1989. Formaldehyde: an experimental multipotential carcinogen. Toxicol. Ind. Health. **5:** 699–730.
6. INTERNATIONAL PROGRAMME ON CHEMICAL SAFETY (IPCS). 1997. Methanol. Environmental Health Criteria 196. WHO. Geneva.
7. GRAYSON, M. 1981. Kirk-Othmer Encyclopedia of Chemical Technology, 3rd edit. **15:** 398–405. John Wiley & Sons. New York.
8. ELVERS, B., S. HAWKINS & G. SCHULZ. 1990. Ullmann's Encyclopedia of Industrial Chemistry, 5th edit. **16A:** 465–486. VCH-Verlag.
9. SRI. 1992. Chemical Economics Handbook: Marketing Research Report on Methanol. SRI International. Menlo Park, CA.
10. CONCAWE. 1995. Alternative fuels in the automotive market. Brussels, CONCAWE, 67 pp (Report No. 2/95, prepared for the Concawe Automotive Emission Management Group by its Technical Coordinator, R.C. Hutcheson).
11. FIEDLER, E., G. GROSSMANN, B. KERSEBOHM, et al. 1990. Methanol. In Ullmann's Encyclopedia of Industrial Chemistry, 5th edit. B. Weinheim, S. Elvers, G. Hawkins & G. Schutz, Eds.: **16A:** 465–486. VCH Verlag. Weinheim, Germany.
12. AXELROD, J. & J. DALY. 1965. Pituitary gland: enzymic formation of methanol from S-adenosylmetionine. Science **158:** 892–893.
13. CEC (COMMISSION OF THE EUROPEAN COMMUNITIES). 1988. Solvent in common use: Health risks to workers. Royal Society of Chemistry. Cambridge. Publication EUR/11553. 1–7, 157–186.
14. LEAF, G. & L.J. ZATMAN. 1952. A study of the conditions under which methanol may exert a toxic hazard in industry. Br. J. Ind. Med. **9:** 19–31.
15. ERIKSEN, S.P. & A.B. KULKARNI. 1963. Methanol in normal human breath. Science **141:** 639–640.
16. LARSSON, B.T. 1965. Gas chromatography of organic volatiles in human breath and saliva. Acta Med. Scand. **19:** 159–164.
17. KROTOSZYNSKI, B.K., G.M. BRUNEAU & H.J. O'NEILL. 1979. Measurement of chemical inhalation exposure in urban populations in the presence of endogenous effluents. J. Anal. Toxicol. **3:** 225–234.
18. JONES, A.W., S. SKAGERBERG, T. YONEKURAT & A. SATO. 1990. Metabolic interaction between endogenous ethanol studied in human volunteers by analysis of breath. Pharmacol. Toxicol. **66:** 62–65.
19. PELLIZZARI, E.D, T.D. HARTWELL, B.S.H. HARRIS III, et al. 1982. Purgeable organic compounds in mother's milk. Bull. Environ. Contam. Toxicol. **28:** 322–328.
20. KAVET, R. & K.M. NAUSS. 1990. The toxicity of inhaled methanol vapors. CRC Crit. Rev. Toxicol. **21:** 21–50.
21. OWENS, L.D., R.G. GILBERT, G.E. GRIEBEL & J.D. MENZIES. 1969. Identification of plant volatiles that stimulates microbial respiration and growth in soil. Phytopathology **59:**1468–1472.
22. HOLZER, G., H. SHANFIELD, A. ZLATKIS, et al. 1977. Collection and analysis of trace emissions from natural sources. J. Chromatogr. **142:** 755–764.
23. GRAEDEL, T.E., D.T HAWKINS & L.D. CLAXTON. 1986. Atmospheric Chemical Compounds: Sources, Occurrence and Bioassay. **557:** 512–514. Academic Press. New York. London.
24. U.S. EPA. 1994. 1992 Toxic release inventory: Public data. US Environmental Protection Agency (EPA 745/R–94-001). Washington, DC.
25. MEDINSKY, M.A. & D.C DORMAN. 1994. Assessing risks of low-level methanol exposure. CIIT Act. **14:** 1–7.
26. GOLD, M.D. & C.E. MOULIF. 1988. Effects of emission standards on methanol vehicle-related ozone, formaldehyde and methanol exposure. Presented at the 81st Meeting of Air Pollution Control Association, Dallas, TX, June 19–24. Pittsburgh, PA.
27. U.S. EPA. 1976. Frequency of organic compounds identified in water. Washington, DC, US Environmental Protection Agency (EPA 600/4–76-062).

28. KIRCHNER, J.G. & J.M. MILLER. 1957. Volatile water-soluble and oil constituents of Valencia orange juice. J. Agric. Food Chem. **5:** 283–291.
29. FRANCOT, P. & P. GEOFFROY. 1956. Le méthanol dans les jus de fruits, les boissons fermentées des alcools et spiriteux. Rev. Ferment. Ind. Aliment. **11:** 279–287.
30. MONTE, W.C. 1984. Aspartame: methanol and the public health. J. Appl. Nutr. **36:** 42–54.
31. GREIZERSTEIN, H.B. 1981. Congener contents of alcoholic beverages. J. Stud. Alcohol **42:** 1030–1037.
32. GILGER, A.P. & A.M. POTTS. 1955. Studies on the visual toxicity of methanol. V. The role of acidosis in experimental methanol poisoning. Am. J. Ophthalmol. **39:** 63–86.
33. NEW ENERGY DEVELOPMENT ORGANIZATION (NEDO). 1987. Toxicological research of methanol as a fuel for power station: summary report on test with monkeys, rats and mice. Tokyo: 1–296.
34. KATOH, M. 1989. New Energy Development Organization data. Presented at the Methanol Vapors and Health Effects Workshop: What we know and what we need to know —Summary Report. Washington, DC, ILSI Risk Science Institute/US Environmental Protection Agency/Health Effects Institute/American Petroleum Institute, A–7.
35. MELLAN, I. 1950. Industrial Solvents: 454. Reinhold. New York.
36. WALSH, B. & M. GRANT. 1985. Public Health Implications of Alcohol Production and Trade (WHO Offset Publication No. 88). World Health Organization. Geneva.
37. BERG, C. 2001. World Ethanol Production 2001. http://www.distill.com/world_ethanol_production.htm.
38. MELLAN, I. 1950. Industrial Solvents: 460–466. Reinhold. New York.
39. ARMSTRONG, S.R. 1999. Ethanol, Brief Report on its Use in Gasoline. Cambridge Environmental, Inc. Cambridge, MA.
40. GROSJEAN, E., D. GROSJEAN, R. GUNAWARDENA & R.A. RASMUSSEN. 1998. Environ. Sci. Technol. **32:** 736.
41. MALCOLM PIRNIE, INC. 1998. Evaluation of the Fate and Transport of Ethanol in the Environment. Prepared for the American Methanol Institute.
42. NATIONAL SCIENCE AND TECHNOLOGY COUNCIL (NSTC). 1997. Interagency Assessment of Oxygenate Fuels. Executive Office of the President.
43. POHORECKY, L.A. & J. BRICK. 1988. Pharmacology of ethanol. Pharmacol. Ther. **36:** 335–427.
44. RITCHIE, J.M. 1980. The aliphatic alcohols. *In* The Pharmacological Basis of Therapeutics. 6th edit.: 376–390. Macmillan. New York.
45. HOLFORD, N.H.G. 1987. Clinical pharmacokinetics of ethanol. Clin. Pharmacokinet. **13:** 273–292.
46. INTERNATIONAL AGENCY FOR RESEARCH ON CANCER. 1988. Monographs on the Evaluation of Carcinogenic Risks to Humans. Vol. **44**. IARC, Lyon, France.

Results of Long-Term Carcinogenicity Bioassays on *Tert*-Amyl-Methyl-Ether (TAME) and Di-Isopropyl-Ether (DIPE) in Rats

FIORELLA BELPOGGI, MORANDO SOFFRITTI, FRANCO MINARDI, LUCIANO BUA, ELISA CATTIN, AND CESARE MALTONI[†]

Cancer Research Center, European Ramazzini Foundation for Oncology and Environmental Sciences, Bologna, Italy

ABSTRACT: *Tert*-amyl-methyl ether (TAME) was administered by gavage in extra virgin olive oil solution at concentrations of 750, 250, or 0 mg/kg bw to groups of 100 male and 100 female Sprague-Dawley rats 8 weeks old at the start of the experiment. Di-isopropyl ether (DIPE) was administered in the same manner at the doses of 1000, 250, or 0 mg/kg body weight to groups of 100 male and 100 female Sprague-Dawley rats. TAME and DIPE were each delivered in 1-mL solution 4 days a week for 78 weeks. Control animals received 1 mL of extra virgin olive oil without TAME or DIPE. At the end of the treatment period, all animals were kept under observation until spontaneous death. Under these test conditions, TAME and DIPE were found to be potential carcinogenic agents for various organs and tissues.

KEYWORDS: oxygenated gasoline additives; *tert*-amyl-methyl-ether; di-isopropyl-ether; carcinogenicity; long-term bioassay; rat

INTRODUCTION

The gasoline oxygenated additives proposed at present include ethers, such as methyl-*tert*-butyl ether (MTBE), ethyl-*tert*-butyl ether (ETBE), *tert*-amyl-methyl ether (TAME), di-isopropyl ether (DIPE), and alcohols such as methyl alcohol, ethyl alcohol, or *tert*-butyl alcohol (TBA). Adding oxygenates to gasoline increases its oxygen content and allegedly reduces emissions of CO and possibly some air toxics, such as ozone-forming hydrocarbons and benzene. However, these oxygenates may increase toxic aldehydes, such as formaldehyde or acetaldehyde. When the use of gasoline-oxygenated additives, particularly MTBE, began in the 1970s, there were no precise data on their impact on the environment and health. No scientific information was available on the carcinogenicity of these compounds.

Only in the middle 1990s were the results of two different experiments on carcinogenicity of MTBE made available. In one study, conducted at the Cancer Research

Address for correspondence: Morando Soffritti, M.D., Cancer Research Center, European Ramazzini Foundation for Oncology and Environmental Sciences, Bentivoglio Castle, 40010 Bentivoglio (BO), Italy. Voice: +39-051-6640460; fax: +39-051-6640223.
crcfr@tin.it.
[†]Deceased.

Center of the Ramazzini Foundation (CRC/RF), MTBE was shown to cause an increase in lymphomas/leukemias (mainly due to lymphoimmunoblastic lymphomas) in female Sprague-Dawley rats and an increase in interstitial cell adenomas of the testis in male rats.[1-3] Another study, sponsored by producers and users of MTBE, found that MTBE caused an increase in hepatocellular adenoma in female CD-1 mice and an increase in renal tubular adenomas and carcinomas in male Fisher 344 rats.[4,5]

In recent years, great concern has arisen regarding the use of MTBE because of groundwater contamination associated with gasoline spills and leaks from underground storage tanks[6] and complaints of people exposed to MTBE from such sources, complaining of unpleasant odor, headaches, and burning of the eyes and throat.[7] Because of MTBE's toxic effects, carcinogenic potential, air pollution, and contamination of the water supply, the petroleum industry has had to propose other oxygenated additives as alternative oxygenated gasoline additives.

In the 1980s, in the context of our research program on the carcinogenicity of fuels, a systematic and integrated project of experimental carcinogenicity bioassay was started on various gasoline additives, namely, the oxygenated additives methyl alcohol, ethyl alcohol, MTBE, ETBE, TAME, DIPE, and the isoparaffin, 2,2,4-trimethyl pentane (TMP). The experiments were performed on Sprague-Dawley rats from the CRC/RF colony on which there is abundant information regarding expected pathology from historical controls. The final results of the experiments on MTBE and ETBE have already been published,[1-3,8] and those on methyl alcohol and ethyl alcohol are reported in this volume. This report outlines the final results of the carcinogenicity bioassays on TAME and DIPE.

Tert-amyl-methyl ether (TAME) is a colorless, flammable liquid. TAME ($C_6H_{14}O$) has a molecular weight of 102.18. Di-isopropyl ether (DIPE) is a colorless, flammable liquid with a sharp, sweet, ether-like odor. DIPE ($C_6H_{14}O$) has a molecular weight of 102.18.

TAME is manufactured from isoamylene and methanol feedstocks. The principal source of isoamylene is the C5-olefin stream from a crude oil-refining process called fluid catalytic cracking. TAME manufacturing provides a refinery with a way to reduce the gasoline vapor pressure, reduce the light olefin content of gasoline, and create a high octane gasoline-blending component. However, similar benefits could also be obtained by using the C5-olefins in the refinery's hydrocarbon alkylation process.[9] DIPE is commercially prepared by the action of sulfuric acid on isopropyl alcohol and also obtained as a by-product in the production of isopropyl alcohol from the propylene fraction of cracked gasoline.[10] The world production capacity of DIPE is unknown but likely to be small. In the United States, it is a permitted additive under U.S. Federal reformulated gasoline regulations.[11]

The information on toxicity, mutagenicity, and carcinogenicity on TAME is extremely limited.[12]

MATERIAL AND METHODS

TAME and DIPE were supplied by SIGMA-ALDRICH, Division of SAF Bulk Chemicals, Milan, Italy, and their purity was higher than 97% and 98%, respectively. The extra virgin oil used as a carrier was provided by Oliaria Toscana (the same oil used in the CRC/RF laboratory for 25 years).

TABLE 1. Long-term carcinogenicity bioassays on *tert*-amyl-methyl ether (TAME) administered by gavage to male (M) and female (F) Sprague-Dawley rats

NUMBER AND PERCENTAGE OF SPRAGUE-DAWLEY RATS BEARING VARIOUS TYPES OF BENIGN AND MALIGNANT TUMORS[a]

Site	Groups											
	I: 750 mg/kg b.w.				II: 250 mg/kg b.w.				III: 0[b] (control)			
	Male		Female		Male		Female		Male		Female	
Histotype	No.	%	No.	%	No.	%	No.	%	No.	%	No.	%
Skin												
Dermatofibroma	0	-	0	-	1	1.0	0	-	1	1.0	0	-
Sebaceous adenoma	0	-	0	-	0	-	0	-	0	-	1	1.0
Squamous cell carcinoma	0	-	0	-	0	-	0	-	1	1.0	0	-
Basocellular carcinoma	0	-	0	-	0	-	1	1.0	0	-	0	-
Subcutaneous tissue												
Fibrolipoma	2	2.0	0	-	0	-	0	-	0	-	0	-
Mammary glands												
Fibroma and fibroadenoma	2	2.0	42(61)	42.0	3	3.0	52(70)	52.0	8(10)	8.0	34(42)	34.0
Lipoma and fibrolipoma	3	3.0	1	1.0	3	3.0	1	1.0	3	3.0	0	-
Fibroangioma	1	1.0	0	-	0	-	0	-	0	-	0	-
Adenocarcinoma	0	-	7(9)	7.0	0	-	14(18)	14.0	0	-	10(15)	10.0
Fibrosarcoma	0	-	0	-	0	-	0	-	0	-	1	1.0
Liposarcoma	0	-	0	-	1	1.0	0	-	1	1.0	2	2.0
Angiosarcoma	1	1.0	0	-	0	-	0	-	0	-	0	-
Zymbal glands												
Acanthoma	0	-	0	-	1	1.0	0	-	0	-	0	-
Sebaceous adenoma	0	-	0	-	0	-	0	-	1	1.0	1	1.0
Carcinoma	1	1.0	0	-	1	1.0	1	1.0	2(3)	2.0	2	2.0
Ear ducts												
Sebaceous adenoma	0	-	0	-	0	-	0	-	1	1.0	0	-
Carcinoma	5	5.0	4	4.0	4	4.0	4(5)	4.0	1	1.0	2	2.0
Nasal cavities												
Carcinoma	0	-	1	1.0	0	-	1	1.0	2	2.0	0	-
Oral cavity, tongue and lips												
Carcinoma	0	-	0	-	0	-	0	-	1	1.0	1	1.0
-Tooth												
Odontoma	0	-	0	-	0	-	0	-	1	1.0	0	-
Pharynx												
Carcinoma	0	-	0	-	0	-	0	-	1	1.0	0	-

— *Continued*

TABLE 1. *Continued*

NUMBER AND PERCENTAGE OF SPRAGUE-DAWLEY RATS BEARING VARIOUS TYPES OF BENIGN AND MALIGNANT TUMORS[a]

					Groups							
	I: 750 mg/kg b.w.				II: 250 mg/kg b.w.				III: 0[b] (control)			
	Male		Female		Male		Female		Male		Female	
Site / Histotype	No.	%	No.	%	No.	%	No.	%	No.	%	No.	%
Lung												
Angioma	1	1.0	0	-	0	-	0	-	0	-	0	-
Leiomyosarcoma	1	1.0	0	-	1	1.0	0	-	0	-	0	-
Stomach												
- Forestomach												
Acanthoma	3	3.0	1	1.0	0	-	2	2.0	2	2.0	1	1.0
- Glandular stomach												
Leiomyosarcoma	0	-	0	-	0	-	1	1.0	0	-	0	-
Intestine												
Glandular polyp	0	-	1	1.0	0	-	0	-	0	-	0	-
Adenomatous polyp	0	-	0	-	0	-	0	-	0	-	1	1.0
Adenocarcinoma	0	-	0	-	1	1.0	0	-	0	-	1	1.0
Leiomyosarcoma	0	-	0	-	1	1.0	1	1.0	0	-	0	-
Salivary glands												
Adenoma	0	-	1	1.0	0	-	0	-	0	-	0	-
Liver												
Cholangioma	0	-	0	-	0	-	0	-	1	1.0	0	-
Hepatocarcinoma	1	1.0	0	-	0	-	0	-	0	-	0	-
Pancreas												
Exocrine adenoma	0	-	0	-	1	1.0	2	2.0	0	-	2	2.0
Islet cell adenoma	2	2.0	0	-	2	2.0	0	-	5	5.0	2	2.0
Islet cell carcinoma	0	-	0	-	0	-	1	1.0	0	-	0	-
Kidneys												
Nephroblastoma	1	1.0	0	-	0	-	0	-	2	2.0	0	-
Adenocarcinoma	0	-	0	-	2	2.0	0	-	0	-	0	-
Pelvis and ureters												
Transitional cell papilloma	0	-	0	-	0	-	1	1.0	0	-	0	-
Testes												
Interstitial cell adenoma	4	4.0			3	3.0			0	-		
Ovaries												
Cystadenoma			0	-			1	1.0			1	1.0
Granulosa cell tumor			0	-			1	1.0			0	-
Adenocarcinoma			0	-			1	1.0			0	-

— *Continued*

TABLE 1. *Continued*

NUMBER AND PERCENTAGE OF SPRAGUE-DAWLEY RATS BEARING VARIOUS TYPES OF BENIGN AND MALIGNANT TUMORS[a]

Site / Histotype	I: 750 mg/kg b.w.				II: 250 mg/kg b.w.				III: 0[b] (control)			
	Male		Female		Male		Female		Male		Female	
	No.	%	No.	%	No.	%	No.	%	No.	%	No.	%
Uterus												
Polyp			13	13.0			14	14.0			10	10.0
Leiomyoma			6	6.0			1	1.0			1	1.0
Squamous cell carcinoma			0	-			1	1.0			0	-
Adenocarcinoma			1	1.0			2	2.0			3	3.0
Leiomyosarcoma			2	2.0			0	-			0	-
Angiosarcoma			2	2.0			0	-			0	-
Uterus & Vagina												
Malignant Schwannoma			1	1.0			1	1.0			3	3.0
Vagina												
Fibroma			0	-			0	-			2	2.0
Benign Schwannoma			1	1.0			0	-			0	-
Squamous cell carcinoma			0	-			0	-			1	1.0
Malignant Schwannoma			1	1.0			0	-			1	1.0
Peritoneum												
Mesothelioma	0	-	1	1.0	0	-	3	3.0	0	-	1	1.0
Pituitary gland												
Adenoma	38	38.0	26	26.0	27	27.0	25	25.0	44	44.0	18	18.0
Adenocarcinoma	0	-	0	-	2	2.0	0	-	0	-	0	-
Thyroid gland												
Follicular adenoma	0	-	0	-	0	-	0	-	0	-	1	1.0
C-cell adenoma	1	1.0	3	3.0	0	-	2	2.0	1	1.0	4	4.0
Parathyroid glands												
Adenoma	0	-	0	-	0	-	1	1.0	0	-	1	1.0
Adrenal glands												
Cortical adenoma	2	2.0	5	5.0	0	-	5	5.0	1	1.0	8(9)	8.0
Pheochromocytoma	7(8)	7.0	9(10)	9.0	6(8)	6.0	5	5.0	6(8)	6.0	9(10)	9.0
Cortical adenocarcinoma	0	-	2	2.0	0	-	5(6)	5.0	0	-	2	2.0
Pheochromoblastoma	0	-	0	-	1	1.0	1(2)	1.0	0	-	5	5.0
Sympathoblastoma	0	-	0	-	1	1.0	0	-	0	-	0	-

— Continued

TABLE 1. Continued

NUMBER AND PERCENTAGE OF SPRAGUE-DAWLEY RATS BEARING VARIOUS TYPES OF BENIGN AND MALIGNANT TUMORS[a]

Site Histotype	I: 750 mg/kg b.w.				II: 250 mg/kg b.w.				III: 0[b] (control)			
	Male		Female		Male		Female		Male		Female	
	No.	%	No.	%	No.	%	No.	%	No.	%	No.	%
Central nervous system												
- Brain												
Multiform glioblastoma	0	-	1	1.0	0	-	0	-	0	-	0	-
Oligodendroglioma	3	3.0	2	2.0	4	4.0	2	2.0	0	-	2	2.0
- Meninges												
Benign meningioma	0	-	0	-	0	-	0	-	0	-	1	1.0
Peripheral nervous system												
- Major peripheral nerves												
Malignant Schwannoma	0	-	0	-	1	1.0	0	-	2	2.0	0	-
Bones												
- Head												
Osteosarcoma	7	7.0	2	2.0	1	1.0	4	4.0	2	2.0	7	7.0
- Other sites												
Osteosarcoma	0	-	0	-	0	-	1	1.0	1	1.0	1	1.0
Soft tissues												
Histiocytoma	0	-	0	-	0	-	1	1.0	0	-	0	-
Leiomyosarcoma	0	-	0	-	0	-	1	1.0	0	-	0	-
Liposarcoma	0	-	1	1.0	0	-	0	-	0	-	0	-
Heart												
Myxoma	0	-	0	-	0	-	3	3.0	0	-	0	-
Malignant Schwannoma	1	1.0	0	-	2	2.0	0	-	0	-	0	-
Spleen												
Fibroangioma	0	-	0	-	0	-	0	-	0	-	1	1.0
Angiosarcoma	0	-	0	-	1	1.0	0	-	0	-	0	-
Pericytosarcoma	0	-	0	-	0	-	0	-	0	-	1	1.0
Hemolymphoreticular tissues[c,d]												
Lymphomas and leukemias	21	21.0	27(28)	27.0	7	7.0	14	14.0	17	17.0	7	7.0

[a] Between brackets the number of tumors (one animal can bear more than one tumor)
[b] Olive oil alone
[c] Including spleen
[d] See table 3

TABLE 2. Long-term carcinogenicity bioassays on *tert*-amyl-methyl ether (TAME) administered by gavage to male (M) and female (F) Sprague-Dawley rats

				TOTAL MALIGNANT TUMORS			
	Daily dose			Malignant tumors			
Group	(mg/kg b.w.	Animals		Tumor-bearing animals		Tumors	
No.	in olive oil)	Sex	No.	No.	%	No.	Per 100 animals
I	750	M	100	37	37.0	42	42.0
		F	100	44	44.0	58	58.0
		M+F	200	81	40.5	100	50.0
II	250	M	100	24	24.0	31	31.0
		F	100	46	46.0	67	67.0
		M+F	200	70	35.0	98	49.0
III	0[a]	M	100	30	30.0	34	34.0
		F	100	37	37.0	58	58.0
		M+F	200	67	33.5	92	46.0

[a] Olive oil alone

During experiments, compounds TAME and DIPE were stored at 4°C. TAME was administered by gavage in 1 mL extra virgin olive oil solution at concentrations of 750, 250, or 0 mg/kg bw to groups of 100 male and 100 female Sprague-Dawley rats 8 weeks old at the start of the experiment. TAME was administered daily, 4 days weekly, for 78 weeks. The animals were then maintained under control conditions until spontaneous death. Control animals received 1 mL of extra virgin olive oil without TAME. The experiment began in February 1995 and ended after 135 weeks of treatment with the death of the last animal at 143 weeks of age. Testing of DIPE was concurrent with that of TAME, and the experimental protocols were identical, apart from the higher dose level of DIPE (1000 mg/kg body weight). The DIPE experiment ended after 163 weeks of treatment with the death of the last animal at 171 weeks of age. Both experiments were performed according to Good Laboratory Practices (GLP) and Standard Operating Procedure (SOP) of the CRC/RF.

The experimental conditions, protocol, and histopathology are described in detail elsewhere in this volume.[13] Multiple tumors of different type and site, of different type in the same site, of the same type in bilateral organs, of the same type in the skin, subcutaneous tissue, and mammary glands, or at distant sites of diffuse tissue (i.e., bones, skeletal muscle, etc.) were plotted as single/independent tumors. Multiple tumors of the same type in the same tissue and organ, including the bilateral organs, were plotted only once. Statistical analysis was performed using the χ^2 test to evaluate differences in tumor incidence between treated and control groups. The Cochrane Armitage test was used to evaluate dose-response relationships.

TABLE 3. Long-term carcinogenicity bioassays on *tert*-amyl-methyl ether (TAME) administered by gavage to male (M) and female (F) Sprague-Dawley rats

HEMOLYMPHORETICULAR NEOPLASIAS AND THEIR DISTRIBUTION BY HISTOCYTOTYPE

Group No.	Daily dose (mg/kg b.w. in olive oil)	Sex	Animals No.	Total[a] No.	Total[a] %	Animals with hemolymphoreticular neoplasias											
						Lymphoblastic lymphoma[b]		Lymphoblastic leukemia		Lymphocytic lymphoma[b]		Lymphoimmuno-blastic lymphoma[b]		Histiocytic sarcoma/ monocytic leukemia		Myeloid leukemia[b]	
						No.	%	No.	%	No.	%	No.	%	No.	%	No.	%
I	750	M	100	21	21.0	0	-	0	-	0	-	17	81.0	4	19.0	0	-
		F	100	27[c]	27.0***♦	1	3.8	0	-	1	3.7	21	77.8	5	18.6	0	-
		M+F	200	48[c]	24.0	1	2.1	0	-	1	2.1	38	79.2	9	18.8	0	-
II	250	M	100	7	7.0	0	-	0	-	0	-	6	85.7	1	14.3	0	-
		F	100	14	14.0♦	2	14.3	0	-	2	14.3	6	42.9	4	28.6	0	-
		M+F	200	21	10.5	2	9.5	0	-	2	9.5	12	57.1	5	23.8	0	-
III	0[d]	M	100	17	17.0	0	-	0	-	0	-	13	76.5	3	17.6	1	5.9
		F	100	7	7.0	3	42.9	0	-	0	-	3	42.9	1	14.3	0	-
		M+F	200	24	12.0	3	12.5	0	-	0	-	16	66.7	4	16.7	1	4.2

[a] Percentages refer to the number of animals at start
[b] Percentages refer to the number of animals bearing hemolymphoreticular neoplasias
[c] One animal bore a lymphoblastic lymphoma and histiocytic sarcoma
[d] Olive oil alone
*** $p < 0.01$ using χ^2 test
♦♦ $p < 0.05$ using Cochrane-Armitage test for dose-response relationship

TABLE 4. Long-term carcinogenicity bioassays on di-isopropyl (DIPE)) administered by gavage to male (M) and female (F) Sprague-Dawley rats[a]

NUMBER AND PERCENTAGE OF SPRAGUE-DAWLEY RATS BEARING VARIOUS TYPES OF BENIGN AND MALIGNANT TUMORS[a]

Site		I: 1,000 mg/kg b.w.				II: 250 mg/kg b.w.				III: 0[b] (control)			
		Male		Female		Male		Female		Male		Female	
Histotype		No.	%	No.	%	No.	%	No.	%	No.	%	No.	%
Skin													
Acanthoma		0	-	0	-	1	1.0	1	1.0	0	-	0	-
Cheratoacanthoma		1	1.0	0	-	0	-	0	-	0	-	0	-
Dermatofibroma		1	1.0	0	-	1	1.0	0	-	1	1.0	0	-
Squamous cell carcinoma		0	-	0	-	0	-	0	-	1	1.0	0	-
Sebaceous adenoma		0	-	0	-	0	-	0	-	0	-	1	1.0
Subcutaneous tissue													
Fibroma		1	1.0	0	-	2	2.0	0	-	0	-	0	-
Angiosarcoma		0	-	0	-	1	1.0	0	-	0	-	0	-
Fibrosarcoma		0	-	0	-	0	-	1	1.0	0	-	0	-
Liposarcoma		0	-	0	-	1	1.0	0	-	0	-	0	-
Mammary glands													
Fibroma and fibroadenoma		4	4.0	42(53)	42.0	4(5)	4.0	44(60)	44.0	8(10)	8.0	34(42)	34.0
Myxoma		0	-	0	-	1	1.0	0	-	0	-	0	-
Lipoma and fibrolipoma		2	2.0	0	-	5	5.0	0	-	3	3.0	0	-
Adenocarcinoma		0	-	7(8)	7.0	0	-	12(15)	12.0	0	-	10(15)	10.0
Fibrosarcoma		0	-	1	1.0	0	-	0	-	0	-	1	1.0
Liposarcoma		0	-	0	-	1	1.0	1	1.0	1	1.0	2	2.0
Angiosarcoma		0	-	0	-	0	-	1	1.0	0	-	0	-
Zymbal glands													
Sebaceous adenoma		1	1.0	0	-	0	-	0	-	1	1.0	1	1.0
Carcinoma		0	-	1	1.0	1	1.0	3	3.0	2(3)	2.0	2	2.0
Ear ducts													
Sebaceous adenoma		0	-	0	-	0	-	0	-	1	1.0	0	-
Carcinoma		4	4.0	4	4.0	9(11)	9.0	3(4)	3.0	1	1.0	2	2.0

— Continued

TABLE 4. Continued

NUMBER AND PERCENTAGE OF SPRAGUE-DAWLEY RATS BEARING VARIOUS TYPES OF BENIGN AND MALIGNANT TUMORS[a]

Site Histotype	I: 1,000 mg/kg b.w.				II: 250 mg/kg b.w.				III: 0[b] (control)			
	Male		Female		Male		Female		Male		Female	
	No.	%	No.	%	No.	%	No.	%	No.	%	No.	%
Nasal cavities												
Carcinoma	0	-	0	-	0	-	0	-	2	2.0	0	-
Oral cavity, tongue and lips												
Carcinoma	0	-	1	1.0	0	-	0	-	1	1.0	1	1.0
- Tooth												
Adamantinoma	0	-	1	1.0	0	-	0	-	0	-	0	-
Odontoma	0	-	0	-	0	-	0	-	1	1.0	0	-
Pharynx												
Carcinoma	0	-	0	-	0	-	0	-	1	1.0	0	-
Lung												
Adenoma	0	-	0	-	1	1.0	0	-	0	-	0	-
Fibroangioma	0	-	0	-	0	-	1	1.0	0	-	0	-
Stomach												
- Forestomach												
Acanthoma	0	-	2	2.0	4	4.0	2	2.0	2	2.0	1	1.0
Lipoma	1	1.0	0	-	0	-	0	-	0	-	0	-
- Glandular stomach												
Adenocarcinoma	0	-	1	1.0	0	-	0	-	0	-	0	-
Intestine												
Adenomatous polyp	0	-	0	-	0	-	0	-	0	-	1	1.0
Leiomyoma	0	-	1	1.0	0	-	0	-	0	-	0	-
Adenocarcinoma	0	-	0	-	0	-	0	-	0	-	1	1.0
Leiomyosarcoma	1	1.0	0	-	0	-	0	-	0	-	0	-
Salivary glands												
Adenocarcinoma	0	-	0	-	1	1.0	0	-	0	-	0	-

— Continued

TABLE 4. *Continued*

NUMBER AND PERCENTAGE OF SPRAGUE-DAWLEY RATS BEARING VARIOUS TYPES OF BENIGN AND MALIGNANT TUMORS[a]

	Groups											
	I: 1,000 mg/kg b.w.				II: 250 mg/kg b.w.				III: 0[b] (control)			
	Male		Female		Male		Female		Male		Female	
Site / Histotype	No.	%	No.	%	No.	%	No.	%	No.	%	No.	%
Liver												
Cholangioma	0	-	1	1.0	0	-	0	-	1	1.0	0	-
Pancreas												
Exocrine adenoma	1	1.0	0	-	0	-	3	3.0	0	-	2	2.0
Islet cell adenoma	3	3.0	2	2.0	9	9.0	2	2.0	5	5.0	2	2.0
Exocrine adenocarcinoma	0	-	0	-	1	1.0	1	1.0	0	-	0	-
Islet cell carcinoma	0	-	0	-	1	1.0	1	1.0	0	-	0	-
Kidneys												
Nephroblastoma	0	-	0	-	0	-	0	-	2	2.0	0	-
Prostate												
Adenoma	0	-			3	3.0			0	-		
Seminal vesicles												
Adenoma	0	-			1	1.0			0	-		
Testes												
Interstitial cell adenoma	2	2.0			2	2.0			0	-		
Ovaries												
Cystadenoma			1	1.0			3(4)	3.0			1	1.0
Granulosa cell tumor			1	1.0			0	-			0	-
Sertoli cell tumor			0	-			3(4)	3.0			0	-
Malignant granulosa cell tumor			0	-			3	3.0			0	-
Uterus												
Polyp			15	15.0			8	8.0			10	10.0
Leiomyoma			4	4.0			2	2.0			1	1.0
Fibroangioma			1	1.0			1	1.0			0	-
Adenocarcinoma			3	3.0			3	3.0			3	3.0
Malignant Schwannoma			1	1.0			0	-			0	-

— *Continued*

TABLE 4. Continued

NUMBER AND PERCENTAGE OF SPRAGUE-DAWLEY RATS BEARING VARIOUS TYPES OF BENIGN AND MALIGNANT TUMORS[a]

Site / Histotype	I: 1,000 mg/kg b.w.				II: 250 mg/kg b.w.				III: 0[b] (control)			
	Male		Female		Male		Female		Male		Female	
	No.	%	No.	%	No.	%	No.	%	No.	%	No.	%
Uterus & Vagina												
Malignant Schwannoma			5	5.0			8	8.0			3	3.0
Vagina												
Fibroma			0	-			0	-			2	2.0
Squamous cell carcinoma			0	-			0	-			1	1.0
Malignant Schwannoma			1	1.0			2	2.0			1	1.0
Peritoneum												
Mesothelioma	0	-	2	2.0	0	-	2	2.0	0	-	1	1.0
Pituitary gland												
Adenoma	43	43.0	24	24.0	50	50.0	38	38.0	44	44.0	18	18.0
Adenocarcinoma	1	1.0	1	1.0	0	-	1	1.0	0	-	0	-
Thyroid gland												
Follicular adenoma	0	-	0	-	0	-	2	2.0	0	-	1	1.0
C-cell adenoma	2	2.0	5	5.0	2	2.0	1	1.0	1	1.0	4	4.0
Follicular adenocarcinoma	0	-	0	-	0	-	2	2.0	0	-	0	-
C-cell carcinoma	0	-	1	1.0	3	3.0	0	-	0	-	0	-
Parathyroid glands												
Adenoma	0	-	1	1.0	0	-	0	-	0	-	1	1.0
Adrenal glands												
Cortical adenoma	0	-	1	1.0	2	2.0	3	3.0	1	1.0	8(9)	8.0
Pheochromocytoma	12	12.0	19(25)	19.0	15(18)	15.0	15(17)	15.0	6(8)	6.0	9(10)	9.0
Cortical adenocarcinoma	0	-	4	4.0	1	1.0	4	4.0	0	-	2	2.0
Pheochromoblastoma	2	2.0	0	-	0	-	2	2.0	0	-	5	5.0

—Continued

TABLE 4. Continued

NUMBER AND PERCENTAGE OF SPRAGUE-DAWLEY RATS BEARING VARIOUS TYPES OF BENIGN AND MALIGNANT TUMORS[a]

Site / Histotype	I: 1,000 mg/kg b.w.				II: 250 mg/kg b.w.				III: 0[b] (control)			
	Male		Female		Male		Female		Male		Female	
	No.	%	No.	%	No.	%	No.	%	No.	%	No.	%
Central nervous system												
- Brain												
Oligodendroglioma	3	3.0	2	2.0	4	4.0	3	3.0	0	-	2	2.0
Glioblastoma	0	-	0	-	0	-	1	1.0	0	-	0	-
- Meninges												
Benign meningioma	1	1.0	0	-	2	2.0	0	-	0	-	1	1.0
Peripheral nervous system												
- Major peripheral nerves												
Malignant Schwannoma	0	-	1	1.0	2	2.0	0	-	2	2.0	0	-
- Ganglia												
Ganglioneuroma	0	-	0	-	1	1.0	0	-	0	-	0	-
Skeletal muscle												
Rhabdomyosarcoma	0	-	1	1.0	0	-	0	-	0	-	0	-
Bones												
- Head												
Osteoma	0	-	0	-	0	-	1	1.0	0	-	0	-
Osteosarcoma	6	6.0	4	4.0	2	2.0	3	3.0	2	2.0	7	7.0
- Other sites												
Osteosarcoma	1	1.0	0	-	0	-	1	1.0	1	1.0	1	1.0
Soft tissues												
Fibroangioma	0	-	0	-	1	1.0	0	-	0	-	0	-
Sarcoma	0	-	0	-	0	-	1	1.0	0	-	0	-
Heart												
Myxoma	0	-	2	2.0	0	-	1	1.0	0	-	0	-
Thymus												
Malignant thymoma[c]	0	-	0	-	1	1.0	0	-	0	-	0	-
Spleen												
Fibroangioma	0	-	1	1.0	0	-	0	-	0	-	1	1.0
Pericytosarcoma	0	-	0	-	0	-	0	-	0	-	1	1.0
Subcutaneous lymph nodes												
Angioma	0	-	1	1.0	0	-	0	-	0	-	0	-
Hemolymphoreticular tissues[d,e]												
Lymphomas and leukemias	31	31.0	41	41.0	27	27.0	28	28.0	17	17.0	7	7.0

[a]Between brackets the number of tumors (one animal can bear more than one tumor). [b]Olive oil alone. [c]In 96% of cases the tumor itself is composed of a mixture in varying proportions of epithelial cells and lymphocytes. In the remaining 4%, only epithelial cells are present. We consider that a tumor composed exclusively of lymphocytes should not be classified as a thymoma but as a lymphoma involving the thymus. [d]Including thymus, spleen and subcutaneous lymph nodes. [e]See TABLES 6.

TABLE 5. Long-term carcinogenicity bioassays on di-isopropyl ether (DIPE)) administered by gavage to male (M) and female (F) Sprague-Dawley rats

TOTAL MALIGNANT TUMORS

Group No.	Daily dose (mg/kg b.w. in olive oil)	Animals Sex	Animals No.	Malignant tumors Tumor-bearing animals No.	%	Tumors No.	Per 100 animals
I	1,000	M	100	38	38.0	49	49.0
		F	100	57	57.0 ***♦♦	85	85.0 ***
		M+F	200	95	47.5	134	67.0
II	250	M	100	46	46.0 * ♦♦	58	58.0 ***
		F	100	55	55.0 **♦♦	90	90.0 ***
		M+F	200	101	50.5	148	74.0
III	0[a]	M	100	30	30.0	34	34.0
		F	100	37	37.0	58	58.0
		M+F	200	67	33.5	92	46.0

[a] Olive oil alone
* $p < 0.05$ using χ^2 test
*** $p < 0.01$ using χ^2 test
♦ $p < 0.01$ using Cochrane-Armitage test for dose-response relationship

TABLE 6. Long-term carcinogenicity bioassays on di-isopropyl ether (DIPE)) administered by gavage to male (M) and female (F) Sprague-Dawley rats

HEMOLYMPHORETICULAR NEOPLASIAS AND THEIR DISTRIBUTION BY HISTOCYTOTYPE

Group No.	Daily dose (mg/kg b.w. in olive oil)	Animals Sex	Animals No.	Total[a] No.	Total[a] %	Lymphoblastic lymphoma[b] No.	Lymphoblastic lymphoma[b] %	Lymphoblastic leukemia No.	Lymphoblastic leukemia %	Lymphocytic lymphoma[b] No.	Lymphocytic lymphoma[b] %	Lymphoimmunoblastic lymphoma[b] No.	Lymphoimmunoblastic lymphoma[b] %	Histiocytic sarcoma/monocytic leukemia[b] No.	Histiocytic sarcoma/monocytic leukemia[b] %	Myeloid leukemia[b] No.	Myeloid leukemia[b] %
I	1,000	M	100	31	31.0 *♦	2	6.5	0	-	0	-	22	71.0	7	22.6	0	-
		F	100	41[c]	41.0 ***♦♦	1	2.4	0	-	1	2.4	33	80.5	8	19.5	0	-
		M+F	200	72[c]	36.0	3	4.2	0	-	1	1.4	55	76.4	15	20.8	0	-
II	250	M	100	27	27.0 ♦	2	7.4	0	-	0	-	20	74.1	5	18.5	0	-
		F	100	28	28.0 ***♦♦	1	3.6	0	-	6	21.4	15	53.6	5	17.9	1	3.6
		M+F	200	55	27.5	3	5.5	0	-	6	10.9	35	63.6	10	18.2	1	1.8
III	0[d]	M	100	17	17.0	0	-	0	-	0	-	13	76.5	3	17.6	1	5.9
		F	100	7	7.0	3	42.9	0	-	0	-	3	42.9	1	14.3	0	-
		M+F	200	24	12.0	3	12.5	0	-	0	-	16	66.7	4	16.7	1	4.2

[a] Percentages refer to the number of animals at start
[b] Percentages refer to the total number of animals bearing hemolymphoreticular neoplasias
[c] Two animals bore a lymphoimmunoblastic lymphoma and histiocytic sarcoma
[d] Olive oil alone
* $p < 0.05$ using χ^2 test
*** $p < 0.01$ using χ^2 test
♦ $p < 0.05$ using Cochrane-Armitage test for dose-response relationship
♦♦ $p < 0.01$ using Cochrane-Armitage test for dose-response relationship

RESULTS

No significant differences were observed in daily water or feed consumption, body weight, behavior, or treatment-related nononcological pathological changes between TAME- or DIPE-treated and control animals. In the period between the 40th and 104th week of age, a decrease in survival was observed in TAME-treated males compared to controls. A decrease in survival was observed in DIPE-treated males versus controls in the period between the 56th and 88th week of age.

The occurrence of benign and malignant tumors in TAME-treated and control animals is shown in TABLE 1. Significant findings in TAME-treated and control animals are summarized as follows: (1) no differences were observed in the total number of malignant tumors (TABLE 2); (2) increases in some sporadic malignant tumors of the gastrointestinal tract occurred in males and females treated with the lower dose compared to the control group; (3) an increased incidence of ear duct carcinomas was noted in treated males and females compared to control animals; (4) an increase in the incidence of interstitial cell adenomas of the testis was observed in treated animals; (5) an increase in the incidence of glial malignant tumors of the brain was noted in males treated at both doses; and (6) an increase in the incidence of hemolymphoreticular neoplasias in males treated at the highest dose and in females treated at both doses (TABLE 3).

The occurrence of benign and malignant tumors in DIPE-treated and control animals is shown in TABLE 4. Significant findings in DIPE-treated and control animals are summarized as follows: (1) an increase in total malignant tumors was observed in males and females of both treated groups (TABLE 5); (2) an increase in the incidence of carcinomas of the ear duct occurred in males treated at 1,000 and 250 mg/kg bw ($P <0.05$) and in treated females; (3) the onset of some interstitial cell adenomas of the testis was noted in the treated group; (4) a slight increase in malignant sarcomas of the uterus and vagina was observed in the treated group; (5) an increase in the incidence of glial malignant tumors of the brain occurred in male- and female-treated animals; and (6) an increase of hemolymphoreticular neoplasias was observed in males and females of the treated groups (TABLE 6).

CONCLUSIONS

TAME and DIPE were demonstrated to be potential carcinogenic agents for various organs and tissues.

The present experiments have some limitations: only two doses were studied, and the number of animals per group, although higher than usually foreseen for long-term bioassays, may not have been sufficient to completely show the carcinogenic potency of the compound tested.

If TAME and DIPE are expected to have increased commercial use, our data show that further carcinogenicity studies are essential before their introduction. Future animal studies should include a wider range of doses, a larger number of animals, and at least two animal models.

ACKNOWLEDGMENTS

This research was partially supported by the Regional Agency for Prevention and Environment (Agenzia Regionale Prevenzione e Ambiente, ARPA) of the Emilia-Romagna Region, Italy

REFERENCES

1. BELPOGGI, F. et al. 1995. Methyl-tertiary-butyl ether (MTBE), a gasoline additive, causes testicular and lymphohaematopoietic cancers in rats. Toxicol. Ind. Health **11:** 119–149.
2. BELPOGGI, F. et al. 1997. Results of long-term experimental studies on the carcinogenicity of methyl-*tert*-butyl ether. Ann. N.Y. Acad. Sci. **837:** 77–95.
3. BELPOGGI, F. et al. 1998. Pathological characterization of testicular tumours and lymphomas-leukaemias, and of their precursors observed in Sprague-Dawley rats exposed to methyl-tertiary-butyl-ether (MTBE). Eur. J. Oncol. **3:** 201–206.
4. BURLEIGH-FLAYER, H.D. et al. 1992. Methyl-tertiary-butyl ether: vapor inhalation oncogenicity study in CD-1 mice. BRRC report 91N0013A. Union Carbide, Bushy Run Research Center, Export, PA.
5. CHUN, J.S. et al. 1992. Methyl-tertiary-butyl ether: vapor inhalation oncogenicity study in Fisher-344 rats. BRRC report 91N0013B. Union Carbide, Bushy Run Research Center, Export, PA.
6. SQUILLACE, P.J. et al. 1996. Preliminary assessment of the occurrence and possible sources of MTBE in groundwater in the United States, 1993-1994. Environ. Sci. Technol. **30:** 227–233.
7. MEHLMAN, M.A. 2001. Methyl-tertiary-butyl-ether (MTBE) misclassified. Am. J. Ind. Med. **39:** 505-508.
8. MALTONI, C. et al. 1999. Comprehensive long-term experimental project of carcinogenicity bioassays on gasoline oxygenated additives: plan and first report of results from the study on ethyl-tertiary butyl ether (ETBE). Eur. J. Oncol. **4:** 493–508.
9. WITHE, R.D. et al. 1995. Health effects of inhaled tertiary amyl methyl ether and ethyl tertiary butyl ether. Toxicol. Lett. (Elsevier Science Ireland) **82/83:** 719–724.
10. PATTY, F.A. 1963. Industrial Hygiene and Toxicology, 2nd edit., revised. Vol. II. Toxicology. Interscience Publishers, Wiley & Sons, Inc. New York.
11. ZOGORSKI, J.S. et al. 1996. Fuel Oxygenate and Water Quality: Current Understanding of Sources, Occurrence in Natural Waters, Environmental Behavior, Fate, and Significance. U.S. Geological Survey. Rapid City, South Dakota.
12. FROINES, J.R. et al. 1998. An Evaluation of the Scientific Peer-Reviewed Research and Literature on the Human Health Effects of MTBE, its Metabolites, Combustion Products and Substitute Compounds. Report to the Legislature of the State of California. Vol. II. :179. Human Health Effects.
13. SOFFRITTI, M. et al. 2002. Results of long-term experimental studies on the carcinogenicity of methyl alcohol and ethyl alcohol in rats. Ann. N.Y. Acad. Sci. **982:** this volume.

Results of Long-Term Experimental Studies on the Carcinogenicity of Formaldehyde and Acetaldehyde in Rats

MORANDO SOFFRITTI, FIORELLA BELPOGGI, LUCA LAMBERTINI, MICHELINA LAURIOLA, MICHELA PADOVANI, AND CESARE MALTONI[†]

Cancer Research Center, European Ramazzini Foundation for Oncology and Environmental Sciences, Bologna, Italy

ABSTRACT: Formaldehyde was administered for 104 weeks in drinking water supplied *ad libitum* at concentrations of 1500, 1000, 500, 100, 50, 10, or 0 mg/L to groups of 50 male and 50 female Sprague-Dawley rats beginning at seven weeks of age. Control animals (100 males and 100 females) received tap water only. Acetaldehyde was administered to 50 male and 50 female Sprague-Dawley rats beginning at six weeks of age at concentrations of 2,500, 1,500, 500, 250, 50, or 0 mg/L. Animals were kept under observation until spontaneous death. Formaldehyde and acetaldehyde were found to produce an increase in total malignant tumors in the treated groups and showed specific carcinogenic effects on various organs and tissues.

KEYWORDS: formaldehyde; acetaldehyde; carcinogenicity; long-term bioassay; rat

INTRODUCTION

Formaldehyde and acetaldehyde are two compounds produced in large amounts, employed almost universally, and diffused ubiquitously. Little is known about their carcinogenicity. Formaldehyde and acetaldehyde develop from combustion of gasoline, methanol, and ethanol fuels. Because of the increased production and use of oxygenated gasolines, particularly those with added methyl *tert*-butyl ether (MTBE) or ethyl *tert*-butyl ether (ETBE), large numbers of people may be exposed to formaldehyde and acetaldehyde daily. This report outlines the final results of the carcinogenicity experiments on formaldehyde and acetaldehyde performed at the Cancer Research Center of the European Ramazzini Foundation (CRC/RF) at the Castle of Bentivoglio.

Formaldehyde (CH_2O) is a colorless gas with a molecular weight of 30, and acetaldehyde (CH_3CHO) is a colorless, mobile liquid with a molecular weight of 44.05 that is miscible with water and most common organic solvents.

Address for correspondence: Morando Soffritti, M.D., Cancer Research Center, European Ramazzini Foundation for Oncology and Environmental Sciences, Bentivoglio Castle, 40010 Bentivoglio (BO), Italy. Voice: +39-051-6640460; fax: +39-051-6640223.
crcfr@tin.it
[†]Deceased.

FORMALDEHYDE

Two methods are used to produce formaldehyde: the silver catalyst and metal oxide catalyst process and the metal oxide (Formox) process.[1,2] By-products are carbon monoxide, dimethyl ether, and small amounts of carbon dioxide and formic acid.

The worldwide production of formaldehyde in 1992 was 12 million tons.[3] The widest use of formaldehyde is in the production of resins with urea, phenol, and melamine. Formaldehyde-based resins are used as adhesives and impregnating resins in the manufacture of particle board, plywood, furniture, and other wood products.[4] Formaldehyde is used as an intermediate for synthesizing other industrial chemical compounds, such as 1,4-butanediol, trimethylopropane, and neopentyl glycol, which are used in the manufacture of polyurethane and polyester plastics, synthetic resin coatings, synthetic lubricating oil, and plasticizers. Formaldehyde itself is used as a preservative and disinfectant. It is used as an antimicrobial agent in many cosmetics products, hair preparations, deodorants, lotions, make-up, mouthwashes, and nail products.[5]

From 1981 to 1983, about 1,500,000 workers in the United States were estimated to have been exposed to formaldehyde during all or part of that period, representing about 0.6% of the population. In factories producing formaldehyde-based resins, a mean concentration of <1 to >10 ppm has been reported. More or less the same concentrations have been reported in wood products and paper factories, and those of textile garments, and other activities.[5]

Ambient levels are generally <1 $\mu g/m^3$ in remote areas[6] and 1–20 $\mu g/m^3$[4,6–9] in urban environments. A major source of formaldehyde is incomplete combustion of hydrocarbon fuels, especially from vehicle emissions. The introduction of MTBE as a gasoline additive increased formaldehyde emissions. A study by the state of Utah Department of Environmental Quality found that during the months of December 1994 and January 1995, when 2.7% oxygen-content-ethanol– and/or MTBE– oxygenated fuel was used, air concentrations of formaldehyde were increased by 300% to 1,800%, and those of acetaldehyde were increased by 470% to 860%.[10]

Indoor levels of formaldehyde are higher than outdoor levels, with the concentrations depending on the sources of formaldehyde that are present, the age of the source materials, ventilation, temperature, and humidity.[6] Cigarette smoke has been reported to contain levels of a few micrograms to several milligrams of formaldehyde per cigarette.[4] The sweetening agent aspartame hydrolyzes in the gastrointestinal tract to become free methyl alcohol, which is metabolized in the liver to formaldehyde, formic acid, and CO_2.[11]

Formaldehyde induces gene mutation in bacteria, fungi, yeast, *Drosophila larvae*, and cultured rodent and human cells. It also causes single-strand breaks in DNA, sister chromatid exchanges, chromosome aberrations, and the transformation of rodent cells in a variety of *in vitro* assays.[12–15] Formaldehyde induces cytogenetic damage in tissues that are locally exposed, either by gavage or inhalation. Sprague-Dawley rats treated orally with formaldehyde and sacrificed 16–30 h after treatment had a greater than fivefold increase in the frequency of micronucleated cells in the stomach, duodenum, ileum, and colon.[16] Male Sprague-Dawley rats exposed by inhalation to formaldehyde vapor for one and eight weeks showed a significant increase in chromosomal aberrations in pulmonary lavage cells.[17] Fisher 344 rats and B6C3F1 mice who received whole-body exposure to formaldehyde vapor for up to 24 months showed nasal cavity malignancies.[18]

In two experiments, male Syrian Golden hamsters exposed to 30 ppm or 10 ppm of formaldehyde for 5 h/day, 1 or 5 days/week for life showed no nasal tumors.[19] Wistar rats who received drinking water containing formaldehyde for up to 24 months at concentrations of 0–82 mg/kg bw per day for males, and 0–109 mg/kg bw per day for females, showed no treatment-related neoplastic changes.[20]

In a lifetime study performed in our laboratory, formaldehyde was administered in drinking water to male and female Sprague-Dawley rats. In this study, male and female breeder rats were given formaldehyde at 2,500 or 0 ppm for 104 weeks. The offspring were initially exposed to 2,500 or 0 ppm formaldehyde *in utero* starting on day 13 of gestation and then received these levels in drinking water for 104 weeks. Preliminary results demonstrated an increase in a variety of malignant and benign tumors of the stomach and intestine in treated animals. It was also noted that, although the incidence of intestinal tract tumors was low, there were no comparable tumors in the control groups in this study; some of these tumors were reported to be uncommon among historical controls.[21]

A comprehensive review of the epidemiological studies conducted on industrial and other professional groups exposed to formaldehyde has been reported.[5] Overall, epidemiological studies may only suggest a causal relationship between formaldehyde exposure and nasopharyngeal cancer. No excess risks have been shown for oropharyngeal, laryngeal, and lung cancer; low or no risks have been shown for lymphatic or hematopoietic cancer.

ACETALDEHYDE

The production capacity for acetaldehyde in the United States in 1989 was 443,000 tons/year.[23] Acetaldehyde is used principally as a chemical intermediate, predominantly of acetic acid, which is primarily used to make vinyl acetate, cellulose acetate, and other acetic esters.[24] Acetaldehyde is also used in silvering mirrors, leather tanning, as a denaturant for alcohol, in fuel mixtures, as a hardener for gelatin fibers, in glue and casein products, as a preservative for fish and fruit, as well as in the paper industry and as a synthetic rubber.[25]

Acetaldehyde is a natural product of combustion and photo-oxidation of hydrocarbons and has been detected at low levels in drinking water, surface water, rainwater, effluents, engine exhaust, and ambient and indoor air samples. It is present in small amounts in alcoholic beverages such as beer, wine and spirits, and in plant juices, essential oils, and roasted coffee.[23,25–26] Cigarette smoke has been reported to contain acetaldehyde at the level of 980 µg/cigarette.[27] The concentration of acetaldehyde in the whole blood of normal fasting human subjects was reported by Lynch *et al*.[28] to be 1.30 µmol/L (57 µg/L). Acetaldehyde has been detected in mothers' milk.[29] Acetaldehyde is formed during the intracellular oxidation of ethanol.[30] It is metabolized to acetic acid by NAD^+-dependent aldehyde dehydrogenases.[31]

Fetal malformations have been found in mice and rats treated with acetaldehyde.[32] Acetaldehyde causes gene mutations in bacteria and gene mutation, sister chromatid exchanges, micronuclei and aneuploidy in cultured mammalian cells, without metabolic activation. *In vivo*, it causes mutation in *Drosophyla melanogaster* but not micronuclei in mouse germ cells. It causes DNA damage in cultured mammalian cells and in mice *in vivo*.[33] Acetaldehyde-DNA adducts have been found in white blood cells from human alcohol abusers.[34]

In long-term carcinogenicity bioassays, WU albino Wistar rats treated by inhalation showed dose-related increases in nasal carcinomas in both sexes.[35] In hamsters, inhalation of acetaldehyde enhanced the incidence of respiratory tract tumors produced by intratracheal instillation of benzo[*a*]pyrene.[36]

To date, the available epidemiological studies are inadequate for evaluation of the carcinogenicity of acetaldehyde to humans.

MATERIALS AND METHODS

Formaldehyde was supplied in aqueous solution by Montedison S.p.A., Italy, at a concentration of $30.0 \pm 0.2\%$. The impurities included iron 0.6 mg/L; lead 0.1 mg/L; sulphur <5.0 mg/L; chlorine <5.0 mg/L; methyl alcohol (stabilizer) 0.3%. Acetaldehyde was supplied by FLUKA Chemica-Biokemica, AG, Switzerland; its purity was higher than 99.0%. During experiments, formaldehyde and acetaldehyde were supplied every two months and were always kept in the dark at a temperature of 14–20°C and 4–5°C, respectively.

Formaldehyde was administered for 104 weeks in the drinking water to 7-week-old Sprague-Dawley rats ($n = 50$ males and $n = 50$ females) and supplied *ad libitum* at concentrations of 1,500, 1,000, 500, 100, 50, or 10 mg/L. One group of 50 males and 50 females received methyl alcohol in 15 mg/L of drinking water. Control animals (100 males and 100 females) received tap water. Acetaldehyde was administered at concentrations of 2,500, 1,500, 500, 250, 50, or 0 mg/L to Sprague-Dawley rats (50 males and 50 females), beginning at age 6 weeks. The studies on formaldehyde and acetaldehyde ended with the death of the last animal at 163 and 161 weeks of age, respectively.

Experiments were performed according to the Good Laboratory Practices (GLP) and the Standard Operating Procedure (SOP) of the CRC/RF. Detailed experimental methods and protocol for counting tumors have been described previously.[37] Statistical analysis was performed using the χ^2 test to evaluate differences in tumor incidence between treated and control groups. The Cochrane Armitage test was used to evaluate dose-response relationships.

RESULTS

Formaldehyde

During the experiment, the intake of liquids decreased in males treated at the highest dose and in females treated at 1,500, 1,000, and 500 mg/L of formaldehyde. No differences were observed in daily feed consumption, body weight, behavior, or survival between treated and control animals. Yellow hair coat was observed in animals exposed to formaldehyde, mainly at the highest concentration. No treatment-related non-oncological pathological changes were detected by gross inspection or histopathological examination.

TABLE 1. Long-term carcinogenicity bioassays on formaldehyde administered with drinking water supplied *ad libitum* to male (M) and female (F) Sprague-Dawley rats

NUMBER AND PERCENTAGE OF MALE AND FEMALE SPRAGUE-DAWLEY RATS BEARING VARIOUS TYPES OF BENIGN AND MALIGNANT TUMORS[a]

Site / Histotype	I: 1,500 mg/l Male No.	%	I: 1,500 mg/l Female No.	%	II: 1,000 mg/l Male No.	%	II: 1,000 mg/l Female No.	%	III: 500 mg/l Male No.	%	III: 500 mg/l Female No.	%	IV: 100 mg/l Male No.	%	IV: 100 mg/l Female No.	%	V: 50 mg/l Male No.	%	V: 50 mg/l Female No.	%	VI: 10 mg/l Male No.	%	VI: 10 mg/l Female No.	%	VII: 0[b] Male No.	%	VII: 0[b] Female No.	%	VIII: 0 (control) Male No.	%	VIII: 0 (control) Female No.	%
Skin																														
Cheratoacanthoma	0	-	1	2.0	0	-	0	-	0	-	0	-	0	-	0	-	0	-	0	-	0	-	0	-	0	-	0	-	0	-
Dermatofibroma	0	-	0	-	1	2.0	0	-	0	-	0	-	0	-	0	-	0	-	0	-	0	-	0	-	0	-	0	-	1	1.0
Squamous cell carcinoma	0	-	0	-	0	-	0	-	0	-	0	-	0	-	0	-	0	-	0	-	0	-	0	-	1	2.0	0	-	0	-
Basocellular carcinoma	1	2.0	0	-	0	-	0	-	0	-	0	-	0	-	0	-	0	-	0	-	0	-	0	-	1	-	0	-	0	-
Subcutaneous tissue																														
Fibroma	0	-	0	-	0	-	0	-	1	-	1	2.0	2	-	0	-	1	2.0	0	-	3	-	0	-	0	-	0	-	0	-
Lipoma	1	2.0	0	-	0	-	0	-	0	-	0	-	0	-	0	-	0	-	0	-	1	2.0	0	-	0	-	2	2.0	0	-
Fibrosarcoma	0	-	0	-	0	-	1	2.0	1	2.0	0	-	0	-	1	2.0	0	-	0	-	0	-	1	2.0	1	2.0	0	-	2	2.0	0	-
Liposarcoma	0	-	0	-	1	2.0	0	-	1	2.0	0	-	0	-	0	-	0	-	0	-	0	-	0	-	0	-	0	-	2	2.0	0	-
Angiosarcoma	0	-	0	-	1	2.0	0	-	0	-	0	-	0	-	0	-	0	-	0	-	0	-	0	-	0	-	0	-	0	-
Mammary glands																														
Fibroma and fibroadenoma	2	4.0	25(28)	50.0	3	6.0	20(23)	40.0	2	4.0	19(20)	38.0	1	2.0	14(20)	28.0	4(5)	8.0	12(14)	24.0	2	4.0	9	18.0	4	8.0	16(20)	32.0	6(8)	6.0	45(61)	45.0
Fibrolipoma	0	-	0	-	0	-	0	-	1	2.0	1	2.0	2.0	-	0	-	0	-	0	-	0	-	0	-	0	-	0	-	0	-	0	-
Lipoma	0	-	0	-	0	-	2	4.0	0	-	0	-	0	-	0	-	0	-	0	-	0	-	2	4.0	0	-	7(9)	14.0	1	1.0	11(12)	11.0
Adenocarcinoma	1	2.0	11(15)	22.0	0	-	9(10)	18.0	0	-	3	-	8(12)	16.0	0	-	4	8.0	0	-	2	4.0	2	-	2.0	-	0	-	0	-
Fibrosarcoma	0	-	0	-	0	-	1	2.0	0	-	2	-	2	4.0	0	-	1	2.0	1	-	2.0	-	1	2.0	0	-	0	-
Liposarcoma	0	-	1	2.0	0	-	0	-	0	-	2	4.0	2.0	-	0	-	0	-	0	-	0	-	0	-	0	-	0	-
Angiosarcoma	0	-	0	-	0	-	0	-	1	2.0	0	-	0	-	0	-	0	-	0	-	0	-	0	-	0	-
Harderian gland																														
Adenocarcinoma	0	-	0	-	0	-	0	-	0	-	0	-	0	-	0	-	1	2.0	1	2.0	0	-	0	-	0	-	0	-
Zymbal glands																														
Sebaceous adenoma	0	-	0	-	0	-	0	-	0	-	0	-	0	-	1	2.0	0	-	0	-	0	-	0	-	1	1.0
Carcinoma	2(3)	4.0	0	-	3	6.0	0	-	4	8.0	1	2.0	1	2.0	0	-	1	-	2	4.0	3	6.0	0	-	3	3.0	1	1.0
Ear ducts																														
Carcinoma	0	-	0	-	3	6.0	3	6.0	0	-	-	2.0	0	-	0	-	1	2.0	2	4.0	1	2.0	1	2.0	1	2.0	7	7.0	3	3.0
Nasal cavities																														
Carcinoma	1	2.0	0	-	1	2.0	1	2.0	0	-	0	-	0	-	0	-	1	2.0	0	-	0	-	0	-	1	2.0	0	-	0	-	0	-
Neuroblastoma	1	-	0	-	1	-	0	-	0	-	0	-	0	-	0	-	0	-	0	-	0	-	0	-	0	-
Oral cavity lips and tongue																														
Acanthoma	0	-	0	-	0	-	0	-	1	2.0	0	-	0	-	0	-	0	-	0	-	0	-	1	2.0	0	-	2	2.0	0	-	0	-
Carcinoma	3	6.0	1	2.0	0	3	6.0	1	2.0	0	-	0	-	1	2.0	0	3	6.0	0	1	4.0	2	-	0	-	1	2.0	0	5	5.0	3	3.0
Lung																														
Adenoma	0	-	0	-	0	-	0	-	0	-	0	-	0	-	2	4.0	0	-	0	-	0	-	1	2.0	0	-	2	2.0	1	1.0
Angioma	0	-	0	-	0	-	0	-	0	-	0	-	0	-	0	-	0	-	0	-	1	2.0	0	-	0	-	0	-	1	1.0	0	-
Adenocarcinoma	0	-	0	-	0	-	0	-	0	-	0	-	0	-	0	-	0	-	1	2.0	0	-	0	-	0	-	1	2.0	0	-	0	-
Esophagus																														
Carcinoma	1	2.0	0	-	0	-	0	-	0	-	0	-	0	-	0	-	0	-	0	-	0	-	0	-	0	-	0	-	0	-
Stomach																														
- Forestomach																														
Acanthoma	1	2.0	0	-	0	-	0	-	0	-	0	-	0	-	0	-	1	2.0	0	-	3	6.0	0	-	0	-	0	-	2	2.0	2	2.0
Squamous cell carcinoma	0	-	0	-	0	-	0	-	0	-	0	-	0	-	0	-	0	-	0	-	0	-	1	2.0	0	-	0	-	0	-	0	-
Leiomyosarcoma	0	-	0	-	0	-	0	-	0	-	0	-	0	-	0	-	0	-	0	-	1	2.0	0	-	0	-	0	-	0	-	0	-

—Contintued

TABLE 1. Continued

NUMBER AND PERCENTAGE OF MALE AND FEMALE SPRAGUE-DAWLEY RATS BEARING VARIOUS TYPES OF BENIGN AND MALIGNANT TUMORS[a]

Site / Histotype	I: 1,500 mg/l Male No. / %	I: 1,500 mg/l Female No. / %	II: 1,000 mg/l Male No. / %	II: 1,000 mg/l Female No. / %	III: 500 mg/l Male No. / %	III: 500 mg/l Female No. / %	IV: 100 mg/l Male No. / %	IV: 100 mg/l Female No. / %	V: 50 mg/l Male No. / %	V: 50 mg/l Female No. / %	VI: 10 mg/l Male No. / %	VI: 10 mg/l Female No. / %	VII: 0[b] Male No. / %	VII: 0[b] Female No. / %	VIII: 0 (control) Male No. / %	VIII: 0 (control) Female No. / %
- Glandular stomach																
Adenomatous polyp	6 / 12.0	0 / -	0 / -	0 / -	0 / -	0 / -	0 / -	0 / -	0 / -	0 / -	0 / -	0 / -	0 / -	0 / -	0 / -	0 / -
Adenocarcinoma	1 / 2.0	2 / 4.0	0 / -	0 / -	0 / -	0 / -	0 / -	0 / -	0 / -	0 / -	1 / 2.0	0 / -	0 / -	0 / -	0 / -	0 / -
Leiomyosarcoma	0 / -	0 / -	1 / 2.0	0 / -	0 / -	0 / -	0 / -	0 / -	0 / -	0 / -	0 / -	0 / -	0 / -	0 / -	0 / -	0 / -
Intestine																
Adenomatous polyp	0 / -	0 / -	0 / -	1 / 2.0	0 / -	0 / -	0 / -	0 / -	0 / -	0 / -	0 / -	0 / -	0 / -	0 / -	0 / -	0 / -
Leiomyoma	0 / -	3 / 6.0	0 / -	0 / -	0 / -	0 / -	0 / -	0 / -	0 / -	1 / 2.0	0 / -	2 / 4.0	0 / -	0 / -	0 / -	0 / -
Adenocarcinoma	3 / 6.0	0 / -	0 / -	0 / -	0 / -	0 / -	0 / -	0 / -	0 / -	1 / 2.0	1 / 2.0	0 / -	0 / -	0 / -	0 / -	0 / -
Leiomyosarcoma	2 / 4.0	0 / -	0 / -	0 / -	0 / -	0 / -	0 / -	0 / -	0 / -	1 / 2.0	0 / -	0 / -	0 / -	0 / -	0 / -	0 / -
Salivary glands																
Adenocarcinoma	0 / -	0 / -	0 / -	1 / 2.0	0 / -	0 / -	0 / -	0 / -	0 / -	0 / -	0 / -	0 / -	0 / -	0 / -	0 / -	0 / -
Liver																
Cholangioma	0 / -	0 / -	0 / -	1 / 2.0	0 / -	0 / -	0 / -	1 / 2.0	1 / 2.0	1 / 2.0	0 / -	0 / -	1 / 2.0	0 / -	1 / 2.0	1 / 1.0
Hepatocarcinoma	2 / 4.0	0 / -	0 / -	1 / 2.0	1 / 2.0	0 / -	0 / -	2 / 4.0	0 / -	0 / -	0 / -	1 / 2.0	1 / 2.0	0 / -	2 / 2.0	0 / -
Cholangiocarcinoma	0 / -	0 / -	0 / -	0 / -	0 / -	0 / -	0 / -	0 / -	0 / -	0 / -	0 / -	1 / 2.0	0 / -	0 / -	1 / -	0 / -
Angiosarcoma	0 / -	0 / -	1 / 2.0	0 / -	1 / 2.0	0 / -	0 / -	0 / -	0 / -	0 / -	1 / 2.0	0 / -	0 / -	0 / -	1 / 2.0	1 / 1.0
Embryonal sarcoma	0 / -	0 / -	0 / -	0 / -	0 / -	0 / -	0 / -	0 / -	0 / -	0 / -	0 / -	0 / -	0 / -	0 / -	0 / -	- / -
Pancreas																
Exocrine adenoma	2 / 4.0	0 / -	0 / -	0 / -	0 / -	0 / -	0 / -	0 / -	0 / -	0 / -	0 / -	1 / 2.0	1 / 2.0	0 / -	0 / -	1 / 1.0
Islet cell adenoma	1 / 2.0	0 / -	3 / 6.0	1 / 2.0	3 / 6.0	2 / 4.0	0 / -	2 / 4.0	1 / 2.0	2 / 4.0	1 / 2.0	1 / 2.0	2 / 4.0	2 / 4.0	5 / 5.0	3 / 3.0
Islet cell carcinoma	1 / 2.0	0 / -	0 / -	0 / -	0 / -	0 / -	0 / -	1 / 2.0	0 / -	0 / -	0 / -	0 / -	0 / -	1 / 2.0	0 / -	0 / -
Kidneys																
Adenoma	0 / -	0 / -	0 / -	0 / -	0 / -	1 / 2.0	0 / -	0 / -	0 / -	0 / -	0 / -	0 / -	0 / -	0 / -	1 / 1.0	0 / -
Lipoma	0 / -	0 / -	0 / -	0 / -	0 / -	0 / -	0 / -	1 / 2.0	0 / -	0 / -	0 / -	0 / -	0 / -	0 / -	0 / -	0 / -
Adenocarcinoma	0 / -	0 / -	0 / -	0 / -	1 / 2.0	0 / -	0 / -	0 / -	0 / -	0 / -	0 / -	1 / 2.0	0 / -	0 / -	0 / -	0 / -
Nephroblastoma	0 / -	0 / -	0 / -	0 / -	0 / -	0 / -	0 / -	0 / -	1 / 2.0	0 / -	1 / 2.0	0 / -	0 / -	0 / -	0 / -	1 / 1.0
Pelvis and ureters																
Transitional cell carcinoma	0 / -	0 / -	0 / -	0 / -	0 / -	0 / -	0 / -	1 / 2.0	0 / -	0 / -	0 / -	0 / -	0 / -	0 / -	0 / -	0 / -
Bladder																
Angioma	0 / -	0 / -	0 / -	1 / 2.0	0 / -	0 / -	0 / -	0 / -	0 / -	0 / -	0 / -	0 / -	0 / -	0 / -	0 / -	0 / -
Prostate																
Adenoma	1 / 2.0		0 / -		0 / -		0 / -		0 / -		0 / -		0 / -		1 / 1.0	
Testes																
Interstitial cell adenoma	9(11) / 18.0		12(21) / 24.0		10(14) / 20.0		6(7) / 12.0		6(11) / 12.0		3(5) / 6.0		3(5) / 6.0		10(12) / 10.0	
Interstitial cell malignant tumor	0 / -		0 / -		1 / 2.0		0 / -		0 / -		0 / -		0 / -		0 / -	
Ovaries																
Cystadenoma		0 / -		0 / -		1 / 2.0		0 / -		1 / 2.0		0 / -		0 / -		0 / -
Granulosa cell tumor		2(4) / 4.0		3(5) / 6.0		2(4) / 4.0		0 / -		0 / -		1(2) / 2.0		2(3) / 4.0		4(7) / 4.0
Granulosa and theca cell tumor		2 / 4.0		0 / -		0 / -		0 / -		0 / -		0 / -		0 / -		2 / 2.0
Sertoli cell tumor		0 / -		1 / 2.0		1 / 2.0		0 / -		0 / -		0 / -		0 / -		0 / -
Benign arrhenoblastoma		0 / -		0 / -		0 / -		0 / -		0 / -		0 / -		0 / -		1 / 1.0
Adenocarcinoma		1 / 2.0		0 / -		0 / -		0 / -		0 / -		0 / -		0 / -		0 / -

— *Continued*

TABLE 1. Continued

NUMBER AND PERCENTAGE OF MALE AND FEMALE SPRAGUE-DAWLEY RATS BEARING VARIOUS TYPES OF BENIGN AND MALIGNANT TUMORS[a]

Site Histotype	I: 1,500 mg/l Male No.	%	Female No.	%	II: 1,000 mg/l Male No.	%	Female No.	%	III: 500 mg/l Male No.	%	Female No.	%	IV: 100 mg/l Male No.	%	Female No.	%	V: 50 mg/l Male No.	%	Female No.	%	VI: 10 mg/l Male No.	%	Female No.	%	VII: 0 Male No.	%	Female No.	%	VIII: 0 (control) Male No.	%	Female No.	%
Uterus																																
Polyp			2	4.0			6	12.0			5	10.0			2	4.0			5	10.0			4	8.0			2	4.0			10	10.0
Leiomyoma			0	-			0	-			0	-			0	-			0	-			1	2.0			0	-			0	-
Fibroangioma			0	-			0	-			0	-			0	-			1	2.0			0	-			1	2.0			0	-
Angioma			0	-			0	-			0	-			1	2.0			0	-			0	-			0	-			0	-
Squamous cell carcinoma			0	-			0	-			1	2.0			0	-			0	-			2	4.0			1	2.0			1	1.0
Adenocarcinoma			3	6.0			3	6.0			3	6.0			8	16.0			2	4.0			0	-			1	2.0			3	3.0
Fibrosarcoma			0	-			0	-			2	4.0			0	-			0	-			0	-			0	-			0	-
Leiomyosarcoma			0	-			0	-			1	2.0			0	-			0	-			0	-			0	-			0	-
Angiosarcoma			1	2.0			0	-			0	-			0	-			0	-			0	-			0	-			0	-
Uterus & Vagina																																
Malignant Schwannoma			0	-			0	-			0	-			0	-			0	-			2	4.0			0	-			3	3.0
Vagina																																
Malignant Schwannoma			0	-			0	-			0	-			0	-			0	-			1	2.0			0	-			0	-
Peritoneum																																
Lipoma	0	-	0	-	0	-	0	-	0	-	0	-	0	-	0	-	0	-	1	2.0	0	-	0	-	0	-	0	-	0	-	0	-
Mesothelioma	1	2.0	1	2.0	0	-	0	-	0	-	0	-	0	-	0	-	0	-	1	2.0	0	-	0	-	0	-	0	-	0	-	1	1.0
Pituitary gland																																
Adenoma	5	10.0	20	40.0	9	18.0	19	38.0	8	16.0	31	62.0	9	18.0	30	60.0	6	12.0	17	34.0	3	6.0	19	38.0	6	12.0	27	54.0	22	22.0	44	44.0
Thyroid gland																																
C-cell adenoma	2	4.0	2	4.0	0	-	0	-	1	2.0	2	4.0	0	-	1	2.0	1	2.0	1	2.0	1	2.0	1	2.0	0	-	2	4.0	9	9.0	2	2.0
Follicular carcinoma	0	-	0	-	2	4.0	0	-	0	-	0	-	0	-	0	-	0	-	0	-	0	-	0	-	0	-	0	-	0	-	0	-
C-cell carcinoma	0	-	1	2.0	0	-	1	2.0	0	-	0	-	1	2.0	0	-	1	2.0	1	2.0	0	-	0	-	0	-	0	-	2	2.0	0	-
Adrenal glands																																
Cortical adenoma	0	-	0	-	0	-	3	6.0	2	4.0	3	6.0	0	-	3	6.0	1	2.0	3	6.0	0	-	2	4.0	0	-	3	6.0	2	2.0	10(11)	10.0
Pheochromocytoma	11(16)	22.0	12(16)	24.0	22(33)	44.0	8(11)	16.0	19(26)	38.0	20(26)	40.0	25(36)	50.0	16(26)	32.0	25(37)	50.0	17(22)	34.0	21(30)	42.0	11(17)	22.0	16(24)	32.0	10(14)	20.0	53(77)	53.0	36(47)	36.0
Cortical adenocarcinoma	0	-	4(7)	8.0	0	-	0	-	0	-	0	-	0	-	0	-	0	-	2	4.0	0	-	0	-	0	-	1	2.0	0	-	4	4.0
Pheochromoblastoma	4(5)	8.0	2	4.0	1	2.0	0	-	8(10)	16.0	1	2.0	3	6.0	1	2.0	0	-	1	2.0	2	4.0	0	-	5(6)	10.0	1	2.0	5(6)	5.0	6	6.0
Central nervous system																																
- Brain																																
Oligodendroglioma	0	-	0	-	1	2.0	1	2.0	1	2.0	0	-	1	2.0	1	2.0	0	-	0	-	1	2.0	2	4.0	0	-	5	10.0	3	3.0	2	2.0
- Meninges																																
Malignant meningioma	0	-	0	-	0	-	0	-	0	-	0	-	0	-	0	-	0	-	1	2.0	0	-	0	-	0	-	2	4.0	2	2.0	0	-
Peripheral nervous system																																
- Major peripheral nerves																																
Malignant Schwannoma	0	-	0	-	0	-	0	-	0	-	0	-	0	-	0	-	0	-	0	-	0	-	0	-	0	-	0	-	0	-	1	1.0
- Ganglia																																
Pheochromocytoma	0	-	0	-	1	2.0	0	-	0	-	0	-	0	-	0	-	0	-	0	-	0	-	0	-	0	-	0	-	0	-	0	-
Bones																																
- Head																																
Osteosarcoma	3	6.0	1	2.0	3	6.0	0	-	2	4.0	0	-	3	6.0	3	6.0	0	-	0	-	1	2.0	2	4.0	1	2.0	2	4.0	4	4.0	0	-
- Other																																
Osteosarcoma	0	-	1	2.0	1	2.0	0	-	0	-	0	-	0	-	0	-	0	-	0	-	0	-	0	-	1	2.0	0	-	2	2.0	0	-

— Continued

TABLE 1. Continued

NUMBER AND PERCENTAGE OF MALE AND FEMALE SPRAGUE-DAWLEY RATS BEARING VARIOUS TYPES OF BENIGN AND MALIGNANT TUMORS[a]

Site Histotype	I: 1,500 mg/l Male No.	I: 1,500 mg/l Male %	I: 1,500 mg/l Female No.	I: 1,500 mg/l Female %	II: 1,000 mg/l Male No.	II: 1,000 mg/l Male %	II: 1,000 mg/l Female No.	II: 1,000 mg/l Female %	III: 500 mg/l Male No.	III: 500 mg/l Male %	III: 500 mg/l Female No.	III: 500 mg/l Female %	IV: 100 mg/l Male No.	IV: 100 mg/l Male %	IV: 100 mg/l Female No.	IV: 100 mg/l Female %	V: 50 mg/l Male No.	V: 50 mg/l Male %	V: 50 mg/l Female No.	V: 50 mg/l Female %	VI: 10 mg/l Male No.	VI: 10 mg/l Male %	VI: 10 mg/l Female No.	VI: 10 mg/l Female %	VII: 0[b] Male No.	VII: 0[b] Male %	VII: 0[b] Female No.	VII: 0[b] Female %	VIII: 0 (control) Male No.	VIII: 0 (control) Male %	VIII: 0 (control) Female No.	VIII: 0 (control) Female %
Soft tissues																																
Fibroangioma	0	-	0	-	0	-	0	-	1	2.0	0	-	0	-	0	-	0	-	0	-	0	-	0	-	0	-	0	-	1	1.0	0	-
Liposarcoma	1	2.0	0	-	0	-	0	-	0	-	0	-	0	-	0	-	0	-	0	-	0	-	0	-	0	-	0	-	0	-	0	-
Angiosarcoma	0	-	0	-	0	-	1	2.0	0	-	0	-	0	-	0	-	0	-	0	-	0	-	0	-	0	-	0	-	0	-	0	-
Heart																																
Malignant Schwannoma	2	4.0	1	2.0	0	-	0	-	0	-	0	-	0	-	0	-	0	-	0	-	0	-	0	-	0	-	0	-	0	-	0	-
Thymus																																
Benign thymoma[c]	0	-	0	-	0	-	2	4.0	0	-	0	-	0	-	0	-	0	-	0	-	0	-	1	2.0	0	-	0	-	0	-	0	-
Malignant thymoma[c]	1	2.0	0	-	0	-	0	-	0	-	0	-	0	-	0	-	0	-	0	-	0	-	0	-	0	-	0	-	0	-	0	-
Spleen																																
Angioma	1	2.0	0	-	0	-	0	-	0	-	0	-	0	-	0	-	0	-	0	-	0	-	0	-	0	-	0	-	0	-	0	-
Angiosarcoma	1	2.0	0	-	0	-	0	-	0	-	0	-	0	-	0	-	0	-	0	-	0	-	0	-	0	-	0	-	0	-	0	-
Subcutaneous lymph nodes																																
Fibroangioma	0	-	0	-	0	-	0	-	0	-	0	-	0	-	0	-	0	-	0	-	0	-	0	-	0	-	0	-	1(2)	1.0	0	-
Pancreatic lymph nodes																																
Fibroangioma	1	2.0	0	-	0	-	0	-	0	-	0	-	0	-	0	-	0	-	0	-	0	-	0	-	0	-	0	-	0	-	0	-
Mesenteric lymph nodes																																
Angioma	0	-	0	-	0	-	0	-	0	-	0	-	0	-	0	-	0	-	0	-	0	-	1	2.0	0	-	0	-	0	-	0	-
Fibroangioma	1	2.0	0	-	0	-	0	-	1	2.0	0	-	1	2.0	0	-	0	-	0	-	0	-	0	-	0	-	0	-	0	-	2	2.0
Hemolymphoreticular tissues[d,e]																																
Histiocytoma	0	-	0	-	0	-	0	-	0	-	0	-	0	-	0	-	0	-	0	-	0	-	0	-	0	-	1	2.0	0	-	0	-
Lymphomas and leukemias	23	46.0	10	20.0	11	22.0	11	22.0	12	24.0	7	14.0	13	26.0	8	16.0	10	20.0	7	14.0	4	8.0	5	10.0	10	20.0	5	10.0	8	8.0	7	7.0

[a] Number in parentheses are the number of total tumours, one animal can bear more than one tumour
[b] 15 mg of methyl alcohol / liter of drinking water
[c] In 96% of cases the tumor itself is composed of a mixture in varying proportions of epithelial cells and lymphocytes. In the remaining 4%, only epithelial cells are present. We consider that a tumor composed exclusively of lymphocytes should not be classified as a thymoma but as lymphoma involving the thymus.
[d] Including thymus, spleen, subcutaneous, pancreatic and mesenteric lymph nodes
[e] See table 3

TABLE 2. Long-term carcinogenicity bioassays on formaldehyde administered with drinking water supplied *ad libitum* to male (M) and female (F) Sprague-Dawley rats

TOTAL MALIGNANT TUMORS

Group No.	Concentration (mg/l)	Animals Sex	Animals No.	Malignant tumors Tumor-bearing animals No.	Malignant tumors Tumor-bearing animals %	Malignant tumors Tumors No.	Malignant tumors Tumors Per 100 animals
I	1,500	M	50	36	72.0 **	56	112.0 **
		F	50	27	54.0	48	96.0 **
		M+F	100	63	63.0	104	104.0
II	1,000	M	50	23	46.0	30	60.0
		F	50	29	58.0	39	78.0 **
		M+F	100	52	52.0	69	69.0
III	500	M	50	24	48.0	36	72.0 *
		F	50	19	38.0	25	50.0
		M+F	100	43	43.0	61	61.0
IV	100	M	50	22	44.0	23	46.0
		F	50	25	50.0	41	82.0 **
		M+F	100	47	47.0	64	64.0
V	50	M	50	12	24.0	15	30.0
		F	50	20	40.0	26	52.0
		M+F	100	32	32.0	41	41.0
VI	10	M	50	14	28.0	19	38.0
		F	50	20	40.0	22	44.0
		M+F	100	34	34.0	41	41.0
VII	0[a]	M	50	21	42.0	29	58.0
		F	50	23	46.0	32	64.0
		M+F	100	44	44.0	61	61.0
VIII	0	M	100	38	38.0	50	50.0
		F	100	43	43.0	49	49.0
		M+F	200	81	40.5	99	49.5

[a] 15 mg of methyl alcohol / liter of drinking water
* $p < 0.05$ using χ^2 test
** $p < 0.01$ using χ^2 test

The occurrence of benign and malignant tumors in animals exposed to formaldehyde is shown in TABLE 1. Differences observed between formaldehyde-treated and control animals are as follows:

(1) The number of total malignant tumors was highest in males and females treated with 1,500 and 1,000 mg/L, in males treated with 500 mg/L, and in females treated with 100 mg/L (TABLE 2). An increase in total malignant

TABLE 3. Long-term carcinogenicity bioassays on formaldehyde administered with drinking water supplied *ad libitum* to male (M) and female (F) Sprague-Dawley rats

HEMOLYMPHORETICULAR NEOPLASIAS

Group No.	Concentration (mg/l)	Animals		Animals with hemolymphoreticular neoplasias		
		Sex	No.	No.	%	Latency[a] (weeks)
I	1,500	M	50	23	46.0 **	98.2
		F	50	10	20.0 *	110.5
		M+F	100	33	33.0	104.4
II	1,000	M	50	11	22.0 *	101.5
		F	50	11	22.0 *	118.0
		M+F	100	22	22.0	109.8
III	500	M	50	12	24.0 *	121.2
		F	50	7	14.0	113.6
		M+F	100	19	19.0	117.4
IV	100	M	50	13	26.0 **	106.3
		F	50	8	16.0	118.6
		M+F	100	21	21.0	112.5
V	50	M	50	10	20.0	111.3
		F	50	7	14.0	104.3
		M+F	100	17	17.0	107.8
VI	10	M	50	4	8.0	118.7
		F	50	5	10.0	88.6
		M+F	100	9	9.0	103.7
VII	0[b]	M	50	10	20.0	105.6
		F	50	5	10.0	106.8
		M+F	100	15	15.0	106.2
VIII	0	M	100	8	8.0	108.7
		F	100	7	7.0	103.0
		M+F	200	15	7.5	105.9

[a] Age at death
[b] 15 mg of methyl alcohol / liter of drinking water
* $p<0.05$ using χ^2 test
** $p<0.01$ using χ^2 test

tumors was also detected in males and females treated with methyl alcohol alone.

(2) The number of mammary malignant tumors increased in females treated with 1,500 ($P < 0.05$), 1,000, and 100 ($P < 0.01$) mg/L and were also increased in females treated with methyl alcohol alone.

(3) Sporadic cases of oncological lesions of the stomach, including two leiomyosarcomas, which have a very rare spontaneous incidence in the colony of Sprague-Dawley rats used, were observed in males treated with formaldehyde at concentrations of 1,000 and 10 mg/L, but were not found in animals treated with methyl alcohol alone or in the control group.

(4) Oncological lesions of the intestine were detected in the treated groups, particularly at the highest dose. Among the lesions observed were leiomyomas and leiomyosarcomas, very rare tumors in the Sprague-Dawley colony of rats used at the CRC/RF.[21] These tumors were not found in animals treated with methyl alcohol alone or in the control group.

(5) The number of testicular interstitial cell adenomas increased in males treated with 1,500, 1,000 ($P < 0.05$), and 500 mg/L of formaldehyde.

(6) The number of hemolymphoreticular neoplasias increased in both sexes treated with 1,500, 1,000, 500, 100, and 50 mg/L, and a slight increase was observed in females treated with 10 mg/L (TABLE 3). This increase shows a dose-response relationship. An increase in hemolymphoreticular neoplasias was also observed in males treated with methyl alcohol alone when compared to nontreated animals.

Acetaldehyde

No significant differences in the daily consumption of beverages and feed, behavior, body weight, or survival were observed between treated and control animals, nor were any treatment-related nononcological pathological changes detected by gross inspection or histopathological examination.

The occurrence of benign and malignant tumors in acetaldehyde-treated animals is shown in TABLE 4. Differences observed between treated and control animals are as follows:

(1) The number of total malignant tumors was increased in all of the treated groups, except for males treated with 250 mg/L (TABLE 5).

(2) The number of malignant mammary tumors increased in all treated females, with the exception of those treated with 250 mg/L. This increase was not dose related. It is noteworthy that some malignant mammary tumors were also observed in treated males, except for those treated with 250 mg/L.

(3) The overall incidence of carcinomas of the Zymbal gland, external ear ducts, nasal sinuses, and oral cavity increased in males and females treated with the highest dose.

(4) Sporadic cases of lung adenoma/adenocarcinoma were found in groups treated with the three highest doses.

(5) Sporadic cases of tumors of the stomach and intestine were observed in the treated groups.

TABLE 4. Long-term carcinogenicity bioassays on acetaldehyde administered with drinking water supplied *ad libitum* to male (M) and female (F) Sprague-Dawley rats

NUMBER AND PERCENTAGE OF MALE AND FEMALE SPRAGUE-DAWLEY RATS BEARING VARIOUS TYPES OF BENIGN AND MALIGNANT TUMORS[a]

Site / Histotype	I: 2,500 mg/l Male No.	%	Female No.	%	II: 1,500 mg/l Male No.	%	Female No.	%	III: 500 mg/l Male No.	%	Female No.	%	IV: 250 mg/l Male No.	%	Female No.	%	V: 50 mg/l Male No.	%	Female No.	%	VI: 0 (control) Male No.	%	Female No.	%
Skin																								
Acanthoma	0	-	0	-	0	-	0	-	0	-	0	-	0	-	0	-	0	-	0	-	1	2.0	0	-
Dermatofibroma	0	-	0	-	2	4.0	0	-	1	2.0	0	-	2	4.0	0	-	0	-	0	-	0	-	0	-
Squamous cell carcinoma	1	2.0	0	-	0	-	0	-	0	-	0	-	0	-	0	-	0	-	0	-	0	-	0	-
Basocellular carcinoma	0	-	0	-	0	-	0	-	0	-	0	-	0	-	0	-	0	-	0	-	0	-	1	2.0
Fibrosarcoma	0	-	1	2.0	0	-	0	-	0	-	0	-	0	-	0	-	0	-	0	-	0	-	0	-
Subcutaneous tissue																								
Fibroma	1	2.0	0	-	0	-	0	-	1	2.0	0	-	0	-	0	-	1	2.0	0	-	0	-	0	-
Lipoma	0	-	0	-	1	2.0	0	-	0	-	0	-	1	2.0	0	-	0	-	0	-	0	-	0	-
Liposarcoma	0	-	1	2.0	0	-	0	-	1	2.0	0	-	0	-	1	2.0	0	-	0	-	2	4.0	0	-
Interscapular brown fat pad																								
Lipoma	0	-	0	-	0	-	1	2.0	0	-	0	-	0	-	0	-	0	-	0	-	0	-	0	-
Liposarcoma	0	-	0	-	0	-	0	-	0	-	0	-	0	-	0	-	0	-	0	-	0	-	0	-
Mammary glands																								
Fibroma and fibroadenoma	3	6.0	22(30)	44.0	1	2.0	29(41)	54.0	1	2.0	25(40)	50.0	1	2.0	26(33)	52.0	1	2.0	17(21)	34.0	3	6.0	14(22)	28.0
Fibrolipoma	0	-	1	2.0	0	-	0	-	1	2.0	0	-	0	-	0	-	1	2.0	0	-	0	-	0	-
Lipoma	0	-	0	-	1	2.0	1	2.0	0	-	0	-	0	-	0	-	1	2.0	1	2.0	0	-	0	-
Adenocarcinoma	1	2.0	6(10)	12.0	0	-	8	16.0	0	-	10(11)	20.0	0	-	3	6.0	2	4.0	9(13)	18.0	0	-	3	6.0
Fibrosarcoma	0	-	0	-	0	-	0	-	2	4.0	0	-	0	-	0	-	1	2.0	0	-	0	-	0	-
Liposarcoma	0	-	1	2.0	1	2.0	0	-	0	-	0	-	0	-	0	-	0	-	0	-	0	-	0	-
Carcinosarcoma	0	-	0	-	0	-	0	-	0	-	0	-	0	-	0	-	0	-	0	-	2	4.0	1	2.0
Harderian gland																								
Adenocarcinoma	0	-	0	-	0	-	0	-	0	-	0	-	0	-	0	-	0	-	0	-	1	2.0	0	-
Zymbal glands																								
Sebaceous adenoma	0	-	1	2.0	0	-	0	-	0	-	0	-	0	-	0	-	0	-	0	-	0	-	0	-
Carcinoma	2	4.0	3	6.0	1	2.0	3	6.0	2	4.0	1	2.0	0	-	1	2.0	0	-	1	2.0	1	2.0	0	-
Ear ducts																								
Carcinoma	6	12.0	7(8)	14.0	3	6.0	3	6.0	3	6.0	0	-	2	4.0	4	8.0	1	2.0	5	10.0	4	8.0	3	6.0
Nasal cavities																								
Carcinoma	2	4.0	0	-	0	-	0	-	0	-	0	-	0	-	0	-	0	-	0	-	0	-	0	-
Oral cavity lips and tongue																								
Carcinoma	2	4.0	2	4.0	0	-	0	-	1	2.0	0	-	1	2.0	0	-	0	-	0	-	0	-	1	2.0
Pharynx																								
Carcinoma	0	-	0	-	0	-	0	-	0	-	1	2.0	0	-	0	-	0	-	0	-	0	-	0	-
Larynx																								
Carcinoma	0	-	0	-	0	-	0	-	0	-	1	2.0	0	-	0	-	0	-	0	-	0	-	0	-

— Continued

TABLE 2. NUMBER AND PERCENTAGE OF MALE AND FEMALE SPRAGUE-DAWLEY RATS BEARING VARIOUS TYPES OF BENIGN AND MALIGNANT TUMORS[a]

Site Histotype	I: 2,500 mg/l Male No.	%	Female No.	%	II: 1,500 mg/l Male No.	%	Female No.	%	III: 500 mg/l Male No.	%	Female No.	%	IV: 250 mg/l Male No.	%	Female No.	%	V: 50 mg/l Male No.	%	Female No.	%	VI: 0 (control) Male No.	%	Female No.	%
Lung																								
Adenoma	1	2.0	0	-	0	-	0	-	0	-	0	-	0	-	0	-	0	-	0	-	0	-	0	-
Adenocarcinoma	0	-	0	-	1	2.0	1	2.0	1	2.0	0	-	0	-	0	-	0	-	0	-	0	-	0	-
Angiosarcoma	0	-	0	-	0	-	1	2.0	0	-	0	-	0	-	0	-	0	-	0	-	0	-	0	-
Pleura																								
Mesothelioma	0	-	0	-	0	-	0	-	0	-	0	-	0	-	0	-	0	-	0	-	1	-	1	2.0
Stomach																								
- Forestomach																								
Acanthoma	2	4.0	2	4.0	1	2.0	2	4.0	0	-	0	-	1	2.0	0	-	2	4.0	0	-	0	-	0	-
Squamous cell carcinoma	1	2.0	0	-	0	-	0	-	0	-	0	-	0	-	0	-	0	-	0	-	0	-	0	-
- Glandular stomach																								
Adenocarcinoma	0	-	0	-	0	-	1	2.0	0	-	0	-	0	-	0	-	0	-	0	-	0	-	0	-
Intestine																								
Fibroma	0	-	1	2.0	0	-	0	-	0	-	0	-	0	-	0	-	0	-	0	-	0	-	0	-
Adenocarcinoma	0	-	0	-	0	-	0	-	0	-	1	2.0	0	-	1	2.0	0	-	0	-	0	-	0	-
Liver																								
Hepatocarcinoma	0	-	0	-	1	2.0	0	-	0	-	0	-	0	-	0	-	1	2.0	0	-	0	-	0	-
Pancreas																								
Exocrine adenoma	0	-	0	-	0	-	0	-	1	2.0	0	-	0	-	0	-	0	-	0	-	0	-	1	2.0
Islet cell adenoma	9	18.0	2	4.0	2	4.0	3	6.0	7	14.0	3	6.0	5	10.0	2	4.0	4	8.0	3	6.0	2	4.0	1	2.0
Kidneys																								
Adenoma	0	-	0	-	0	-	1	2.0	0	-	0	-	0	-	0	-	0	-	0	-	0	-	0	-
Fibrolipoma	0	-	0	-	0	-	0	-	0	-	1	2.0	0	-	0	-	0	-	0	-	0	-	0	-
Bladder																								
Transitional cell carcinoma	0	-			0	-			0	-			0	-			0	-			0	-		
Seminal vesicles																								
Adenoma	0	-			0	-			0	-			1	2.0			1	2.0			0	-		
Testes																								
Interstitial cell tumor	4(5)	8.0			3(4)	6.0			4(5)	8.0			1	2.0			6(7)	12.0			2	4.0		
Ovaries																								
Cystadenoma			0	-			0	-			0	-			1	2.0			0	-			0	-
Granulosa cell tumor			0	-			1	2.0			0	-			0	-			0	-			0	-
Fibroangioma			0	-			0	-			1	2.0			0	-			0	-			0	-
Adenocarcinoma			0	-			0	-			0	-			1	2.0			0	-			1	2.0
Granulosa cell malignant tumors			0	-			0	-			0	-			1	2.0			1	2.0			0	-
Uterus																								
Polyp			6	12.0			6	12.0			7	14.0			8	16.0			8	16.0			7	14.0
Squamous cell carcinoma			1	2.0			1	2.0			0	-			0	-			1	2.0			0	-
Adenocarcinoma			1	2.0			2	4.0			1	2.0			5	10.0			0	-			0	-
Fibrosarcoma			0	-			0	-			0	-			0	-			1	2.0			0	-
Angiosarcoma			0	-			0	-			0	-			0	-			0	-			1	2.0

—Continued

TABLE 4. Continued

NUMBER AND PERCENTAGE OF MALE AND FEMALE SPRAGUE-DAWLEY RATS BEARING VARIOUS TYPES OF BENIGN AND MALIGNANT TUMORS[a]

Site / Histotype	I: 2,500 mg/l				II: 1,500 mg/l				III: 500 mg/l				IV: 250 mg/l				V: 50 mg/l				VI: 0 (control)			
	Male		Female		Male		Female		Male		Female		Male		Female		Male		Female		Male		Female	
	No.	%	No.	%	No.	%	No.	%	No.	%	No.	%	No.	%	No.	%	No.	%	No.	%	No.	%	No.	%
Uterus & Vagina																								
Malignant Schwannoma			0	-			0	-			1	2.0			0	-			1	2.0			2	4.0
Vagina																								
Fibrosarcoma			0	-			0	-			1	2.0			0	-			0	-			0	-
Peritoneum																								
Fibroangioma	0	-	0	-	0	-	0	-	0	-	0	-	0	-	0	-	0	-	0	-	0	-	1	2.0
Liposarcoma	0	-	0	-	0	-	0	-	0	-	0	-	0	-	0	-	0	-	0	-	0	-	1	2.0
Mesothelioma	0	-	1	2.0	1	2.0	0	-	0	-	0	-	0	-	0	-	0	-	0	-	0	-	0	-
Angiosarcoma	0	-	0	-	0	-	0	-	0	-	0	-	0	-	0	-	0	-	0	-	0	-	0	-
Pituitary gland																								
Adenoma	0	-	0	-	0	-	0	-	0	-	0	-	0	-	1	2.0	0	-	0	-	0	-	0	-
Thyroid gland																								
Adenoma	13	26.0	21	42.0	16	32.0	22	44.0	14	28.0	19	38.0	19	38.0	24	48.0	19	38.0	19	38.0	14	28.0	14	28.0
C-cell adenoma	1	2.0	3	6.0	3	6.0	4	8.0	3	6.0	3	6.0	2	4.0	2	4.0	3	6.0	3	6.0	2	4.0	0	-
Follicular carcinoma	0	-	0	-	0	-	0	-	0	-	0	-	0	-	1	2.0	0	-	1	2.0	0	-	0	-
C-cell carcinoma	1	2.0	0	-	0	-	0	-	0	-	0	-	0	-	0	-	0	-	0	-	0	-	0	-
Adrenal glands																								
Cortical adenoma	0	-	4	8.0	1	2.0	7(8)	14.0	1	2.0	2	4.0	0	-	4(5)	8.0	1	2.0	2(3)	4.0	0	-	5(6)	10.0
Pheochromocytoma	17(25)	34.0	8(14)	16.0	15(21)	30.0	12(16)	24.0	13(19)	26.0	12(15)	24.0	20(27)	40.0	11(15)	22.0	22(33)	44.0	10(13)	20.0	17(26)	34.0	10(14)	20.0
Fibroangioma	0	-	0	-	0	-	0	-	0	-	0	-	0	-	0	-	0	-	0	-	1	2.0	0	-
Cortical adenocarcinoma	0	-	0	-	0	-	0	-	0	-	1(2)	2.0	0	-	0	-	0	-	3(5)	6.0	0	-	0	-
Pheochromoblastoma	0	-	2	4.0	1	2.0	0	-	2	4.0	0	-	1	2.0	0	-	0	-	3(5)	6.0	0	-	0	-
Central nervous system																								
- Brain																								
Oligodendroglioma	0	-	0	-	1	2.0	0	-	0	-	0	-	0	-	2	4.0	0	-	0	-	1	2.0	0	-
- Meninges																								
Malignant meningioma	2	4.0	0	-	0	-	1	2.0	0	-	0	-	0	-	0	-	0	-	0	-	0	-	1	2.0
Peripheral nervous system																								
- Major peripheral nerves																								
Neuroblastoma	0	-	0	-	0	-	0	-	0	-	0	-	0	-	1	2.0	0	-	0	-	0	-	0	-
Malignant Schwannoma	0	-	0	-	0	-	0	-	0	-	0	-	0	-	0	-	0	-	0	-	0	-	1	2.0

— Continued

TABLE 4. Continued

NUMBER AND PERCENTAGE OF MALE AND FEMALE SPRAGUE-DAWLEY RATS BEARING VARIOUS TYPES OF BENIGN AND MALIGNANT TUMORS[a]

Site Histotype	Groups																							
	I: 2,500 mg/l				II: 1,500 mg/l				III: 500 mg/l				IV: 250 mg/l				V: 50 mg/l				VI: 0 (control)			
	Male		Female		Male		Female		Male		Female		Male		Female		Male		Female		Male		Female	
	No.	%	No.	%	No.	%	No.	%	No.	%	No.	%	No.	%	No.	%	No.	%	No.	%	No.	%	No.	%
Bones																								
- Head																								
Osteosarcoma	7	14.0	2	4.0	0	-	2	4.0	2	4.0	0	-	1	2.0	2	4.0	5	10.0	2	4.0	0	-	2	4.0
- Other																								
Osteosarcoma	0	-	0	-	1	2.0	0	-	0	-	1	2.0	0	-	0	-	2	4.0	0	-	1	2.0	0	-
Soft tissues																								
Lipoma	0	-	1	2.0	0	-	0	-	0	-	0	-	0	-	0	-	0	-	0	-	0	-	0	-
Fibrosarcoma	0	-	0	-	0	-	1	2.0	0	-	0	-	0	-	0	-	0	-	0	-	0	-	0	-
Angiosarcoma	0	-	0	-	0	-	0	-	0	-	0	-	1	2.0	0	-	0	-	0	-	0	-	0	-
Malignant Schwannoma	0	-	0	-	0	-	0	-	0	-	0	-	0	-	0	-	0	-	0	-	1	2.0	1	2.0
Spleen																								
Fibroma	0	-	0	-	0	-	0	-	1	2.0	0	-	0	-	0	-	0	-	0	-	0	-	0	-
Fibroangioma	0	-	0	-	2	4.0	0	-	0	-	0	-	0	-	0	-	0	-	0	-	0	-	0	-
Subcutaneous lymph nodes																								
Fibroangioma	1	2.0	0	-	0	-	0	-	0	-	0	-	0	-	0	-	0	-	0	-	0	-	0	-
Hemolymphoreticular tissues[b,c]																								
Lymphomas and leukemias	8	16.0	7	14.0	15(16)	30.0	3	6.0	9	18.0	4	8.0	10	20.0	8	16.0	14	28.0	5	10.0	6	12.0	2	4.0

[a] Numbers in parentheses indicate the total number of tumours; one animal can bear more than one tumor
[b] Including spleen, mediastinal, subcutaneous and mesenteric lymph nodes
[c] See table 6

TABLE 5. Long-term carcinogenicity bioassays on acetaldehyde administered with drinking water supplied *ad libitum* to male (M) and female (F) Sprague-Dawley rats

TOTAL MALIGNANT TUMORS

Group No.	Concentration (mg/l)	Animals		Malignant tumors			
				Tumor-bearing animals		Tumors	
		Sex	No.	No.	%	No.	Per 100 animals
I	2,500	M	50	23	46.0	33	**66.0** *
		F	50	21	42.0	39	**78.0** *
		M+F	100	44	44.0	72	**72.0**
II	1,500	M	50	21	42.0	27	**54.0**
		F	50	20	40.0	29	**58.0**
		M+F	100	41	41.0	56	**56.0**
III	500	M	50	20	40.0	23	**46.0**
		F	50	19	38.0	25	**50.0**
		M+F	100	39	39.0	48	**48.0**
IV	250	M	50	15	30.0	17	**34.0**
		F	50	26	52.0	33	**66.0**
		M+F	100	41	41.0	50	**50.0**
V	50	M	50	20	40.0	26	**52.0**
		F	50	24	48.0	41	**82.0** *
		M+F	100	44	44.0	67	**67.0**
VI	0	M	50	14	28.0	17	34.0
		F	50	20	40.0	23	46.0
		M+F	100	34	34.0	40	40.0

* $p<0.05$ using χ^2 test

(6) An increased incidence of interstitial cell adenomas of the testis was observed in all treated males except for those in group IV.

(7) An increased incidence of malignant uterine adenocarcinomas was detected in females treated with 250 mg/L of acetaldehyde.

(8) An increased incidence of cranial osteosarcomas occurred in males treated with the highest ($P <0.05$) and the lowest doses.

(9) An increased incidence of hemolymphoreticular neoplasias was observed to a varying degree in all treated groups (TABLE 6).

CONCLUSION

Formaldehyde, administered with drinking water, was shown to be carcinogenic based on an increased incidence of total malignant tumors and oncological lesions varying in site and histotype. Tumors included malignant mammary tumors, oncological lesions of the stomach and intestine, testicular interstitiae cell adenomas, and hemolymphoreticular neoplasias.

TABLE 6. Long-term carcinogenicity bioassays on acetaldehyde administered with drinking water supplied *ad libitum* to male (M) and female (F) Sprague-Dawley rats

HEMOLYMPHORETICULAR NEOPLASIAS

Group No.	Concentration (mg/l)	Animals Sex	Animals No.	Animals with hemolymphoreticular neoplasias No	%[a]	Latency[b] (weeks)
I	2,500	M	50	8	16.0	105.2
		F	50	7	14.0	102.1
		M+F	100	15	15.0	103.7
II	1,500	M	50	15 [c]	30.0	103.7
		F	50	3	6.0	118.7
		M+F	100	18	18.0	111.2
III	500	M	50	9	18.0	82.2
		F	50	4	8.0	60.0
		M+F	100	13	13.0	71.1
IV	250	M	50	10	20.0	92.2
		F	50	8	16.0	91.4
		M+F	100	18	18.0	91.8
V	50	M	50	14	28.0	100.8
		F	50	5	10.0	104.2
		M+F	100	19	19.0	102.5
VI	0	M	50	6	12.0	105.7
		F	50	2	4.0	121.0
		M+F	100	8	8.0	113.4

[a] Percentages refer to the number of animals at start
[b] Age at death
[c] One animal bore a lymphoimmunoblastic lymphoma and histiocytic sarcoma

The results reported on acetaldehyde administered in drinking water indicate a carcinogenic effect of the compound on different organs and tissues, even though this effect is often not dose related.

Because of the extreme rarity of lesions of the stomach and intestine in nontreated animals,[21] their onset in formaldehyde-treated animals cannot be considered accidental and should not be underestimated. The production and use of oxygenated compounds as alternative fuels or as additives in reformulated gasolines was started with the intention of reducing the deleterious effects of vehicular fuels on the environmental quality. On the basis of these new data on the carcinogenicity of formaldehyde and acetaldehyde, and on what was already long known,[18,35] we think that this strategy should be carefully revaluated through an up-to-date analysis of risks and benefits.

ACKNOWLEDGMENT

This research has been partially supported by the Regional Agency for Prevention and Environment (Agenzia Regionale Prevenzione e Ambiente, ARPA) of the Emilia-Romagna Region, Italy.

REFERENCES

1. REUSS, G., W. DISTELDORF, O. GRUNDLER & A. HILT. 1988. Formaldehyde. *In* Ullmann's Encyclopedia of Industrial Chemistry, 5th rev. edit. W. Gerhartz, Y.S. Yamamoto, B. Elvers, J.F. Rounsaville & G. Schulz, Eds. **A11:** 619–651. VCH Publishers. New York.
2. GERBERICH, H.R. & G.C. SEAMAN. 1994. Formaldehyde. *In* Kirk-Othmer Encyclopedia of Chemical Technology. J.I. Kroschwitz & M. Howe-Grant, Eds. **11:** 929–951. John Wiley & Sons. New York.
3. SMITH, R. 1993. Environmental economics and the new paradigm. Chem. Ind. Newsl. Nov.–Dec.: 8.
4. WORLD HEALTH ORGANIZATION. 1989. Formaldehyde. *In* Environmental Health Criteria 89. International Program on Chemical Safety. Geneva.
5. INTERNATIONAL AGENCY FOR RESEARCH ON CANCER. 1995. Wood dust and formaldehyde. *In* Monographs on the Evaluation of Carcinogenic Risks to Humans **62:** 217–362. IARC. Lyon.
6. PREUSS, P.W., R.L. DAILEY & E.S. LEHMAN. 1985. Exposure to formaldehyde. *In* Formaldehyde. Analytical Chemistry and Toxicology. V. Turoski, Ed.: Adv. Chem. Ser. **210:** 247–259.
7. GAMMAGE, R.G. & C.C. TRAVIS. 1989. Formaldehyde exposure and risk in mobile homes. *In* The Risk Assessment of Environmental and Human Health Hazard: A Textbook of Case Studies. D.J. Paustenbach, Ed.: 601–611. John Wiley & Sons. New York.
8. UNITED STATES NATIONAL RESEARCH COUNCIL. 1980. Formaldehyde: An Assessment of Its Health Effects. National Academy Press. Washington, D.C.
9. UNITED STATES NATIONAL RESEARCH COUNCIL. 1981. Health effects of formaldehyde. *In* Formaldehyde and Other Aldehydes. National Academy Press.: 175–220, 306–340. Washington, D.C.
10. OLSON, R.N. 1998. A Study of the Effects of Oxygenated Gasoline on Particulate Concentrations in Salt Lake and Utah Counties during the Winter Season, 1994–95. Air Monitoring Center Study 126-95, State of Utah, Division of Air Quality http://members.mint.net/troberts/julian/DADW120.html
11. MEDINSKY, M.A. & D.C. DORMAN. 1994. Assessing risks of low-level methanol exposure. CIIT Act **14:** 1–7.
12. AUERBACH, C., M. MOUTSCHEN-DAHMEN & J. MOUTSCHEN. 1977. Genetic and cytogenetical effects of formaldehyde and related compounds. Mutat. Res. **39:** 317–362.
13. INTERNATIONAL AGENCY FOR RESEARCH ON CANCER. 1982. Formaldehyde. *In* Monographs on the Evaluation of the Carcinogenic Risk of Chemicals to Humans, Some Industrial Chemicals and Dyestuffs. **29:** 354–389. IARC. Lyon.
14. SWEMBERG, J.A., C.S. BARROW, C.J. BOREIKO, *et al.* 1983. Non linear biological responses to formaldehyde and their implications for carcinogenic risk assessment. Carcinogenesis **4:** 945–952.
15. CONSENSUS WORKSHOP ON FORMALDEHYDE. 1984. Report on the Consensus Workshop on Formaldehyde. Environ. Health Perspect. **58:** 323–381.
16. MIGLIORE, L., L. VENTURA, R. BARALE, *et al.* 1989. Micronuclei and nuclear anomalies induced in the gastro-intestinal epithelium of rats treated with formaldehyde. Mutagenesis **4:** 327–334.
17. DALLAS, C.E., M.J. SCOTT, J.B. WARD, *et al.* 1992. Cytogenetic analysis of pulmonary lavage and bone marrow cells of rats repeated formaldehyde inhalation. J. Appl. Toxicol. **12:** 199–203.
18. KERNS, W.D., K.L. PAVKOV, K.L. DONOFRIO, *et al.* 1983. Carcinogenicity of formaldehyde in rats and mice after long-term inhalation exposure. Cancer Res. **43:** 4382–4392.
19. DALBEY, C.E. 1982. Formaldehyde and tumors in hamster respiratory tract. Toxicology **24:** 9–14.
20. TILL, H.P., R.A. WOUTERSEN, V.J. FERON, *et al.* 1989. Two-year drinking-water study of formaldehyde in rats. Food Chem. Toxicol. **27:** 77–87.
21. SOFFRITTI, M., C. MALTONI, F. MAFFEI & R. BIAGI. 1989. Formaldehyde: an experimental multipotential carcinogen. Toxicol. Ind. Health **5:** 699–730.

22. HAGERMEYER, H.J. 1978. Acetaldehyde. In Kirk-Othmer Encyclopedia of Chemical Technology, 3rd edit. M. Grayson, Ed. **1:** 97–112. John Wiley & Sons. New York.
23. HAGEMEYER, H.J. 1991. Acetaldehyde. In Kirk-Othmer Encyclopedia of Chemical Technology, 4th edit. J.I. Kroschwitz & M. Howe-Grant, Eds. **1:** 94–109. John Wiley & Sons. New York.
24. INTERNATIONAL AGENCY FOR RESEARCH ON CANCER. 1979. Acetaldehyde In Monographs on the Evaluation of the Carcinogenic Risk of Chemicals to Humans: Some Monomers, Plastics and Synthetic Elastomers, and Acrolein. **19:** 341–366. IARC. Lyon.
25. UNITED STATES NATIONAL LIBRARY OF MEDICINE. 1998. Hazardous Substances Data Bank (HSDB) Database, Bethesda, MD [Record No. 230].
26. JIRA, R., R.J. LAIB & H.M. BOLT. 1985. Acetaldehyde. In Encyclopedia of Industrials Chemistry. W. Gerhartz & Y.S. Yamamoto, Eds. **A1:** 31–44. VCH Publishers. Deerfield Beach, FL.
27. HOFFMANN, D., K.D. BRUNNEMANN & G.B. GORI. 1975. On the carcinogenicity of marijuana smoke. Rev. Adv. Phytochem. **9:** 63–81.
28. LYNCH, C., C.K. LIM, M. THOMAS, et al. 1983. Assay of blood and tissue aldehydes by HPLC analysis of their 2,4-dinitrophenylhydrazine adducts. Clin. Chim. Acta **130:** 117–122.
29. PELLIZZARI, E.D., T.D. HARTWELL, B.S.H. HARRIS, et al. 1982. Purgeable organic compounds in mother's milk. Bull. Environ. Contam. Toxicol. **28:** 322–328.
30. ERIKSSON, C.J. 1983. Human blood acetaldehyde concentration during ethanol oxidation. Pharmacol. Biochem. Behav. **18 (**Suppl. 1): 141–150.
31. BRIEN, J.F. & C.W. LOOMIS. 1983. Pharmacology of acetaldehyde. Can. J. Physiol. Pharmacol. **61:** 1–22.
32. INTERNATIONAL AGENCY FOR RESEARCH ON CANCER. 1985. Acetaldehyde In Monographs on the Evaluation of the Carcinogenic Risk of Chemicals to Humans: Allyl Compounds, Aldehydes, Epoxides and Peroxides. **36:** 99–132. IARC. Lyon.
33. INTERNATIONAL AGENCY FOR RESEARCH ON CANCER. 1999. Acetaldehyde In: Monographs on the Evaluation of the Carcinogenic Risk to Humans, Re-Evaluation of Some Organic Chemicals, Hydrazine and Hydrogen Peroxide (Part Two). **71:** 319-335. IARC. Lyon.
34. FANG, J.L. & C.E. VACA. 1997. Detection of DNA adducts of acetaldehyde in peripheral white blood cells of alcohol abusers. Carcinogenesis **18:** 627–632.
35. WOUTERSEN, R.A., L.M. APPELMAN, A. VAN GARDENER-HOETMER, et al. 1986. Inhalation toxicity of acetaldehyde in rats. III. Cacinogenicity study. Toxicology **41:** 213–231.
36. FERON, V.J., H.P. TIL, F. DE VRIJER, et al. 1991. Aldehydes: occurrence, carcinogenic potential, mechanism of action and risk assessment. Mutat. Res. **259:** 363–385.
37. SOFFRITTI, M., F. BELPOGGI, D. CEVOLANI, et al. 2002. Results of long-term experimental studies on the carcinogenicity of methyl alcohol and ethyl alcohol in rats. Ann. N.Y. Acad. Sci. **982:** this volume.

Results of Long-Term Carcinogenicity Bioassay on Vinyl Acetate Monomer in Sprague-Dawley Rats

FRANCO MINARDI, FIORELLA BELPOGGI, MORANDO SOFFRITTI, ADRIANO CILIBERTI, MICHELINA LAURIOLA, ELISA CATTIN, AND CESARE MALTONI[†]

Cancer Research Center, European Ramazzini Foundation for Oncology and Environmental Sciences, Bologna, Italy

ABSTRACT: Vinyl acetate monomer (VAM) was administered in drinking water supplied *ad libitum* at doses of 5,000, 1,000, and 0 ppm (v/v) to 17-week-old Sprague-Dawley rats (breeders) and to 12-day embryos (offspring). Treatment lasted for 104 weeks; thereafter, animals were kept under control conditions until spontaneous death. VAM was found to cause an increase in total malignant tumors and in carcinomas and/or precursor lesions of the oral cavity, lips, tongue, esophagus, and forestomach. Based on these data, VAM must be considered a multipotent carcinogen.

KEYWORDS: vinyl acetate monomer; carcinogenicity; long-term bioassay; rat

INTRODUCTION

Vinyl acetate monomer (VAM) is an important compound in the plastics industry. VAM ($C_4H_6O_2$) has a molecular weight of 86.09. Industrial production of VAM started in the United States in 1928.[1] VAM is produced mainly by two processes: (1) In a process used since the 1920s, acetylene and acetic acid are reacted in the vapor phase over a catalyst bed,[2] (2) In another process, largely used since the 1970s, ethylene is reacted with acetic acid in the presence of oxygen.[3] The world production of VAM is over 2.5 million tons per year.[4]

The only commercial use of VAM is in the production of polymers (polyvinyl acetate, polyvinyl alcohol, polyvinyl acetals) and copolymers (ethylene-vinyl acetate and polyvinyl-acetate chloride).[3]

Polyvinyl acetate is mainly used in adhesives for paper, wood, glass, metals, and porcelain. It is also used in latex water paint, for paper coating, for textile and leather finishing, as a base for inks and lacquers, in heat-sealing films, in shatterproof photographic bulbs, as an emulsifying agent in cosmetics, pesticide formulations, and pharmaceuticals, and as a food additive.[5,6] Polyvinyl acetate is used as a component

Address for correspondence: Morando Soffritti, M.D., Cancer Research Center, European Ramazzini Foundation for Oncology and Environmental Sciences, Bentivoglio Castle, 40010 Bentivoglio (BO), Italy. Voice: +39-051-6640460; fax: +39-051-6640223.
crcfr@tin.it
[†]Deceased.

Ann. N.Y. Acad. Sci. 982: 106–122 (2002). © 2002 New York Academy of Sciences.

in the production of chewing gum. The amount of polymer used in the U.S. is about 5% of the final product; in some European countries, the amount is higher.

Polyvinyl alcohol is the most highly produced synthetic, water-soluble plastic in the world, used in sizing for textile warp and yarn, in laminating adhesives, photosensitive films, and cements, and as a binder and emulsifying agent.[5,6] Polyvinyl acetals are produced by the condensation of polyvinyl alcohol with an aldehyde. Commonly used aldehydes are formaldehyde, acetaldehyde, and butyraldehyde. Polyvinyl formal, polyvinyl acetals and polyvinyl butyrals are used in adhesives, paints, lacquers, and films. Polyvinyl butyral is also used in sheet form as an interlayer in safety glasses and shatter-resistant acrylic protection in aircraft.[5] Ethylene-vinyl acetate copolymers improve the adhesive properties of hot-melt and pressure-sensitive adhesives. They are also used in medical tubing, milk packaging, and beer-dispensing equipment. Plastic containers with barrier layers of ethylene-vinyl alcohol copolymers are replacing many glass and metal containers for packaging food.[5,6] Polyvinyl chloride-acetate copolymers, compounded with plasticizers, are used for cable and wire coverings, in chemical plants and in protective garments.[6]

VAM is not known to occur in nature. In the workplace, it may be present wherever its polymers are produced, used, and stocked. Concentrations of $0.25-2$ mg/m^3 have been measured[7] in air where vinyl acetate manufacturing or processing facilities are located. In areas near chemical waste disposal sites, concentrations of 0.5 µg/m^3 have been detected.[8] VAM has been found at concentrations of 50 mg/L in wastewater effluents from a polyvinyl acetate plant.[9] VAM was among the volatile chemicals released from food packaging during heating in microwave ovens. A concentration of $0.002-0.14$ µg/cm^2 has been detected.[10] VAM has also been detected in cigarette smoke, at concentrations of 400 ng/cigarette.[11]

VAM has an irritative effect on the upper respiratory system in humans. After subchronic and chronic exposure by inhalation, in experimental animals it causes hyperplasia and metaplasia of the respiratory epithelium, bronchitis, and bronchiolitis.[1] Experimental studies on reproductive and prenatal effects have shown that VAM causes parental toxicity (including decreased fertility), developmental toxicity, and minor skeletal alterations.[1]

Rats exposed to VAM exhaled acetaldehyde as a result of hydrolysis by esterase.[12,13] Acetaldehyde is known to be carcinogenic in experimental animals,[14] and its carcinogenic potential was clearly demonstrated in a study performed in our laboratory, the results of which are reported in another paper in this volume.

The experimental studies on rodents conducted until 1997 proved in one way or another inadequate to evaluate the carcinogenic potential of VAM. To date, seven carcinogenicity studies have been published in the scientific literature. In the first study, 96 male and female Sprague-Dawley rats were exposed to 2500 ppm VAM by inhalation for 52 weeks. Early mortality was high: only 49 animals survived for 26 or more weeks. No tumors related to VAM were reported during 135 weeks.[15-17] Because of the poor survival rate, this study was inadequate for detecting the carcinogenic potential of the monomer.

In a second study, 60 male and 60 female Sprague-Dawley rats were exposed to $0-600$ ppm vinyl acetate for about 104 weeks. A slight increase in benign and malignant nasal cavity tumors was found.[18]

VAM was administered at doses of $0-2500$ mg/L in drinking water for 100 weeks, to 20 male and 20 female Fischer F344 rats. An increase in liver neoplastic nodules, in uterine adenocarcinomas and polyps, and in thyroid C-cell adenomas was observed.[19] The number of the animals tested was small and the histopathological examination was limited to gross lesions and major organs only.

Male ($n = 72$) and female ($n = 144$) Sprague-Dawley rats received 0–5000 mg/L VAM in drinking water. Treatment began 10 weeks before mating and was continued for an additional four weeks for males and throughout mating, gestation, and lactation for females. Sixty male and 60 female F_1 pups were administered 0–5000 mg/L vinyl acetate in drinking water for 104 weeks. No treatment-related increase in tumor incidence was observed.[20]

Fifty male and 50 female F344 rats received 0–10,000 ppm vinyl acetate (98% pure) for 104 weeks. Statistically significant increases in preneoplastic changes and squamous cell neoplasms were observed at several sites in the upper digestive tract, but only at the 10,000 ppm dose (unpublished data).[21]

The last two studies were performed on mice exposed to VAM by inhalation. Swiss mice exposed to 0–600 ppm vinyl acetate for about 104 weeks showed no treatment-related increase in tumor incidence.[18] BDF1 mice exposed to 0–10,000 ppm vinyl acetate for 104 weeks showed statistically significant increases in preneoplastic changes and squamous cell neoplasms at several sites in the upper digestive tract, but only at the 10,000 ppm dose (unpublished data).[21]

In a cohort study aimed at identifying the specific exposure associated with an excess of lung cancer risk in humans in a synthetic chemical plant, 19 chemicals were studied: the subgroup with undifferentiated large-cell lung cancer had slightly higher cumulative exposure to VAM.[22] In a nested case-control study on a cohort of 29,139 men employed in two U.S. facilities, who died of lymphomas or leukemias, no significant association was reported.[23]

VAM showed genotoxic effects in both human and rodent cells.[1] *In vitro*, VAM produced a dose-related, statistically significant increase of sister chromatid exchanges and chromosomal aberrations in human lymphocytes and whole blood and an increase of sister chromatid exchanges in ovarian cells of Chinese hamsters.[1]

In the 1980s, a research project on VAM consisting of three experiments conducted with the same protocol, using Sprague-Dawley and Wistar rats and Swiss mice, was started at the Cancer Research Center of the Ramazzini Foundation (CRC/RF). The results of the experiment on Swiss mice have been published,[24] and those on Wistar rats are in publication.[25] Results of the experiment on Sprague-Dawley rats are reported herein for the first time.

MATERIALS AND METHODS

VAM, purity >99%, was supplied by an Italian chemical plant. The impurities were: benzene 30–45 ppm; methyl and ethyl acetate 50 ppm; crotonaldehyde 6–16 ppm; acetaldehyde 2–11 ppm; acetone 330–500 ppm.

VAM was administered in drinking water supplied *ad libitum* at concentrations of 5000, 1000, or 0 ppm to 17-week-old male and female Sprague-Dawley rats (breeders) and 12-day embryos (offspring). Control animals received tap water without VAM. After 104 weeks of treatment, all animals received regular tap water.

The experimental protocol, including method of tumor reporting and statistical methods, are reported in detail elsewhere in this volume.[26]

RESULTS

There were no substantial differences between treated animals and controls in mean body weight, survival, behavior, or treatment–related nononcological pathological changes.

The occurrence of benign and malignant tumors is shown in TABLE 1. Differences observed between treated and control animals were:

(1) an increase in total malignant tumors per 100 animals in male breeders and in male and female offspring of the VAM-treated group (TABLE 2);

(2) an increased incidence of squamous cell carcinomas of the oral cavity and lips in female breeders treated at two dose levels, and in male and female offspring treated at 5000 ppm (TABLE 3);

(3) an increased incidence of squamous cell carcinomas of the tongue in female breeders and in male and female offspring exposed at 5000 ppm; an increased incidence of squamous cell dysplasias was observed in all treated female breeders and offspring and in male offspring treated at 5000 ppm (TABLE 4);

(4) a dose-related increase in the incidence of squamous cell dysplasia of the esophagus in male offspring and female breeders, in male offspring treated at 5000 ppm, and in the female offspring of both treated groups; one case of squamous cell carcinoma occurred in a male offspring treated at 5000 ppm (TABLE 5);

(5) an increased incidence of squamous cell carcinomas of the forestomach in male and female breeders treated at the higher dose and also in male breeders treated at the lower dose; a dose-related increase in incidence occurred in male and female offspring; the same trend was observed for squamous dysplasias (TABLE 6);

(6) when squamous cell dysplasias and carcinomas of the upper gastrointestinal tract were considered as a whole, a highly significant dose-related increase was observed in male and female breeders and offspring (TABLE 7).

CONCLUSIONS

VAM caused an increase in total malignant tumors and tumors at several body sites. The increase in squamous cell carcinomas of the oral cavity and lips, tongue, esophagus and forestomach is of particular significance for two reasons: 1) the same oncological lesions were found in Swiss mice and Wistar rats during the experiments performed in our laboratory following the same experimental protocols; and 2) because the sites of these tumors were tissues most directly exposed to VAM.

The results of this experiment, together with those of the experiment on Swiss mice[24] and on Wistar rats,[25] show that VAM is a multipotent carcinogen. As reported in another paper of this volume, most tumors arose after 112 weeks of age in the present study. Had we stopped our experiment at 112 weeks of age it is unlikely that we would have found the multipotent carcinogenic activity of VAM.

Based on these experimental findings, regulatory measures must be undertaken to prevent the carcinogenic risk of VAM among workers exposed and consumers of goods containing the monomer. Of particular concern is the current use of VAM-based polymers for containers of food and beverages, as VAM has been found to migrate from plastic material into wine[27] and water.[28] Based on this, the use of VAM-based polymers for the production of chewing gum must also be of concern, as one cannot exclude migration into saliva and other biological fluids.

TABLE 1. Long-term carcinogenicity bioassay on vinyl acetate monomer (VAM) in drinking water supplied *ad libitum* to male (M) and female (F) Sprague-Dawley rats

NUMBER AND PERCENTAGE OF MALE AND FEMALE SPRAGUE-DAWLEY RATS BEARING VARIOUS TYPES OF BENIGN AND MALIGNANT TUMORS (a)

Site	I: 5,000 ppm								II: 1,000 ppm								III: 0 (control)							
	Breeders				Offspring				Breeders				Offspring				Breeders				Offspring			
	Male		Female		Male		Female		Male		Female		Male		Female		Male		Female		Male		Female	
	No.	%	No.	%	No.	%	No.	%	No.	%	No.	%	No.	%	No.	%	No.	%	No.	%	No.	%	No.	%
Skin																								
Acanthoma	0	-	0	-	0	-	1	1.8	0	-	0	-	0	-	0	-	0	-	0	-	0	-	0	-
Dermatofibroma	2(3)	15.4	0	-	2	3.8	0	-	0	-	0	-	1	1.2	0	-	0	-	0	-	2	1.9	1	1.0
Squamous cell carcinoma	0	-	0	-	0	-	0	-	0	-	1	2.7	1	1.2	0	-	0	-	0	-	0	-	0	-
Carcinoma	0	-	0	-	0	-	0	-	0	-	0	-	0	-	0	-	0	-	0	-	0	-	1	1.0
Basocellular carcinoma	0	-	0	-	1	1.9	0	-	0	-	0	-	0	-	0	-	0	-	0	-	0	-	0	-
Siringocarcinoma	0	-	0	-	0	-	0	-	0	-	0	-	1	1.2	0	-	0	-	0	-	0	-	0	-
Sebaceous adenocarcinoma	0	-	0	-	0	-	0	-	1	7.7	0	-	0	-	0	-	0	-	0	-	0	-	0	-
Fibrosarcoma	0	-	0	-	1	1.9	0	-	0	-	0	-	0	-	0	-	0	-	0	-	0	-	0	-
Subcutaneous tissue																								
Fibroma	0	-	0	-	2	3.8	0	-	0	-	0	-	0	-	0	-	0	-	0	-	1	0.9	0	-
Lipoma and fibrolipoma	0	-	0	-	0	-	0	-	1	7.7	0	-	0	-	0	-	0	-	0	-	2	1.9	0	-
Fibrosarcoma	0	-	0	-	0	-	0	-	0	-	0	-	0	-	0	-	1	7.1	0	-	0	-	0	-
Liposarcoma	1	7.7	0	-	1	1.9	0	-	1	7.7	0	-	1	1.2	1	1.1	0	-	0	-	3	2.8	0	-
Mammary glands																								
Fibroma & fibroadenoma	0	-	15(22)	40.5	8	15.1	29(42)	50.9	2	15.4	20(32)	54.1	2	2.4	50(62)	57.5	1	7.1	22(28)	59.5	9(10)	8.4	69(98)	69.7
Lipoma & fibrolipoma	0	-	0	-	1	1.9	0	-	1	7.7	0	-	3(4)	3.6	2	2.3	2	14.3	0	-	5	4.7	1	1.0
Adenocarcinoma	0	-	6	16.2	0	-	13(16)	22.8	0	-	7	18.9	1	1.2	12(19)	13.8	0	-	7(9)	18.9	3	2.8	17(21)	17.2
Carcinosarcoma	0	-	0	-	0	-	0	-	0	-	1	2.7	0	-	0	-	0	-	0	-	0	-	1	1.0
Fibrosarcoma	0	-	0	-	0	-	1	1.8	0	-	0	-	0	-	3	3.4	0	-	3(4)	8.1	0	-	4	4.0
Liposarcoma	1	7.7	0	-	0	-	2	3.5	0	-	0	-	2	2.4	2	2.3	0	-	1	2.7	2	1.9	2	2.0
Rhabdomyosarcoma	0	-	0	-	0	-	0	-	1	7.7	1	2.7	0	-	0	-	0	-	0	-	0	-	0	-
Harderian glands																								
Adenocarcinoma	0	-	0	-	0	-	0	-	0	-	0	-	1	1.2	0	-	0	-	0	-	0	-	0	-
Zymbal glands																								
Sebaceous adenoma	0	-	0	-	0	-	0	-	0	-	0	-	0	-	0	-	0	-	1	2.7	0	-	0	-
Carcinoma	0	-	0	-	1	1.9	0	-	1	7.7	0	-	3	3.6	1	1.1	1	7.1	0	-	2(3)	1.9	0	-
Ear ducts																								
Carcinoma	0	-	0	-	2	3.8	0	-	0	-	0	-	0	-	0	-	0	-	0	-	0	-	0	-

—Continued

TABLE 1. Continued

NUMBER AND PERCENTAGE OF MALE AND FEMALE SPRAGUE-DAWLEY RATS BEARING VARIOUS TYPES OF BENIGN AND MALIGNANT TUMORS (a)

	Groups																							
	I: 5,000 ppm								II: 1,000 ppm								III: 0 (control)							
	Breeders				Offspring				Breeders				Offspring				Breeders				Offspring			
	Male		Female		Male		Female		Male		Female		Male		Female		Male		Female		Male		Female	
Site	No.	%	No.	%	No.	%	No.	%	No.	%	No.	%	No.	%	No.	%	No.	%	No.	%	No.	%	No.	%
Nasal cavities																								
Carcinoma	0	-	0	-	0	-	0	-	0	-	0	-	0	-	0	-	1	7.1	0	-	0	-	1	1.0
Olfactory neuroblastoma	0	-	0	-	0	-	0	-	0	-	0	-	0	-	0	-	0	-	0	-	1	0.9	0	-
Oral cavity & lips [b]																								
Acanthoma	0	-	0	-	1	1.9	1	1.8	0	-	1	2.7	0	-	0	-	1	7.1	1	2.7	1	0.9	0	-
Carcinoma	0	-	2	5.4	13	24.5	9	15.8	0	-	1	2.7	0	-	0	-	0	-	0	-	2	1.9	1	1.0
Fibrosarcoma	0	-	0	-	0	-	0	-	0	-	0	-	0	-	0	-	0	-	0	-	1	0.9	0	-
Tongue [c]																								
Granulosa cell tumor (Abrikosoff's tumor)																								
Carcinoma	0	-	1	2.7	1	1.9	2	3.5	0	-	1	2.7	0	-	0	-	0	-	0	-	0	-	0	-
Pharynx																								
Carcinoma	0	-	0	-	1	1.9	1	1.8	0	-	0	-	0	-	0	-	0	-	0	-	0	-	0	-
Lung																								
Adenoma	0	-	0	-	0	-	0	-	0	-	0	-	0	-	0	-	0	-	0	-	0	-	1	1.0
Fibroangioma	0	-	0	-	0	-	0	-	0	-	0	-	0	-	0	-	0	-	0	-	1	0.9	0	-
Adenocarcinoma	0	-	0	-	0	-	0	-	0	-	0	-	0	-	1	1.1	0	-	0	-	0	-	0	-
Pleura																								
Mesothelioma	0	-	0	-	1	1.9	0	-	0	-	0	-	0	-	0	-	0	-	0	-	0	-	0	-
Oesophagus [d]																								
Carcinoma	0	-	0	-	1	1.9	0	-	0	-	0	-	0	-	0	-	0	-	0	-	0	-	0	-
Stomach																								
- Forestomach [e]																								
Acanthoma	1	7.7	1	2.7	6	11.3	2	3.5	3	23.1	0	-	7	8.4	5	5.7	2	14.3	1	2.7	9	8.4	8	8.1
Carcinoma	1	7.7	3	8.1	7	13.2	4	7.0	1	7.7	0	-	6	7.2	3	3.4	0	-	0	-	0	-	0	-
Intestine																								
Leiomyoma	0	-	0	-	0	-	0	-	0	-	0	-	0	-	0	-	0	-	0	-	0	-	1	1.0
Adenocarcinoma	0	-	0	-	0	-	0	-	0	-	1	2.7	0	-	0	-	0	-	0	-	0	-	0	-
Fibrosarcoma	0	-	1	2.7	0	-	0	-	0	-	0	-	0	-	0	-	0	-	0	-	0	-	0	-

— Continued

TABLE 1. Continued

NUMBER AND PERCENTAGE OF MALE AND FEMALE SPRAGUE-DAWLEY RATS BEARING VARIOUS TYPES OF BENIGN AND MALIGNANT TUMORS (a)

	\multicolumn{8}{c}{Groups}																		
	\multicolumn{8}{c}{I: 5,000 ppm}	\multicolumn{8}{c}{II: 1,000 ppm}	\multicolumn{8}{c}{III: 0 (control)}								
	\multicolumn{4}{c}{Breeders}	\multicolumn{4}{c}{Offspring}	\multicolumn{4}{c}{Breeders}	\multicolumn{4}{c}{Offspring}	\multicolumn{4}{c}{Breeders}	\multicolumn{4}{c}{Offspring}				
	\multicolumn{2}{c}{Male}	\multicolumn{2}{c}{Female}	\multicolumn{2}{c}{Male}	\multicolumn{2}{c}{Female}	\multicolumn{2}{c}{Male}	\multicolumn{2}{c}{Female}	\multicolumn{2}{c}{Male}	\multicolumn{2}{c}{Female}	\multicolumn{2}{c}{Male}	\multicolumn{2}{c}{Female}	\multicolumn{2}{c}{Male}	\multicolumn{2}{c}{Female}
Site	No.	%	No.	%	No.	%	No.	%	No.	%	No.	%	No.	%	No.	%	No.	%	No.	%	No.	%	No.	%
Liver																								
Cholangioma	0	-	0	-	1	1.9	1	1.8	0	-	0	-	0	-	1	1.1	0	-	0	-	1	0.9	1	1.0
Angioma	0	-	0	-	0	-	0	-	0	-	0	-	0	-	0	-	1	7.1	0	-	0	-	0	-
Hepatocarcinoma	0	-	2	5.4	3	5.7	1	1.8	0	-	1	2.7	1	1.2	2	2.3	1	-	0	-	5	4.7	5	5.1
Angiosarcoma	0	-	0	-	0	-	0	-	0	-	0	-	0	-	0	-	0	-	0	-	1	0.9	1	1.0
Pancreas																								
Exocrine adenoma	0	-	0	-	1	1.9	0	-	3	23.1	1	2.7	5	6.0	2	2.3	1	7.1	0	-	0	-	1	1.0
Islet cell adenoma	4	30.8	0	-	2	3.8	1	1.8	1	7.7	1	2.7	9	10.8	2	2.3	0	-	2	5.4	10	9.3	3	3.0
Exocrine adenocarcinoma	0	-	0	-	0	-	0	-	0	-	0	-	0	-	1	1.1	0	-	0	-	0	-	0	-
Islet cell carcinoma	1	7.7	1	2.7	0	-	2	3.5	0	-	0	-	1	1.2	2	2.3	0	-	0	-	2	1.9	0	-
Kidneys																								
Adenoma	0	-	0	-	1	1.9	0	-	0	-	0	-	0	-	0	-	0	-	1	2.7	0	-	0	-
Adenocarcinoma	0	-	0	-	1	1.9	0	-	0	-	0	-	0	-	0	-	0	-	0	-	2	1.9	0	-
Liposarcoma	0	-	0	-	0	-	0	-	0	-	0	-	0	-	1	1.1	0	-	0	-	0	-	1	1.0
Angiosarcoma	0	-	0	-	0	-	0	-	0	-	0	-	1	1.2	0	-	0	-	0	-	0	-	0	-
Pelvis and ureters																								
Transitional cell papilloma	0	-	0	-	0	-	0	-	0	-	0	-	1	1.2	0	-	0	-	0	-	0	-	0	-
Bladder																								
Leiomyosarcoma	0	-	0	-	0	-	0	-	0	-	0	-	0	-	1	1.1	0	-	0	-	0	-	0	-
Testes																								
Leydig cell tumor	0	-			4(6)	7.5			0	-			2(3)	2.4			0	-			5(7)	4.7		
Malignant Leydig cell tuomor	1	7.7			0	-			0	-			0	-			0	-			0	-		
Ovaries																								
Luteoma			0	-			0	-			0	-			1(2)	1.1			0	-			0	-
Granulosa cell tumor			3(4)	8.1			4(5)	7.0			3(4)	8.1			1	1.1			0	-			2(3)	2.0
Granulosa and theca cell tumor			1	2.7			0	-			0	-			2(3)	2.3			1	2.7			2	2.0
Sertoli cell tumor			0	-			1(2)	1.8			0	-			0	-			0	-			0	-
Fibroangioma			1	2.7			0	-			0	-			0	-			0	-			1	1.0
Granulosa cell malignant tumor			1	2.7			0	-			1	2.7			0	-			0	-			0	-

— Continued

TABLE 1. *Continued*

NUMBER AND PERCENTAGE OF MALE AND FEMALE SPRAGUE-DAWLEY RATS BEARING VARIOUS TYPES OF BENIGN AND MALIGNANT TUMORS (a)

	I: 5,000 ppm								II: 1,000 ppm								III: 0 (control)							
	Breeders				Offspring				Breeders				Offspring				Breeders				Offspring			
	Male		Female		Male		Female		Male		Female		Male		Female		Male		Female		Male		Female	
Site	No.	%	No.	%	No.	%	No.	%	No.	%	No.	%	No.	%	No.	%	No.	%	No.	%	No.	%	No.	%
Uterus																								
Polyps			4	10.8			6	10.5			3	8.1			6	6.9			2	5.4			9	9.1
Fibroma			0	-			0	-			0	-			0	-			1	2.7			0	-
Leiomyoma			0	-			0	-			0	-			0	-			1	2.7			0	-
Squamous cell carcinoma			0	-			0	-			0	-			0	-			0	-			1	1.0
Adenocarcinoma			2	5.4			4	7.0			2	5.4			4	4.6			3	8.1			6	6.1
Carcinosarcoma			1	2.7			0	-			0	-			0	-			0	-			0	-
Malignant Schwannoma			0	-			0	-			0	-			1	1.1			0	-			0	-
Uterus and vagina																								
Malignant Schwannoma			0	-			1	1.8			0	-			2	2.3			2	5.4			1	1.0
Peritoneum																								
Mesothelioma	0	-	0	-	0	-	0	-	1	-	0	-	1	1.2	1	1.1	0	-	0	-	0	-	0	-
Liposarcoma	0	-	0	-	0	-	0	-	0	-	0	-	0	-	0	-	1	-	0	-	1	0.9	0	-
Pituitary gland																								
Adenoma	5	38.5	26	70.3	18	34.0	39	68.4	3	23.1	22	59.5	27	32.5	55	63.2	6	42.9	24	64.9	27	25.2	69	69.7
Thyroid gland																								
Follicular adenoma	0	-	0	-	1	1.9	0	-	0	-	0	-	0	-	0	-	0	-	0	-	0	-	0	-
C-cell adenoma	0	-	0	-	0	-	0	-	0	-	0	-	0	-	0	-	1	7.1	0	-	0	-	0	-
Follicular carcinoma	0	-	0	-	0	-	0	-	1	-	0	-	1	1.2	1	1.1	0	-	0	-	0	-	1	1.0
C-cell carcinoma	0	-	1	2.7	3	5.7	1	1.8	0	-	0	-	0	-	1	1.1	1	7.1	0	-	2	1.9	0	-
Adrenal glands																								
Cortical adenoma	0	-	1	2.7	2(3)	3.8	3(4)	5.3	0	-	1(2)	2.7	0	-	6	6.9	1	7.1	2	5.4	1	0.9	13	13.1
Pheochromocytoma	7(11)	53.8	11(15)	29.7	23(35)	43.4	22(28)	38.6	6(8)	46.2	11(18)	29.7	46(66)	55.4	29(46)	33.3	8(11)	57.1		27.0	63(102)	58.9	37(46)	37.4
Cortical adenocarcinoma	0	-	3	8.1	0	-	5(6)	8.8	0	-	3	8.1	0	-	2	2.3	0	-	0	-	0	-	2	2.0
Pheochromoblastoma	1	7.7	0	-	9(12)	17.0	3	5.3	1	7.7	0	-	7(9)	8.4	6(9)	6.9	3	21.4	3	8.1	11(18)	10.3	1	1.0
Central nervous system																								
- Brain																								
Oligodendroglioma	0	-	0	-	0	-	2	3.5	0	-	1	2.7	0	-	0	-	0	-	0	-	3	2.7	2	2.8
Ependymoma	0	-	0	-	0	-	0	-	0	-	0	-	1	1.2	0	-	0	-	0	-	0	-	0	-
- Meninges																								
Benign meningioma	0	-	1	2.7	0	-	0	-	0	-	1	2.7	0	-	0	-	1	7.1	0	-	0	-	0	-
Malignant Schwannoma	0	-	0	-	0	-	0	-	0	-	1	2.7	0	-	0	-	0	-	0	-	0	-	0	-

— *Continued*

TABLE 1. Continued

NUMBER AND PERCENTAGE OF MALE AND FEMALE SPRAGUE-DAWLEY RATS BEARING VARIOUS TYPES OF BENIGN AND MALIGNANT TUMORS (a)

	Groups																							
	I: 5,000 ppm								II: 1,000 ppm								III: 0 (control)							
	Breeders				Offspring				Breeders				Offspring				Breeders				Offspring			
	Male		Female		Male		Female		Male		Female		Male		Female		Male		Female		Male		Female	
Site	No.	%	No.	%	No.	%	No.	%	No.	%	No.	%	No.	%	No.	%	No.	%	No.	%	No.	%	No.	%
Peripheral nervous system																								
- Major peripheral nerves																								
Benign Schwannoma	0	-	0	-	-	-	0	-	0	-	0	-	0	-	0	-	0	-	0	-	1	0.9	0	-
Malignant Schwannoma	0	-	0	-	-	-	0	-	0	-	0	-	-	-	1	1.2	0	-	0	-	0	-	0	-
Bones																								
- Head																								
Osteosarcoma	0	-	1	2.7	1	1.9	0	-	0	-	0	-	0	-	1	1.1	0	-	0	-	0	-	1	1.0
- Other																								
Osteosarcoma	0	-	0	-	1	1.9	0	-	0	-	1	2.7	3	3.6	1	1.1	0	-	0	-	0	-	2	2.0
Soft tissues																								
Lipoma	0	-	1	2.7	0	-	0	-	0	-	0	-	1	1.2	0	-	0	-	0	-	1	0.9	2	2.0
Liposarcoma	0	-	0	-	1	1.9	0	-	0	-	0	-	0	-	0	-	0	-	0	-	1	0.9	0	-
Thymus																								
Malignant thymoma [f]	1	7.7	0	-	1	1.9	0	-	0	-	0	-	0	-	0	-	0	-	0	-	0	-	0	-
Spleen																								
Fibroangioma	1	7.7	0	-	1	1.9	0	-	0	-	0	-	0	-	1	1.1	0	-	0	-	0	-	0	-
Mesenteric lymph nodes																								
Fibroangioma	0	-	0	-	0	-	0	-	0	-	0	-	1	1.2	1	1.1	1	7.1	0	-	0	-	0	-
Hemolymphoreticular tissues [g]																								
Lymphomas and leukemias	1	7.7	2	5.4	6	11.3	7	12.3	4	30.8	10	27.0	20	24.1	19	21.8	0	-	8	21.6	13	12.1	11	11.1

[a] Number in brackets indicate the total number of tumors; one animal can bear more than one tumor
[b] See table 3
[c] See table 4
[d] See table 5
[e] See table 6
[f] In 96 percent of cases, the tumor itself is composed of a mixture in varying proportions of epithelial cells and lymphocytes. In the remaining 4 percent, only epithelial cells are present. We consider that a tumor composed exclusively of lymphocytes would not be classified as a thymoma but as a lymphoma involving the thymus
[g] Including thymus, spleen and mesenteric lymph nodes

TABLE 2. Long-term carcinogenicity bioassay on vinyl acetate monomer (VAM) in drinking water supplied ad libitum to male (M) and female (F) Sprague-Dawley rats

TOTAL MALIGNANT TUMORS

Group dose (ppm, v/v)	Animals				Malignant tumors				Tumors	
	Age	Sex	No.	Tumor-bearing animals		Malignant tumors				
				No.	%	%	No.	Per 100 animals		
I 5,000	Breeders (17 weeks old)	M	13	8	61.5		11	84.6 **		
		F	37	18	48.6		27	73.0		
		M+F	50	26	52.0		38	76.0		
	Offspring (Embryos)	M	53	31		58.5 *	59		111.3 **	
		F	57	32		56.1	62		108.8 **	
		M+F	110	63		57.3	121		110.0	
II 1,000	Breeders (17 weeks old)	M	13	7	53.8		9	69.2 *		
		F	37	20	54.1		33	89.2		
		M+F	50	27	54.0		42	84.0		
	Offspring (Embryos)	M	83	38		45.8	55		66.3	
		F	87	42		48.3	79		90.8 *	
		M+F	170	80		47.1	134		78.8	
III 0 [a]	Breeders (17 weeks old)	M	14	5	35.7		7	50.0		
		F	37	22	59.5		34	91.9		
		M+F	51	27	52.9		41	80.4		
	Offspring (Embryos)	M	107	43		40.2	63		58.9	
		F	99	43		43.4	66		66.7	
		M+F	206	86		41.7	129		62.6	

[a] Drinking water alone. * $p < 0.05$ using χ^2 test. ** $p < 0.01$ using χ^2 test.

TABLE 3. Long-term carcinogenicity bioassay on vinyl acetate monomer (VAM) in drinking water supplied *ad libitum* to male (M) and female (F) Sprague-Dawley rats

ONCOLOGICAL LESIONS OF THE ORAL CAVITY AND LIPS

Group/dose (ppm, v/v)	Animals			Animals with oncological lesions								Total	
				Acanthomas		SqDy		SqCa					
	Age	Sex	No.	No.	%	No.	%	No.	%			No.	Per 100 animals
I (5,000)	Breeders	M	13	0	-	0	-	0	-			0	-
		F	37	0	-	0	-	2	5.4			2	5.4
		M+F	50	0	-	0	-	2	4.0			2	4.0
	Offspring	M	53	1	1.9	3	5.7	13	24.5 **			17	32.1 **
		F	57	1	1.8	0	-	9	15.8 **			10	17.5 **♦♦
		M+F	110	2	1.8	3	2.7	22	20.0			27	24.5
II (1,000)	Breeders	M	13	0	-	0	-	0	-			0	-
		F	37	1	2.7	0	-	1	2.7			2	5.4
		M+F	50	1	2.0	0	-	1	2.0			2	4.0
	Offspring	M	83	0	-	0	-	0	-			0	-
		F	87	0	-	2	2.3	0	-			2	2.3 ♦♦
		M+F	170	0	-	2	1.2	0	-			2	1.2
III 0[a]	Breeders	M	14	1	7.1	0	-	0	-			1	7.1
		F	37	1	2.7	0	-	0	-			1	2.7
		M+F	51	2	3.9	0	-	0	-			2	3.9
	Offspring	M	107	1	0.9	1	0.9	2	1.9			4	3.7
		F	99	0	-	0	-	1	1.0			1	1.0
		M+F	206	1	0.5	1	0.5	3	1.5			5	2.4

[a]Drinking water alone. * $p < 0.01$ using χ^2 test. ♦♦ $p < 0.01$ using Cochrane-Armitage test for dose-response relationship.

TABLE 4. Long-term carcinogenicity bioassay on vinyl acetate monomer (VAM) in drinking water supplied *ad libitum* to male (M) and female (F) Sprague-Dawley rats

ONCOLOGICAL LESIONS OF THE TONGUE

Group/dose (ppm, v/v)	Age	Animals Sex	Animals No.	Acanthomas No.	Acanthomas %	SqDy No.	SqDy %	SqCa No.	SqCa %	%	Total No.	Total Per 100 animals
I (5,000)	Breeders	M	13	0	-	0	-	0	-	-	0	-
		F	37	0	-	7	18.9 *	1	2.7	-	8	21.6 ***♦♦♦
		M+F	50	0	-	7	14.0	1	2.0	-	8	16.0
	Offspring	M	53	0	-	3	5.7	1	1.9	-	4	7.5 *
		F	57	0	-	9	15.8 ***♦♦♦	2	3.5	-	11	19.3 ***♦♦♦
		M+F	110	0	-	12	10.9	3	2.7	-	15	13.6
II (1,000)	Breeders	M	13	0	-	0	-	0	-	-	0	-
		F	37	0	-	3	8.1	1	2.7	-	4	10.8 ♦♦
		M+F	50	0	-	3	6.0	1	2.0	-	4	8.0
	Offspring	M	83	0	-	0	-	0	-	-	0	-
		F	87	0	-	2	2.3 ♦♦	0	-	-	2	2.3 ♦♦
		M+F	170	0	-	2	1.2	0	-	-	2	1.2
III 0 [a]	Breeders	M	14	0	-	0	-	0	-	-	0	-
		F	37	0	-	0	-	0	-	-	0	-
		M+F	51	0	-	0	-	0	-	-	0	-
	Offspring	M	107	0	-	0	-	0	-	-	0	-
		F	99	0	-	1	1.0	0	-	-	1	1.0
		M+F	206	0	-	1	0.5	0	-	-	1	0.5

[a] Drinking water alone. * $p < 0.05$ using χ^2 test. ** $p < 0.01$ using χ^2 test. ♦♦ $p < 0.01$ using Cochrane-Armitage test for dose-response relationship.

TABLE 5. Long-term carcinogenicity bioassay on vinyl acetate monomer (VAM) in drinking water supplied *ad libitum* to male (M) and female (F) Sprague-Dawley rats

ONCOLOGICAL LESIONS OF THE ESOPHAGUS

Group/dose (ppm, v/v)	Animals Age	Animals Sex	Animals No.	Acanthomas No.	Acanthomas %	SqDy No.	SqDy %	SqCa No.	SqCa %	Total No.	Total Per 100 animals
I (5,000)	Breeders	M	13	0	-	2	15.4	0	-	2	15.4
		F	37	0	-	8	21.6 ✦✦✦	0	-	8	21.6 ✦✦✦
		M+F	50	0	-	10	20.0	0	-	10	20.0
	Offspring	M	53	0	-	19	35.8 **	1	1.9	20	37.7 **
		F	57	0	-	23	40.4 **✦✦	0	-	23	40.4 **✦✦
		M+F	110	0	-	42	38.2	1	0.9	43	39.1
II (1,000)	Breeders	M	13	0	-	1	7.7	0	-	1	7.7
		F	37	0	-	2	5.4 ✦✦	0	-	2	5.4 ✦✦
		M+F	50	0	-	3	6.0	0	-	3	6.0
	Offspring	M	83	0	-	0	-	0	-	0	-
		F	87	0	-	4	4.6 ✦✦	0	-	4	4.6 ✦✦
		M+F	170	0	-	4	2.4	0	-	4	2.4
III 0 [a]	Breeders	M	14	0	-	0	-	0	-	0	-
		F	37	0	-	1	2.7	0	-	1	2.7
		M+F	51	0	-	1	2.0	0	-	1	2.0
	Offspring	M	107	0	-	0	-	0	-	0	-
		F	99	0	-	0	-	0	-	0	-
		M+F	206	0	-	0	-	0	-	0	-

[a]Drinking water alone. *$p < 0.05$ using χ^2 test. **$p < 0.01$ using χ^2 test. ✦✦$p < 0.01$ using Cochrane-Armitage test for dose-response relationship.

TABLE 6. Long-term carcinogenicity bioassay on vinyl acetate monomer (VAM) in drinking water supplied *ad libitum* to male (M) and female (F) Sprague-Dawley rats

ONCOLOGICAL LESIONS OF THE FORESTOMACH

Group/ dose (ppm, v/v)	Animals			Animals with oncological lesions									Total	
				Acanthomas		SqDy			SqCa					Per 100
	Age	Sex	No.	No.	%	No.	%		No.	%		No.		animals
I (5,000)	Breeders	M	13	1	7.7	4	30.8		1	7.7		6		46.2
		F	37	1	2.7	11	29.7 *		3	8.1		15		40.5 **
		M+F	50	2	4.0	15	30.0		4	8.0		21		42.0
	Offspring	M	53	6	11.3	13	24.5 ♦♦♦♦		7	13.2 **		26		49.1 ****♦♦♦♦
		F	57	2	3.5	14	24.6 ♦♦♦♦		4	7.0 *		20		35.1 ****♦♦♦
		M+F	110	8	7.3	27	24.5		11	10.0		46		41.8
II (1,000)	Breeders	M	13	3	23.1	3	23.1		1	7.7		7		53.8
		F	37	0	-	2	5.4		0	-		2		5.4
		M+F	50	3	6.0	5	10.0		1	2.0		9		18.0
	Offspring	M	83	7	8.4	16	19.3 ♦♦♦♦		6	7.2 *		29		34.9 ****♦♦♦♦
		F	87	5	5.7	14	16.1 ♦♦♦		3	3.4		22		25.3 ****♦♦♦
		M+F	170	12	7.1	30	17.6		9	5.3		51		30.0
III 0 [a]	Breeders	M	14	2	14.3	1	7.1		0	-		3		21.4
		F	37	1	2.7	3	8.1		0	-		4		10.8
		M+F	51	3	5.9	4	7.8		0	-		7		13.7
	Offspring	M	107	9	8.4	4	3.7		0	-		13		12.1
		F	99	8	8.1	4	4.0		0	-		12		12.1
		M+F	206	17	8.3	8	3.9		0	-		25		12.1

[a] Drinking water alone. *$p < 0.05$ using χ^2 test. **$p < 0.01$ using χ^2 test. ♦♦$p < 0.01$ using Cochrane-Armitage test for dose-response relationship.

TABLE 7. Long-term carcinogenicity bioassay on vinyl acetate monomer (VAM) in drinking water supplied *ad libitum* to male (M) and female (F) Sprague-Dawley rats

UPPER GIT SQUAMOUS CELL CARCINOMAS (SqCa) PLUS THEIR PRECURSOR (SqDy)

Group No.	Dose (ppm, v/v)	Animals Age	Sex	No.	SqDy + SqCa No. per 100 animals
I	5,000	Breeders (17 weeks old)	M	13	53.8 ****
			F	37	86.5 ****
			M+F	50	78.0
		Offspring (Embryos)	M	53	113.2 ****
			F	57	107.0 ****
			M+F	110	110.0
II	1,000	Breeders (17 weeks old)	M	13	38.5 ****
			F	37	24.3 ***
			M+F	50	28.0
		Offspring (Embryos)	M	83	26.5 ****
			F	87	28.7 ****
			M+F	170	27.6
III	0 [a]	Breeders (17 weeks old)	M	14	7.1
			F	37	10.8
			M+F	51	9.8
		Offspring (Embryos)	M	107	6.5
			F	99	6.1
			M+F	206	6.3

[a] Drinking water alone. *$p < 0.05$ using χ^2 test. **$p < 0.01$ using χ^2 test. ♦♦$p < 0.01$ using Cochrane-Armitage test for dose-response relationship.

ACKNOWLEDGMENTS

This research has been partially supported by the Regional Agency for Prevention and Environment (Agenzia Regionale Prevenzione e Ambiente, ARPA) of the Emilia-Romagna Region, Italy.

REFERENCES

1. INTERNATIONAL AGENCY FOR RESEARCH ON CANCER. 1995. Monographs on the evaluation of the carcinogenic risks to humans, Vol. 63. Dry cleaning, some chlorinated solvents and other industrial chemicals. IARC. Lyon, France.
2. ROSCHER, G., E. HOFMANN, K. ARMSTRONG ADEY, et al. 1983. Vinyl acetate. In Ullmann's Encyclopaedia of Industrial Chemistry, 4th edit. W. Gerhartz, C. Mayer, D. Moegling, et al., Eds. **23:** 597–618. VCH Publishers. New York.
3. DANIELS, W. 1983. Vinyl acetate monomer. In Kirk-Othmer Encyclopedia of Chemical Technology, 3rd edit. H.F. Mark, D. F. Othmer & C.G. Overberger, et al., Eds. **23:** 817–848. John Wiley & Sons. New York.
4. ENVIRONMENTAL CHEMICALS DATA AND INFORMATION NETWORK. 1993. Vinyl acetate. Ispra, JRC-CEC, last update: 02.09.1993.
5. SAX, N.I. & R.W. LEWIS. 1987. Hawley's Condensed Chemical Dictionary. 11th edit. **491:** 945–946. Van Nostrand Reinhold. New York.
6. ROSATO, D.V. 1993. Rosato's Plastics Encyclopedia and Dictionary. **573:** 235–236. New York.
7. UNITED STATES AGENCY FOR TOXIC SUBSTANCES AND DISEASE REGISTRY. 1992. Toxicological profile for vinyl acetate. Clement International Corp. Contract No. 205-88-0608. United States Department of Health and Human Services. Washington DC.
8. PELLIZZARI, E.D. 1982. Analysis for organic vapor emissions near industrial and chemical waste disposal sites. Environ. Sci. Technol. **16:** 781–785.
9. STEPANYAN, I.S., G.M. PADARYAN, L.K. AIRAPETYAN, et al. 1970. Ionization-chromatographic method for determining some components of waste waters from the plant "Polivinilatsetat." Prom. Arm. **9:** 76–78 (in Russian).
10. MCNEAL, T.P. & H.C. HOLLIFIELD. 1993. Determination of volatile chemicals released from microwave-heat-susceptor food packaging. J. Assoc. Off. Anal. Chem. Int. **76:** 1268–1275.
11. GUERIN, J.C. 1970. Chemical composition of cigarette smoke. In Banbury Report: A Safe cigarette? G.B. Gori & F.G. Bock, Eds. **3:** 191–204. CSH Press. Cold Spring Harbor, NY.
12. SIMON, P., V.G. FILSER & H.M. BOLT. 1985. Metabolism and pharmacokinetics of vinyl acetate. Arch. Toxicol. **57:** 19-23.
13. FILOV, V.A. 1959. The fate of complex esters of vinyl alcohol and fatty acid in the body. Gig. Tr. Zabol. **G3G:** 42–46.
14. INTERNATIONAL AGENCY FOR RESEARCH ON CANCER. 1985. Monographs on the evaluation of the carcinogenic risk of chemicals to humans, Vol. 36. Allyl compounds, aldehydes, epoxides and peroxides. IARC. Lyon.
15. MALTONI, C. & G. LEFEMINE. 1974. Carcinogenicity bioassays of vinyl chloride. I. Research plan and early results. Environ. Res. **7:** 387–405.
16. MALTONI, C., G. LEFEMINE, P. CHIECO, et al. 1974. Vinyl chloride carcinogenesis: Current results and perspectives. Med. Lav. **65:** 421–444.
17. MALTONI, C. & G. LEFEMINE. 1975. Carcinogenicity bioassays of vinyl chloride: current results. Ann. N.Y. Acad. Sci. **246:** 195–218.
18. BOGDANFFY, M.S., H.C. DREEF-VAN DER MEULEN, R.B. BEEMS, et al. 1994. Chronic toxicity and oncogenicity inhalation study with vinyl acetate in the rat and mouse. Fundam. Appl. Toxicol. **23:** 215–229.
19. LIJINSKI, W. & M.D. REUBER. 1983. Chronic toxicity studies of vinyl acetate in Fischer rats. Toxicol. Appl. Pharmacol. **68:** 43–53.

20. BOGDANFFY, M.S., T.R. TYLER, M.B. VINEGAR, et al. 1994. Chronic toxicity and oncogenicity study with vinyl acetate in the rat: in utero exposure in drinking water. Fundam. Appl. Toxicol. **23:** 206–214.
21. U.S. ENVIRONMENTAL PROTECTION AGENCY. 1997. Carcinogenesis study of vinyl acetate (drinking water study) in rats and mice, with cover letter from Japan Bioassay Research Center dated 1/31/97. EPA/OTS, FYI-OTS-0297-1286. Washington, DC.
22. WAXWEILER, R.J., A.H. SMITH, H. FALK, et al. 1981. Excess lung cancer in a synthetic chemical plant. Environ. Health Perspect. **41:** 159–165.
23. OTT, M.G., M.J. TETA & H.L. GREENBERG. 1989. Lymphatic and hematopoietic tissue cancer in a chemical manufacturing environment. Am. J. Ind. Med. **16:** 631–643.
24. MALTONI, C., A. CILIBERTI, G. LEFEMINE, et al. 1997. Results of a long-term experimental study on the carcinogenicity of vinyl acetate monomer in mice. *In* Preventive Strategies for Living in a Chemical World. E. Bingham & D.P. Rall, Eds. Ann. N.Y. Acad. Sci. **837:** 209–238.
25. SOFFRITTI, M., F. BELPOGGI, F. MINARDI, et al. 2002. Results of a long-term carcinogenicity bioassay on vinyl acetate monomer in Wistar rats. In press.
26. SOFFRITTI, M., F. BELPOGGI, D. CEVOLANI, et al. 2002. Results of long-term experimental studies on the carcinogenicity of methyl alcohol and ethyl alcohol in rats. Ann. N.Y. Acad. Sci. **982:** 46–69.
27. NOBLE, A.C., R.A. FLAT & R.R. FORREY. 1980. Wine headspace analysis. Reproducibility and application to varietal classification. J. Agric. Food Chem. **28:** 346–353.
28. MANTEL, A. & S. RAUZY. 1983. Dosages des monomers et solvents legers dans les matériaux plastiques en contact avec l'eau et dans l'eau en contact avec ces materiaux. Rev. Fr. Sci. Eau **2:** 255–256.

Results of Long-Term Experimental Studies on the Carcinogenicity of Ethylene-bis-Dithiocarbamate (Mancozeb) in Rats

FIORELLA BELPOGGI, MORANDO SOFFRITTI, MARINA GUARINO, LUCA LAMBERTINI, DANIELA CEVOLANI, AND CESARE MALTONI[†]

Cancer Research Center, European Ramazzini Foundation for Oncology and Environmental Sciences, Bologna, Italy

ABSTRACT: Mancozeb, an ethylene-bis-dithiocarbamate (EBDC), has been one of the most commonly used fungicides in commercial use for several decades. Nevertheless, up to now, no adequate published experimental studies on the carcinogenicity of Mancozeb have been published. Because of the importance of the compound and of the number of people potentially exposed (workers engaged in the production and use of the fungicide, people living in agricultural areas where the compound is sprayed, and people consuming polluted products), a long-term experimental study of Mancozeb was begun at the Cancer Research Center of the Ramazzini Foundation. Groups of 150 male and female Sprague-Dawley rats, 8 weeks old at the start of the treatment, were administered Mancozeb at the concentration of 1000, 500, 100, 10, and 0 ppm in feed supplied *ad libitum* for 104 weeks. At the end of the treatment, animals were kept under controlled conditions until spontaneous death. Mancozeb caused an increase in (1) total malignant tumors, (2) malignant mammary tumors, (3) Zymbal gland and ear duct carcinomas, (4) hepatocarcinomas, (5) malignant tumors of the pancreas, (6) malignant tumors of the thyroid gland, (7) osteosarcomas of the bones of the head, and (8) hemolymphoreticular neoplasias. On the basis of these data, Mancozeb must be considered a multipotent carcinogenic agent.

KEYWORDS: ethylene-bis-dithiocarbamate; Mancozeb; carcinogenicity; long-term bioassay; rat

INTRODUCTION

The issue of long-term toxicity and carcinogenicity of pesticides has been one of the major problems of public health in the last two decades. Unfortunately, only a small percentage of these compounds are tested for toxicity or carcinogenicity prior to their production, distribution, and general use. There is extensive scientific documentation to show that pesticides, especially during production, application, and the

---

Address for correspondence: Morando Soffritti, M.D., Cancer Research Center, European Ramazzini Foundation for Oncology and Environmental Sciences, Bentivoglio Castle, 40010 Bentivoglio (BO), Italy. Voice: +39-051-6640460; fax: +39-051-6640223.
 crcfr@tin.it
[†]Deceased.

disposal of containers, may be released into the general environment and be present at measurable concentrations in the atmosphere, soil, surface and ground waters, and as residues in food.

Ethylene-bis-dithiocarbamate (Mancozeb) is one of the most widely used commercial fungicides worldwide and is used in agriculture in Italy, particularly in the Emilia Romagna Region. In spite of its widespread use, few data exist on the carcinogenicity of this compound. This experiment was designed to evaluate the carcinogenic effects of Mancozeb because of the importance of the compound, the consumption in our country and region, and the number of people exposed that include workers engaged in the production and use of the compound, citizens living in agricultural areas where the compound is sprayed, and citizens consuming polluted products.

Mancozeb $(C_4H_6MnN_2S_4)_a(Zn)_y$ is a fungicide that belongs to the ethylene(bis) dithiocarbamate family. It is a zinc ion coordination product with a manganese ethylene-1,2-bis-dithiocarbamate polymer and has a molecular weight of 265.3 + 65.4. Mancozeb and similar substances were first marketed in 1944, and their usage increased steadily. After the introduction of systemic fungicides, use of bis-dithiocarbamates declined temporarily, but as resistance to systemic fungicides developed, the use of bis-dithiocarbamates, mainly Mancozeb, increased, and it is now one of the most widely used fungicide products.[1] Mancozeb acts by enzyme activity inhibition. It is synthesized[2,3] from carbamate radicals that have reacted with carbon disulfide to give dialkyldithiocarbamates. The reaction with diamines gives dithiocarbamate groups. The addition of zinc chloride to a suspension of dithiocarbamate, named Maneb, yields a product named Mancozeb, which is superior to Maneb in fungicide activity. The world production of Mancozeb in 1983 was about 30,000 tons.[4]

Because of its efficacy against a broad spectrum of fungi and their associated plant diseases,[2,3] Mancozeb is widely used as a fungicide in agriculture on a variety of horticultural crops and wheat. It is also used in industry as a slimicide in water-cooling systems; in sugar, pulp, and paper manufacturing; as a vulcanization accelerator and antioxidant in the rubber industry; and as a scavenger in wastewater treatment because of its chelating properties.[3]

Mancozeb is not known to occur as a natural product. It has a negligible vapor pressure and a low potential to volatilize into the environment.[5,6] It can be found associated with air-borne particulates or as spray drift.[5] Mancozeb has low solubility in water but hydrolyzes rapidly over a wide range of pH.[5] It is hydrolyzed within 1 day and has a field half-life of 1 to 7 days.[5] It has many metabolites, of which the most important is ethylenethiourea (ETU),[7] which has a high water solubility.[5] Mancozeb has low soil persistence due to its fast hydrolysis and has moderate adsorption capacity, whereas ETU is potentially mobile through the soil and can result in groundwater pollution.[5] Residues are regularly detected in fruit and vegetables, and it has been shown that a significant percentage of ETU is produced during cooking of contaminated vegetables.[8]

Oral administration to rats showed an $LD_{50} > 800$ mg/kg of body weight.[9] Administered by inhalation or ingestion at different dose levels for 13 weeks to rats and mice, Mancozeb altered thyroid hormone levels and produced diffuse thyroid follicular epithelial hyperplasia.[1]

No adequate study exists to evaluate reproductive effects over at least two generations.[10] In a teratogenicity study, pregnant Sprague-Dawley rats, treated by gavage at the 11th day of pregnancy with a single administration of Mancozeb at doses of 0–1320 mg/kg b.w., showed an incidence of offspring with gross malformations of 25% in the group treated with the highest dose.[11] In some studies, Mancozeb has been found to cause chromosome aberrations and sister chromatid exchanges in workers occupationally exposed.[12]

Although Mancozeb has been commercially produced for almost 60 years, the carcinogenicity studies on this fungicide are, in our opinion, still inadequate. To date, Mancozeb carcinogenicity experiments have been conducted on rats and mice.

Charles River Crl:CDBR rats received 0–750 ppm of a technical product of Mancozeb (83.8% pure) for two years. Thyroid follicular cell adenomas and/or carcinomas were noted in both sexes at the highest dose level.[13] Sprague-Dawley (CD) rats were given 0–400 ppm of the same mixture. No evidence of tumorigenicity was observed.[13] In two experiments, groups of Charles River CD-1 mice were given 0–1000 ppm dietary concentrations for 78 weeks. No different incidence of liver nodules and liver masses was observed between treated and control animals; overall, there was no evidence of carcinogenicity.[13]

Studies on the carcinogenicity of ETU, the Mancozeb contaminant, degradation product, and metabolite, were conducted on rats and mice. Charles River rats were treated for 78 weeks with feed containing ETU at concentrations of 0–350 ppm and kept alive for an additional 26 weeks on a controlled diet. At the higher dose, the incidence of thyroid carcinomas was 17% in males and 8% in females. No thyroid tumor was observed in control animals in either sex.[1]

Charles River rats were treated for 104 weeks with feed containing ETU at concentrations of 0–500 ppm. The incidence of thyroid carcinoma was 62% in rats treated with 500 ppm, 16% in those treated with 250 ppm, and between 1% and 2% in animals treated with the lowest doses and in the control group.[1]

Combined perinatal and two-year adult dietary exposure to ETU caused Fischer 344/N rats of each sex to have an increased incidence of thyroid neoplasms and a marginal increase in Zymbal gland neoplasms and mononuclear cell leukemias. In a two-year adult-only dietary exposure of Fischer rats 344/N to ETU, an increased incidence of thyroid follicular cell neoplasms occurred in males and females.[14]

Combined perinatal and two-year adult dietary exposure to ETU caused male and female B6C3F$_1$ mice to have an increased incidence in thyroid follicular cell neoplasms, hepatocellular neoplasms, and adenomas of the pars distalis of the pituitary gland.[14]

It must be noted that all the experiments on Mancozeb technical products and ETU were truncated at 104 weeks. The possibility of tumor development in later life was not evaluated.

MATERIALS AND METHODS

Mancozeb was supplied by the Rohm and Haas Company, Philadelphia, PA, USA. Its purity was 85%, as active ingredient. The general protocol of the experiment has been described in detail in this volume.[15] Mancozeb was added at concentrations of 1000, 500, 100, 10, or 0 ppm to the standard "Corticella" diet, used for 30 years at the laboratories of the Cancer Research Center of the Ramazzini Foundation (CRC/RF), and prepared and supplied monthly by the "Laboratorio Dottori

Piccioni." Mancozeb-treated feed was administered *ad libitum* to Sprague-Dawley rats (75/sex/group), 8 weeks old at the start of the experiments. Control animals received the same feed without Mancozeb. Treated animals were given Mancozeb-treated food for 104 weeks after which time they received standard food. All animals were kept under observation until spontaneous death.

RESULTS

There were no differences between the various groups in mean daily water or food consumption, body weight, survival, or behavior. No treatment-related nononcological pathological changes were detected by gross inspection during the experiment or by histopathological examination.

The occurrence of benign and malignant tumors among male and female rats in treated and control groups is shown in TABLE 1. Differences observed among treated animals and controls were:

(1) increases in total malignant tumors in males and in females in all of the Mancozeb-treated groups—in males, the increase was dose-related (TABLE 2);

(2) increases in total malignant mammary tumors in all Mancozeb-treated females—the differences were statistically significant at the dose levels of 1,000 and 100 ppm;

(3) increased incidence in Zymbal gland and ear duct carcinomas in males treated at 1000 ppm; an increased incidence of carcinomas was also observed in nasal and oral cavities, tongue, lips, pharynx, and larynx. Overall, the incidence of head and neck carcinomas increased in males (dose-related) and in females of all treated groups (TABLE 3);

(4) increased incidence of hepatocarcinomas in males at the highest dose (1000 ppm);

(5) increased incidence of malignant tumors of the pancreas in males and females treated with 1000, 500, and 100 ppm—one case of islet cell carcinoma was observed in a female treated with 10 ppm;

(6) increased incidence of malignant tumors of the thyroid gland in males treated with 1000 and 500 ppm, and in females treated with 1000, 100, and 10 ppm (TABLE 4);

(7) increased incidence of osteosarcomas in bones of the head in males treated with 1000, 500, 100, and 10 ppm and in females treated with 1000, 500, and 10 ppm—the differences were statistically significant in males and females treated at 10 ppm;

(8) increases in hemolymphoreticular neoplasias (lymphomas and leukemias) in males and females treated with 1000, 500, 100, and 10 ppm of Mancozeb (TABLE 5).

CONCLUSIONS

Animals treated with Mancozeb in food from age 8 weeks through age 104 weeks and followed until spontaneous death showed a significant increase in total tumors and in tumors of specific type that were often sex specific. Mancozeb was shown to be carcinogenic on the basis of the number of total malignant tumors and the tumors

TABLE 1. Long-term carcinogenicity bioassay on Mancozeb administered with feed supplied *ad libitum* to male (M) and female (F) Sprague-Dawley rats

NUMBER AND PERCENTAGE OF MALE AND FEMALE SPRAGUE-DAWLEY RATS BEARING VARIOUS TYPES OF BENIGN AND MALIGNANT TUMORS[a]

Site / Histotype	I: 1,000 ppm Male No.	%	I: 1,000 ppm Female No.	%	II: 500 ppm Male No.	%	II: 500 ppm Female No.	%	III: 100 ppm Male No.	%	III: 100 ppm Female No.	%	IV: 10 ppm Male No.	%	IV: 10 ppm Female No.	%	V: 0 (control) Male No.	%	V: 0 (control) Female No.	%
Skin																				
Papilloma	1	1.3	0	-	0	-	0	-	0	-	0	-	0	-	0	-	0	-	0	-
Trichoepithelioma	0	-	0	-	0	-	0	-	0	-	0	-	0	-	0	-	1	1.3	0	-
Squamous cell carcinoma	1	1.3	0	-	0	-	0	-	2	2.7	0	-	0	-	0	-	0	-	0	-
Dermatofibrosarcoma	1	1.3	0	-	0	-	0	-	0	-	0	-	0	-	0	-	0	-	0	-
Subcutaneous tissue																				
Fibroma	1	1.3	0	-	1	1.3	0	-	1	1.3	1	1.3	1	1.3	0	-	0	-	0	-
Lipoma	2	2.7	0	-	0	-	1	1.3	1	1.3	0	-	0	-	0	-	1	1.3	1	1.3
Fibrosarcoma	0	-	0	-	0	-	0	-	0	-	1	1.3	0	-	0	-	0	-	0	-
Liposarcoma	1	1.3	0	-	1	1.3	0	-	1	1.3	0	-	1	1.3	0	-	4	5.3	0	-
Mammary glands																				
Fibroma & fibroadenoma	3	4.0	43(72)	57.3	3	4.0	32(51)	42.7	5	6.7	30(44)	40.0	4	5.3	41(64)	54.7	2	2.7	35(54)	46.7
Lipoma	2(3)	2.7	0	-	3	4.0	0	-	2	2.7	1	1.3	2	2.7	0	-	3(4)	4.0	0	-
Adenocarcinoma	1	1.3	5(7)	6.7	0	-	8(10)	10.7	0	-	9(13)	12.0	0	-	5	6.7	1	1.3	3	4.0
Fibrosarcoma	1	1.3	2	2.7	0	-	0	-	1	1.3	0	-	0	-	0	-	0	-	0	-
Liposarcoma	1	1.3	2	2.7	1	1.3	0	-	1	1.3	0	-	1	1.3	1	1.3	2	2.7	0	-
Zymbal glands[b]																				
Sebaceous adenoma	0	-	0	-	1	1.3	0	-	0	-	0	-	0	-	0	-	0	-	0	-
Carcinoma	12	16.0	5(6)	6.7	6	8.0	6	8.0	4(5)	5.3	4	5.3	1	1.3	6	8.0	1	1.3	1(2)	1.3
Ear ducts[b]																				
Acanthoma	0	-	0	-	0	-	0	-	0	-	0	-	0	-	0	-	1	1.3	0	-
Carcinoma	10	13.3	11	14.7	7	9.3	11(12)	14.7	5(6)	6.7	7(8)	9.3	8(10)	10.7	10	13.3	2	2.7	6	8.0
Nasal cavities[b]																				
Carcinoma	1	1.3	0	-	1	1.3	1	1.3	1	1.3	0	-	1	1.3	0	-	0	-	0	-
Olfactory neuroblastoma	0	-	0	-	1	1.3	1	1.3	0	-	0	-	1	1.3	0	-	0	-	0	-
Oral cavity, tongue & lips																				
Carcinoma	2	2.7	4	5.3	3	4.0	0	-	1	1.3	1	1.3	0	-	3	4.0	2	2.7	0	-
Fibrosarcoma	0	-	0	-	0	-	5	6.7	0	-	3	4.0	0	-	0	-	1	1.3	1	1.3
Pharynx[b]																				
Carcinoma	0	-	1	1.3	0	-	0	-	0	-	1	1.3	1	1.3	1	1.3	0	-	0	-
Larynx[b]																				
Carcinoma	0	-	0	-	1	1.3	0	-	1	1.3	0	-	0	-	4	5.3	0	-	0	-
Lung																				
Adenoma	0	-	0	-	1	1.3	0	-	0	-	0	-	0	-	0	-	0	-	0	-
Adenocarcinoma	0	-	0	-	0	-	0	-	0	-	0	-	0	-	1	1.3	0	-	1	1.3
Fibrosarcoma	0	-	1	1.3	0	-	0	-	0	-	0	-	0	-	0	-	0	-	0	-

— Continued

TABLE 1. Continued

NUMBER AND PERCENTAGE OF MALE AND FEMALE SPRAGUE-DAWLEY RATS BEARING VARIOUS TYPES OF BENIGN AND MALIGNANT TUMORS[a]

Site		I: 1,000 ppm				II: 500 ppm				III: 100 ppm				IV: 10 ppm				V: 0 (control)			
Histotype		Male		Female		Male		Female		Male		Female		Male		Female		Male		Female	
		No.	%	No.	%	No.	%	No.	%	No.	%	No.	%	No.	%	No.	%	No.	%	No.	%
Stomach																					
- Forestomach																					
Acanthoma		2	2.7	4	5.3	2	2.7	3	4.0	3	4.0	2	2.7	4	5.3	3	4.0	5	6.7	8	10.7
Adenoma		0	-	0	-	0	-	0	-	1	1.3	0	-	0	-	0	-	0	-	0	-
Squamous cell carcinoma		1	1.3	0	-	1	1.3	0	-	2	2.7	0	-	0	-	1	1.3	0	-	0	-
Intestine																					
Adenomatous polyp		0	-	1	1.3	0	-	0	-	1	1.3	0	-	1	1.3	0	-	0	-	0	-
Adenocarcinoma		0	-	0	-	0	-	0	-	0	-	0	-	0	-	0	-	0	-	1	1.3
Leiomyosarcoma		0	-	1	1.3	0	-	0	-	1	1.3	0	-	0	-	0	-	1	1.3	0	-
Salivary glands																					
Adenocarcinoma		0	-	0	-	1	1.3	0	-	0	-	0	-	0	-	0	-	0	-	0	-
Liver																					
Cholangioma		1	1.3	1	1.3	0	-	0	-	0	-	0	-	0	-	0	-	0	-	0	-
Hepatocarcinoma		4	5.3	0	-	1	1.3	0	-	1	1.3	2	2.7	0	-	0	-	0	-	1	1.3
Angiosarcoma		0	-	0	-	0	-	0	-	0	-	0	-	0	-	0	-	1	1.3	0	-
Pancreas																					
Exocrine adenoma		0	-	0	-	4	5.3	2	2.7	0	-	0	-	0	-	0	-	3	4.0	1	1.3
Islet cell adenoma		10	13.3	5	6.7	10	13.3	7	9.3	10	13.3	6	8.0	8	10.7	5	6.7	11	14.7	6	8.0
Exocrine adenocarcinoma		1	1.3	1	1.3	0	-	0	-	0	-	0	-	0	-	0	-	0	-	0	-
Islet cell carcinoma		3	4.0	0	-	2	2.7	2	2.7	1	1.3	3	4.0	0	-	1	1.3	0	-	0	-
Kidneys																					
Adenoma		0	-	0	-	0	-	0	-	0	-	0	-	0	-	0	-	1	1.3	0	-
Nephroblastoma		0	-	0	-	0	-	1	1.3	1	1.3	0	-	0	-	0	-	0	-	0	-
Adenocarcinoma		0	-	0	-	1	1.3	1	1.3	0	-	1(2)	1.3	0	-	0	-	0	-	0	-
Liposarcoma		0	-	0	-	0	-	0	-	0	-	0	-	1	1.3	0	-	0	-	0	-
Bladder																					
Papilloma		0	-	0	-	0	-	0	-	0	-	0	-	0	-	0	-	0	-	1	1.3
Leiomyosarcoma		0	-	0	-	0	-	0	-	0	-	0	-	1	1.3	0	-	0	-	0	-
Prostate																					
Adenocarcinoma		0	-			0	-			1	1.3			0	-			0	-		
Testes																					
Interstitial cell adenoma		7(8)	9.3			10	13.3			6(8)	8.0			6(7)	8.0			7(8)	9.3		
Ovaries																					
Cystadenoma				3	4.0			0	-			3	4.0			0	-			0	-
Granulosa cell tumor				1	1.3			1	1.3			0	-			0	-			0	-
Sertoli cell tumor				2	2.7			4(5)	5.3			5	6.7			1	1.3			1	1.3
Granulosa cell malignant tumor				0	-			0	-			0	-			1	1.3			0	-
Malignant thecoma				0	-			0	-			0	-			1	1.3			0	-

— *Continued*

TABLE 1. Continued

NUMBER AND PERCENTAGE OF MALE AND FEMALE SPRAGUE-DAWLEY RATS BEARING VARIOUS TYPES OF BENIGN AND MALIGNANT TUMORS[a]

Site Histotype	I: 1,000 ppm				II: 500 ppm				III: 100 ppm				IV: 10 ppm				V: 0 (control)			
	Male		Female		Male		Female		Male		Female		Male		Female		Male		Female	
	No.	%	No.	%	No.	%	No.	%	No.	%	No.	%	No.	%	No.	%	No.	%	No.	%
Uterus																				
Polyp			11	14.7			11	14.7			16	21.3			14	18.7			7	9.3
Leiomyoma			3	4.0			1	1.3			0	-			1	1.3			2	2.7
Fibroangioma			0	-			2	2.7			0	-			0	-			0	-
Squamous cell carcinoma			0	-			1	1.3			0	-			0	-			0	-
Adenocarcinoma			4	5.3			5	6.7			3	4.0			3	4.0			4	5.3
Fibrosarcoma			0	-			0	-			0	-			0	-			1	1.3
Leiomyosarcoma			4	5.3			3	4.0			0	-			1	1.3			1	1.3
Malignant Schwannoma			0	-			0	-			0	-			1	1.3			0	-
Uterus and vagina																				
Malignant Schwannoma			2	2.7			1	1.3			2	2.7			2	2.7			1	1.3
Vagina																				
Granular cell tumor (Abrikosoff's tumor)			0	-			1	1.3			0	-			0	-			0	-
Malignant Schwannoma			0	-			0	-			0	-			2	2.7			0	-
Peritoneum																				
Lipoma	0	-	0	-	0	-	1	1.3	0	-	0	-	0	-	0	-	0	-	0	-
Fibroangioma	0	-	1	1.3	0	-	0	-	0	-	0	-	0	-	0	-	0	-	0	-
Mesothelioma	1	1.3	0	-	0	-	1	1.3	0	-	0	-	0	-	1	1.3	0	-	0	-
Liposarcoma	1	1.3	0	-	0	-	0	-	0	-	0	-	0	-	0	-	0	-	0	-
Pituitary gland																				
Adenoma	22	29.3	47	62.7	14	18.7	36	48.0	22	29.3	38	50.7	26	34.7	37	49.3	23	30.7	40	53.3
Thyroid gland[c]																				
Follicular adenoma	10	13.3	9	12.0	3	4.0	0	-	1	1.3	3	4.0	0	-	2	2.7	0	-	1	1.3
C-cell adenoma	4	5.3	7[d]	9.3	4[d]	5.3	1	1.3	3	4.0	5	6.7	1	1.3	5	6.7	2[d]	2.7	4	5.3
Follicular carcinoma	7[d]	9.3	12	16.0	2	2.7	0	-	0	-	0	-	0	-	1	1.3	0	-	0	-
C-cell carcinoma	2	2.7	2	2.7	0	-	3	4.0	1	1.3	1	1.3	0	-	0	-	0	-	0	-
Parathyroid glands																				
Adenoma	0	-	0	-	0	-	0	-	0	-	0	-	2	2.7	1	1.3	0	-	1	1.3
Adrenal glands																				
Cortical adenoma	0	-	1	1.3	0	-	4	5.3	0	-	3	4.0	0	-	5(6)	6.7	0	-	3	4.0
Pheochromocytoma	15(21)	20.0	5	6.7	24(29)	32.0	13(16)	17.3	16(24)	21.3	12(14)	16.0	19(22)	25.3	13(16)	17.3	22(36)	29.3	14(16)	18.7
Cortical adenocarcinoma	1	1.3	2	2.7	0	-	0	-	1	1.3	1	1.3	0	-	1	1.3	0	-	0	-
Pheochromoblastoma	2	2.7	2	2.7	3	4.0	1	1.3	0	-	1	1.3	3	4.0	3(4)	4.0	0	-	2	2.7
Central nervous system																				
- Brain																				
Oligodendroglioma	1	1.3	0	-	1	1.3	1	1.3	2	2.7	0	-	0	-	1	1.3	0	-	2	2.7
- Meninges																				
Benign meningioma	1	1.3	0	-	0	-	1	1.3	0	-	0	-	2	2.7	0	-	0	-	0	-
Fibrosarcoma	0	-	0	-	0	-	0	-	0	-	0	-	0	-	0	-	0	-	1	1.3
Malignant meningioma	0	-	0	-	0	-	0	-	2	2.7	0	-	0	-	0	-	0	-	0	-

— Continued

TABLE 1. Continued

NUMBER AND PERCENTAGE OF MALE AND FEMALE SPRAGUE-DAWLEY RATS BEARING VARIOUS TYPES OF BENIGN AND MALIGNANT TUMORS[a]

Site Histotype	I: 1,000 ppm				II: 500 ppm				III: 100 ppm				IV: 10 ppm				V: 0 (control)			
	Male		Female		Male		Female		Male		Female		Male		Female		Male		Female	
	No.	%	No.	%	No.	%	No.	%	No.	%	No.	%	No.	%	No.	%	No.	%	No.	%
Peripheral nervous system																				
- Ganglia																				
Pheochromocytoma	0	-	0	-	0	-	0	-	0	-	0	-	1	1.3	1	1.3	1	1.3	0	-
Bones																				
- Head																				
Chondroma	0	-	0	-	0	-	0	-	2	2.7	0	-	0	-	0	-	0	-	0	-
Osteoma	0	-	0	-	0	-	1	1.3	0	-	0	-	0	-	0	-	0	-	0	-
Chondrosarcoma	0	-	0	-	0	-	0	-	0	-	0	-	0	-	1(2)	1.3	0	-	0	-
Osteosarcoma	8	10.7	3	4.0	5	6.7	4	5.3	8	10.7	1	1.3	13	17.3	8	10.7	2	2.7	1	1.3
- Other																				
Osteosarcoma	1	1.3	1	1.3	1	1.3	0	-	0	-	0	-	0	-	1	1.3	2	2.7	0	-
Soft tissues																				
Liposarcoma	1	1.3	0	-	1	1.3	0	-	0	-	0	-	0	-	0	-	1	1.3	0	-
Angiosarcoma	1	1.3	0	-	0	-	0	-	0	-	0	-	0	-	0	-	0	-	0	-
Heart																				
Myxosarcoma	1	1.3	0	-	0	-	0	-	0	-	0	-	0	-	0	-	0	-	0	-
Malignant Schwannoma	1	1.3	0	-	0	-	0	-	0	-	0	-	0	-	0	-	0	-	1	1.3
Thymus																				
Malignant thymoma[e]	0	-	0	-	0	-	0	-	0	-	1	1.3	0	-	0	-	2	2.7	0	-
Spleen																				
Fibrosarcoma	1	1.3	0	-	1	1.3	0	-	0	-	0	-	0	-	0	-	0	-	0	-
Angiosarcoma	1	1.3	0	-	0	-	1	1.3	1	1.3	0	-	0	-	0	-	0	-	0	-
Mesenteric lymph nodes																				
Fibroangioma	0	-	0	-	1	1.3	0	-	0	-	1	1.3	0	-	0	-	0	-	1	1.3
Hemolymphoreticular tissues[f,g]																				
Lymphomas & leukemias	30	40.0	16	21.3	35(37)	46.7	21	28.0	32	42.7	27	36.0	22	29.3	20	26.7	16	21.3	11	14.7

[a] Numbers between brackets indicate total number of tumors; one animal can bear more than one tumour
[b] See Table 3
[c] See Table 4
[d] One is bearing bilateral tumors
[e] In 96% of cases the tumor itself is composed of a mixture in varying proportions of epithelial cells and lymphocytes. In the remaining 4%, only epithelial cells and lymphocytes are present. We consider that a tumor composed exclusively of lymphocytes should not be classified as a thymoma but as a lymphoma involving the thymus
[f] Including thymus, spleen and mesenteric lymph nodes
[g] See table 5

TABLE 2. Long-term carcinogenicity bioassay on Mancozeb administered with feed supplied *ad libitum* to male (M) and female (F) Sprague-Dawley rats

TOTAL MALIGNANT TUMORS

Group No.	Concentration (ppm)	Animals Sex	Animals No.	Malignant tumors Tumor-bearing animals No.	Malignant tumors Tumor-bearing animals %	Malignant tumors Tumors No.	Malignant tumors Tumors Per 100 animals
I	1,000	M	75	59	78.7 **♦♦	99	132.0 **
		F	75	43	57.3	84	112.0 **
		M+F	150	102	68.0	183	122.0
II	500	M	75	55	73.3 **♦♦	78	104.0 **
		F	75	46	61.3 *	78	104.0 **
		M+F	150	101	67.3	156	104.0
III	100	M	75	51	68.0 **♦♦	72	96.0 **
		F	75	42	56.0	74	98.7 **
		M+F	150	93	62.0	146	97.3
IV	10	M	75	39	52.0 ♦♦	57	76.0 *
		F	75	50	66.7 **	82	109.3 **
		M+F	150	89	59.3	139	92.7
V	0	M	75	29	38.7	38	50.7
		F	75	31	41.3	40	53.3
		M+F	150	60	40.0	78	52.0

*$p < 0.05$ using χ^2 test. **$p < 0.01$ using χ^2 test. ♦♦$p < 0.01$ using Cochrane-Armitage test for dose-response relationship.

TABLE 3. Long-term carcinogenicity bioassay on Mancozeb administered with feed supplied *ad libitum* to male (M) and female (F) Sprague-Dawley rats

CARCINOMAS OF THE HEAD AND THE NECK

Group No.	Concentration (ppm)	Animals Sex	Animals No.	Animals with carcinomas[a] Zymbal glands No.	%	Ear ducts No.	%	Nasal cavities No.	%	Oral cavity, tongue and lips No.	%	Pharynx No.	%	Larynx No.	%	Total No.	%
I	1,000	M	75	12	16.0***	10	13.3*	1	1.3	2	2.7	0	-	0	-	25	33.3*** ♦♦
		F	75	5 (1)	6.7	11	14.7	0	-	4	5.3	1	1.3	0	-	21	28.0***
		M+F	150	17	11.3	21	14.0	1	0.7	6	8.0	1	0.7	0	-	46	30.7
II	500	M	75	6	8.0	7	9.3	2[b]	2.7	3	4.0	0	-	1	1.3	19	25.3*** ♦♦
		F	75	6	8.0	11 (1)	14.7	2[b]	2.7	0	-	0	-	0	-	19	25.3*
		M+F	150	12	8.0	18	12.0	4	2.7	3	2.0	0	-	1	0.7	38	25.3
III	100	M	75	4 (1)	5.3	5 (1)	6.7	1	1.3	1	1.3	0	-	1	1.3	12	16.0 ♦♦
		F	75	4	5.3	7 (1)	9.3	0	-	1	1.3	1	1.3	0	-	13	17.3
		M+F	150	8	5.3	12	8.0	1	0.7	2	1.3	1	0.7	1	0.7	25	16.7
IV	10	M	75	1	1.3	8 (2)	10.7	2[b]	2.7	0	-	1	1.3	0	-	12	16.0 ♦♦
		F	75	6	8.0	10	13.3	0	-	3	4.0	1	1.3	4	5.3	24	32.0***
		M+F	150	7	4.7	18	12.0	2	1.3	3	4.0	2	1.3	4	2.7	36	24.0
V	0	M	75	1	1.3	2	2.7	0	-	2	2.7	0	-	0	-	5	6.7
		F	75	1 (1)	1.3	6	8.0	0	-	0	-	0	-	0	-	7	9.3
		M+F	150	2	1.3	8	5.3	0	-	2	1.3	0	-	0	-	12	8.0

[a]Between brackets the number of animals with bilateral tumors. [b]1 olfactory neuroblastoma. *$p < 0.05$ using χ^2 test. **$p < 0.01$ using χ^2 test. ♦♦$p < 0.01$ using Cochrane-Armitage test for dose-response relationship.

TABLE 4. Long-term carcinogenicity bioassay on Mancozeb administered with feed supplied *ad libitum* to male (M) and female (F) Sprague-Dawley rats

ONCOLOGICAL LESIONS OF THE THYROID GLAND

Group No.	Concentration (ppm)	Animals Sex	Animals No.	Follicular adenomas No.	Follicular adenomas %	C-cell adenomas No.	C-cell adenomas %	Tumor-bearing animals Follicular carcinomas No.	Follicular carcinomas %	C-cell carcinomas No.	C-cell carcinomas %	Total No.	Total %
I	1,000	M	75	10	**13.3**[***]	4	**5.3**	6	**8.0**[*]	2	2.7	22	**29.3**[***]
		F	75	9	**12.0**[*]	7	**9.3**	12	**16.0**[***]	2	2.7	30	**40.0**[***]
		M+F	150	19	**12.7**	11	**7.3**	18	**12.0**	4	2.7	52	**34.7**
II	500	M	75	3	4.0	4	**5.3**	2	2.7	0	-	9	**12.0**
		F	75	0	-	1	1.3	0	-	0	-	1	1.3
		M+F	150	3	2.0	5	3.3	2	1.3	0	-	10	6.7
III	100	M	75	1	1.3	3	4.0	0	-	1	1.3	5	6.7
		F	75	3	4.0	5	6.7	0	-	1	1.3	9	**12.0**
		M+F	150	4	2.7	8	**5.3**	0	-	2	1.3	14	9.3
IV	10	M	75	0	-	1	1.3	0	-	0	-	1	1.3
		F	75	2	2.7	5	6.7	1	**1.3**	0	-	8	**10.7**
		M+F	150	2	1.3	6	4.0	1	0.7	0	-	9	6.0
V	0	M	75	0	-	2	2.7	0	-	0	-	2	2.7
		F	75	1	1.3	4	**5.3**	0	-	0	-	5	6.7
		M+F	150	1	0.7	6	4.0	0	-	0	-	7	4.7

[*] $p < 0.05$ using χ^2 test. [**] $p < 0.01$ using χ^2 test.

TABLE 5. Long-term carcinogenicity bioassay on Mancozeb administered with feed supplied *ad libitum* to male (M) and female (F) Sprague-Dawley rats

HEMOLYMPHORETICULAR NEOPLASIAS AND THEIR DISTRIBUTION BY HISTOCYTOTYPE

Group No.	Concentration (ppm)	Animals			Animals with hemolymphoreticular neoplasias						
					Total[a]		Lymphoimmuno-blastic lymphoma[b]		Other lymphomas and leukemias[b]		
		Sex	No.		No.	%	No.	%	No.	%	
I	1,000	M	75		30	40.0 *	17	56.7	13	43.3	
		F	75		16	21.3	11	68.8	5	31.3	
		M+F	150		46	30.7	28	60.9	18	39.1	
II	500	M	75		35[c]	46.7 **	24	68.6	13	37.1	
		F	75		21	28.0	14	66.7	7	33.3	
		M+F	150		56	37.3	38	67.9	20	35.7	
III	100	M	75		32	42.7 **	17	53.1	15	46.9	
		F	75		27	36.0 **	19	70.4	8	29.6	
		M+F	150		59	39.3	36	61.0	23	39.0	
IV	10	M	75		22	29.3	16	72.7	6	27.3	
		F	75		20	26.7	15	75.0	5	25.0	
		M+F	150		42	28.0	31	73.8	11	26.2	
V	0	M	75		16	21.3	14	87.5	2	12.5	
		F	75		11	14.7	8	72.7	3	27.3	
		M+F	150		27	18.0	22	81.5	5	18.5	

[a]Percentages refer to the number of animals at start. [b]Percentages refer to the number of animals bearing hemolymphoreticular neoplasias. [c]Two animals bore two different hemolymphoreticular neoplasias. *$p < 0.05$ using χ^2 test. **$p < 0.01$ using χ^2 test.

at various sites that included malignant mammary tumors, Zymbal gland and ear duct carcinomas, hepatocarcinomas, malignant tumors of the pancreas, malignant tumors of the thyroid gland, osteosarcomas of the bones of the head, and hemolymphoreticular neoplasias.

In the case of follicular tumors of the thyroid gland, a hormonal mechanism may be envisioned (stimulation of the thyrotropic hormone due to the lowering of thyroid hormones produced by ethylenethiourea); this mechanism cannot apply to tumors of other sites.

Our results indicate that Mancozeb should be considered a multipotent carcinogenic agent capable of producing tumors of many types in various sites in treated animals. The results of our study are not consistent with those produced in other laboratories. It should be noted that, in contrast with our study, which continued until the time of spontaneous death of the animals, the bioassays on rats conducted in other laboratories were truncated after 104 weeks of treatment. In our study, most tumors arose after 112 weeks of age. Had we stopped our experiment at 112 weeks of age, it is unlikely that we would have observed the multipotent carcinogenic activity of Mancozeb.[16]

ACKNOWLEDGMENTS

This research has been partially supported by the Regional Agency for Prevention and Environment (Agenzia Regionale Prevenzione e Ambiente, ARPA) of the Emilia-Romagna Region, Italy.

REFERENCES

1. U.S. ENVIRONMENTAL PROTECTION AGENCY. 1987. Guidance for the reregistration of pesticide products containing Mancozeb as the active ingredient. April 1987. EPA Office of Pesticide Programs. Washington, DC.
2. INTERNATIONAL AGENCY FOR RESEARCH ON CANCER. 1991. Monographs on the evaluation of the carcinogenic risks to humans. Vol. 53. Insecticides. IARC. Lyon.
3. INTERNATIONAL PROGRAMME ON CHEMICAL SAFETY (IPCS). 1988. Dithiocarbamate Pesticides, Ethylenethiourea, and Propylenethiourea: A General Introduction. Environmental Health Criteria 78. WHO. Geneva.
4. BATTELLE'S AGROCHEMICAL DATA BANK. 1983.
5. U.S. ENVIRONMENTAL PROTECTION AGENCY (EPA). 1984. Health and environmental effects profile for mancozeb. Report No.: EPA/600/X-84/129. Washington D.C.
6. LIGOCKI, M.P. & J.F. PANKOW. 1989. Measurements of the gas/particle distribution of atmospheric organic compounds. Environ. Sci. & Technol. **23:** 75–83.
7. HOUETO, P., G. BINDOULA & J.R. HOFFMAN. 1995. Ethylenebisdithiocarbamates and ethylenethiourea: Possible human health hazards. Environ. Health Perspect. **103:** 568–573.
8. NEWSOME, W.H. & G.W. LAVER. 1973. Effects of boiling on the formation of ethylenethiourea in zineb-treated foods. Bull. Environ. Contam. Toxicol. **10:** 151–154.
9. WARTING, C.R. 1979. The Pesticide Manual. A World Compendium, 324. VI edit. British Crop Protection Council. Croydon.
10. LU, M.H. & G.L. KENNEDY, JR. 1986. Teratogenic evaluation of mancozeb in the rat following inhalation exposure. Toxicol. Appl. Pharmacol. **84:** 355–368.
11. LARSSON, K.S., C. ARNANDER, E. CEKANOVA & M. KJELLBERG. 1976. Studies of teratogenic effects of the dithiocarbamates maneb, mancozeb, and propineb. Teratology **14:** 171–184.

12. JABLOICKA, J., H. POLAKOVA, J. KARELOVA & M. VARGOVA. 1989. Analysis of chromosome aberrations and sister-chromatid exchanges in peripheral blood lymphocytes of workers with occupational exposure to the mancozeb-containing fungicide Novozir Mn80. Mutat. Res. **224:** 143–146.
13. WHO. 1993. Pesticide residues in food. 1993. Joint FAO/WHO Meeting on Pesticide Residues. Part II: Mancozeb 257-289. WHO/PCS/94.4. Available from the International Programme on Chemical Safety. WHO, Geneva, CH.
14. NATIONAL TOXICOLOGY PROGRAM. 1992. Toxicology and Carcinogenesis Studies of Ethylene Thiourea (CAS No. 96-45-7) in F344/N Rats and B6C3F1 Mice (Feed Studies). NTP Technical Report 388. Bethesda, MD.
15. SOFFRITTI, M., F. BELPOGGI, D. CEVOLANI, *et al.* 2002. Results of long-term experimental studies on the carcinogenicity of methyl alcohol and ethyl alcohol in rats. Ann. N.Y. Acad. Sci. **982:** 46–69.
16. SOFFRITTI, M., F. BELPOGGI, F. MINARDI & C. MALTONI. 2002. Ramazzini foundation cancer program: history and major projects, life-span carcinogenicity bioassay design. Ann. N. Y. Acad. Sci. **982:** 26–45.

Carcinogenic Effects of Benzene: Cesare Maltoni's Contributions

MYRON A. MEHLMAN

Department of Environmental Medicine, The Mount Sinai Medical Center, New York, New York 10029, USA

ABSTRACT: Cesare Maltoni's contributions to understanding, identifying, and characterizing widely used commercial chemicals in experimental animals are among the most important methods developed in the history of toxicology and serve to protect working men and women, the general population, and our environment from hazardous substances. Maltoni developed experimental methods that have reached the "platinum standard" for protection of public health. Benzene was among the 400 or more chemicals that Maltoni and his associates tested for carcinogenicity. In 1976, Maltoni reported that benzene is a potent experimental carcinogen. Maltoni's experiments clearly demonstrated that benzene is carcinogenic in Sprague-Dawley rats, Wistar rats, Swiss mice, and RF/J mice when administered by inhalation or ingestion. Benzene caused carcinomas of the Zymbal gland, oral cavity, nasal cavities; cancers of the skin, forestomach, mammary glands, and lungs; angiosarcomas and hepatomas of the liver; and hemolymphoreticular cancers. Thus, benzene was shown to be a multipotential carcinogen that produced cancers in several species of animals by various routes of administration. On November 2, 1977, *Chemical Week* reported that Maltoni provided a "bombshell" when he demonstrated the "first direct link" between benzene and cancer. In this paper, I shall summarize early experiments and human studies and reports; Maltoni's experimental contribution to understanding the carcinogenicity of benzene in humans and animals; earlier knowledge concerning benzene toxicity; and benzene standards and permissible exposure levels.

KEYWORDS: benzene; carcinogenesis; gasoline

MALTONI'S REPORT ON EARLIER STUDIES IN EXPERIMENTAL ANIMALS: LONG-TERM CARCINOGENICITY DATA ON BENZENE BEFORE 1976

The available data on long-term carcinogenicity bioassays of benzene available up to 1976, summarized by Maltoni[1] (TABLE 1), are scanty and insubstantial.

IARC Monograph No. 7, 1974,[6] concluded: "Benzene has been tested only in mice by subcutaneous injection and skin application. The data reported do not permit the conclusion that carcinogenic activity has been demonstrated."

Address for correspondence: Myron A. Mehlman, Ph.D., 7 Bouvant Dr., Princeton, NJ 08540. Voice: 609-683-4750; fax: 609-683-0838.
mehlman@rcn.com

TABLE 1. Long-term carcinogenicity bioassays on benzene: available date until 1976[a]

	Animals				Treatment and other experimental		
Authors	Species	Strain	Sex	No.	details	Results	Observations
Lignac, 1933[2]	mouse	albino	M,F	33 T	SQ injection of 0.001 mL of benzene in 0.1 mL olive oil once/week for 17–21 weeks	8 leukemias (from 4–8 months from beginning of treatment)	no control group
Kirschbaum & Strong, 1942[3]	mouse	F	?	30 T 212 C		29 leukemias (30%) from 200–300 days old	leukemia increase in treated animals not statistically significant
Amiel, 1960[4]	mouse	DBZ2	M	30 T	SQ injection of 0.01 mL benzene in 0.1 mL olive oil for life span	0	maximum survival = 6 animals for 730 days
		C3H		30 T		0	
		C57BL6		30 T		0	
		AKR	M	30 T		16 leukemias (between 7th and 16th month of treatment; 8 animals dead before 9th month of treatment	major survival of animals in control group
				35 C		30 leukemias	
Hiraki et al., 1963[5]	mouse	Swiss	M,F	10 T	SQ injection of 0.0001 mL benzene in 0.01 mL olive oil for 10 weeks	5 SQ sarcomas at autopsy performed between the 162nd and 253rd day from start of treatment	2 animals dead within first 8 weeks of treatment no control group
Many	mice rats rabbits	various	M,F	very many	skin applications	no effects	nonsystematic and non-ad hoc planned experiments

[a]Source: Maltoni, 1983.[1]
Abbreviations: T = treated; C = control; SQ = subcutaneous.

In November 1977, early results indicating the carcinogenic effect of benzene were published (Maltoni & Scarinato, 1977).[7] These data were published as definitive (Maltoni & Scarinato, 1979)[8] in studies that showed benzene produces Zymbal gland carcinomas in Sprague-Dawley rats at the two dose levels studied with a dose-response relationship. These tumors are usually seen spontaneously or in animals treated with olive oil alone at a rate below 1%. Benzene also produced a dose-related increase in hemolymphoreticular neoplasia (leukemia) and in mammary carcinomas. Some rarely occurring tumors were found in animals treated with the higher dose and included two skin carcinomas, one subcutaneous angiosarcoma, and one hepatoma. The incidence of all malignant tumors was greater in the treated group.

EARLIER STUDIES IN HUMANS AS REPORTED BY MALTONI AND OTHERS

Maltoni et al.[9] (TABLE 2) summarized reports of hemolymphoreticular neoplasias (leukemias) in individuals exposed to benzene.

Ensuing reports have shown that acute myelogenous leukemia (AML) is the most frequently reported form of leukemia related to benzene exposure. Many other types of leukemia and lymphoma have been found in persons exposed to benzene. The majority of cases have been found in Italy, Turkey, and France, especially among shoemakers handling rubber cement. TABLES 3 and 4 summarize available data on evaluation of leukemia risk in benzene-exposed populations.

TABLE 2. History of benzene carcinogenicity[a]

Year	Report
1897	Santeson[10] observes aplastic anemia and established that benzene is a powerful bone marrow poison
1897	LeNoir[11] observes leukemia in benzene workers
1916	Selling[12] establishes that benzene is a bone marrow poison in rabbit bioassays
1933	Ligna[2] first reports acute leukemia following benzene intoxication
1939	Greenberg et al.[13] report several hundred workers with benzene poisoning in the rotogravure printing industry
1939	Hunter[14] and Mallory et al.[15] report 89 cases of benzene poisoning including several leukemia deaths Hunter[14] observes: "No level of benzene greater than zero is safe over a long period of time."
1960s	Vigliani & Saita[16] report leukemias among workers (mainly shoemakers) heavily exposed to benzene in Italy
1970s	Aksoy et al.[17,18] report leukemias among workers (mainly shoemakers) heavily exposed to benzene in Turkey

[a]Source: Maltoni et al., 1989.[9]

TABLE 3. Case reports of leukemias and related disorders observed in individuals exposed to benzene reported in scientific literature up to 1977[a]

Type	No. of cases	No. of reports
Acute myelogenous leukemia	58	28
Erythroleukemia	16	10
Acute monocytic leukemia	3	2
Chronic myelogenous leukemia	27	7
Myelofibrosis and myeloid metaplasia	7	5
Thrombocythemia	1	1
Acute lymphoblastic leukemia	8	4
Chronic lymphocytic leukemia	9	7
Lymphomas and related disorders	14	7

[a]Source: Mehlman, 1983.[19]

TABLE 4. Available data on the evaluation of leukemia risk in populations exposed to benzene compared to populations not having benzene exposure[a]

Authors	Country	Leukemia risk
Vigliana & Saita, 1964[16]	Italy	20 times greater
Girard et al., 1968;[20] 1970;[21,22] 1971[23,24]	France	statistically significantly greater
Aksoy et al., 1974[18]	Turkey	more than double
Ishimaru et al., 1971[25]	Japan	2–3 times greater
Infante et al., 1977[26]	USA	10 times greater

[a]Source: Mehlman, 1983.[19]

BENZENE-CAUSED CANCERS IN ANIMALS AND MALTONI'S CONTRIBUTION

In numerous studies, Maltoni (Maltoni & Scarnato, 1979;[8] Maltoni, 1982,[27] 1983;[1] Maltoni et al. 1983,[28] 1987,[29] 1989[9]) demonstrated that benzene caused tumors in rats and mice including cancers of the zymbal gland, oral cavity, lung, skin, nasal cavity, forestomach, Harderian gland, mammary gland, ovary, and uterus; lymphoma; hemolymphoreticular neoplasia; and all types of leukemias (TABLE 5). Huff et al. (1989[30]) expanded these studies using a broader dose range and reported numerous cancers occurring at a lower dosage in various organs and tissues (TABLE 6).

While there is a prevalence of acute myelogenous leukemia, it is certain that all forms of leukemia are caused by benzene exposure. In 1971, Ishimaru et al.[25] reported leukemia in adult survivors of Hiroshima and Nagasaki. This report noted that the risk of leukemia was 2.5 times greater among individuals who, in addition to radiation, had also been exposed occupationally to benzene.

TABLE 5. Cancers caused by benzene exposure in rats and mice[a]

Zymbal gland	Mammary gland
Oral cavity	Hepatic
Nasal cavity	Angiosarcoma of liver
Skin	Hemolymphoreticular neoplasia
Forestomach	Lung

[a]Source: Maltoni et al., 1989.[9]

TABLE 6. Cancers caused by benzene exposure in rats and mice[a]

Zymbal gland	Preputial gland
Oral cavity	Forestomach
Skin	Ovary
Lymphoma	Liver
Lung	Uterus
Harderian gland	Leukemia
Mammary gland	

[a]Source: Huff et al., 1989.[30]

TABLE 7. Types of cancers from benzene exposure in humans[a]

Acute myelogenous leukemia	Chronic lymphocytic leukemia
Acute lymphocytic leukemia	Hairy cell leukemia
Acute erythroleukemic leukemia	Hodgkin's lymphoma
Acute myelomonocytic leukemia	Non-Hodgkin's lymphoma
Acute promyelocytic leukemia	Lymphosarcoma
Acute undifferentiated leukemia	Multiple myeloma
Chronic myelogenous leukemia	Reticulum cell sarcoma

[a]Sources: Aksoy et al., 1989;[31] Bond et al., 1986;[32] Decouflé et al., 1983;[33] Delore et al., 1928;[34] Goguel et al., 1967;[35] Goldstein, 1977;[36] Hayes et al., 1997;[37] Infante et al., 1985,[38] 1995;[39] McMichael et al., 1974,[40] 1975,[41] 1976;[42] Rinsky, 1981;[43] 1987;[44] Savitz et al., 1997;[45] Schwartz, 1987;[46] Travis et al., 1994;[47] Vianna & Polan, 1979;[48] Vigliani et al., 1976;[49] Wong, 1987;[50] Yin et al., 1987,[51] 1989,[52] 1994,[53] 1996.[54]

HUMAN LEUKEMIAS AND CANCER CAUSED BY BENZENE

TABLE 7 lists the types of cancers in humans caused from exposure to benzene. In 1989, Yin et al.[52] reported significant increases in human cancers from exposure to benzene. Benzene caused leukemia and cancers of the lung, liver, lymphosarcoma, stomach, esophagus, nasopharnyx, and intestine (TABLE 8). Benzene is also known to cause urothelial cancers in workers in the petroleum refining, rubber, printing, and

TABLE 8. Excess human cancers in benzene workers[a]

Leukemia	Stomach
Lung	Esophagus
Liver	Nasopharynx
Lymphosarcoma	Intestine

[a]Source: Yin et al., 1989.[52]
NOTE: Based on 8460 workers (15,642 males, 12,817 females) from 233 factories and 28,257 control workers (16,621 males, 12,966 females) from 83 factories. Lowest average extimated level of exposure for leukemia, 6.5 mg/m^3.

TABLE 9. Acute myelogenous leukemia deaths from exposure to benzene[a]

Case	ppm-years
1	45.4
2	1.5
3	25.4
4	300.1
5	23.6

[a]Sources: Ott et al., 1978;[60] Bond et al., 1986.[32]
NOTE: Low level: <2 ppm TWA; medium level: 2–9 ppm TWA; high level: 25 ppm TWA.

shoe production industries (Steineck, 1990;[55] Silverman, 1988;[56] 1989;[57,58] Vineis, 1985[59]).

LOW-LEVEL BENZENE EXPOSURE AND LEUKEMIA

Exposure to low levels of benzene, which causes leukemia, is often expressed as a cumulative dose (measured in ppm-years), which is the average exposure over an eight-hour workday multiplied by the time of exposure. For example, an average exposure to 1 ppm benzene for 10 years in a facility would result in 10 ppm-years of cumulative benzene dose. A summary of acute myelogenous leukemia deaths of Dow Chemical workers from exposure to benzene is shown in TABLE 9. Leukemia cases 2, 3, and 5 had very low cumulative benzene exposures of 1.5, 25.4, and 23.6 ppm-years, respectively. An additional study by Bond et al. (1986[32]) of Dow Chemical workers showed a significantly elevated risk of myelogenous leukemia in workers exposed to an average concentration of benzene of 5.5 ppm.

Wong (1987[50]) showed that Case No. 11 (chronic myelogenous leukemia) and Case No. 8 (multiple myeloma) were exposed to average benzene concentrations of 0.5 ppm for 1.2 years and 2.3 years, respectively (TABLE 10).

Reports by Rinsky et al. (1987[61]) and Yin et al. (1989[52]) (TABLES 11 and 12, respectively) show leukemias and multiple myeloma in workers exposed to benzene for reasonably short periods of time.

TABLE 10. Deaths from lymphatic and hematopoietic cancers[a]

Case	Cancer	ppm-years
4	chronic lymphocytic leukemia	43.6
6	chronic myelogenous leukemia	10.0
11	chronic myelogenous leukemia	0.6
8	myeloma	1.2
16	chronic lymphocytic leukemia	—

[a]Source: Wong, 1987.[50]
NOTE: Low level: <1 ppm TWA; medium level: 1–10 ppm TWA; high level: 11–50 ppm TWA. RR = 3.93 ($P < 0.02$) for leukemia and aleukemia; RR = 4.12 ($P < 0.06$) for non-Hodgkin's lymphopoietic cancers.

TABLE 11. Deaths from leukemia and multiple myeloma in workers exposed to benzene[a]

Case	Cancer	ppm-years
1	monocytic leukemia	49.99
2	chronic myelogenous leukemia	0.1
8	myelogenous leukemia	10.16
10	multiple myeloma	19.50
11	multiple myeloma	0.11
13	multiple myeloma	7.75

[a]Source: Rinsky et al., 1987.[61]
NOTE: SMR = 5.6 ($P < 0.001$) for leukemia.

TABLE 12. Leukemias in benzene-exposed workers[a]

Case	Disease	ppm-years
3	acute monoocytic leukemia (AMoL)	89.1
7	acute myelocytic leukemia (AML)	96.0
8	acute myeloctyic leukemia	27.9
9	chronic myeloctyic leukemia (CML)	11.7
11	acute myelocytic leukemia	34.2
15	chronic myelocytic leukemia	16.2
16	acute myelocytic leukemia	28.8
18	acute myelocytic leukemia	12.4
26	acute monoocytic leukemia	29.8
27	acute myelocytic leukemia	10.2

[a]Source: Yin et al., 1989.[52]
NOTE: SMRs for AML = 4.96 ($P < 0.001$); AMoL = 4.54 ($P < 0.05$); CML = 4.24 ($P < 0.01$).

TABLE 13. Distribution of leukemias in Chinese benzine workers[a]

Type of leukemia	Cell-specific[39] SMR	P value
Acute lymphocytic leukemia	1.94	
Acute myelogenous leukemia	4.96	<0.001
Acute monocytic leukemia	4.54	<0.05
Acute myelomonocytic leukemia	5.56	<0.05
Chronic myelogenous leukemia	4.24	<0.01

[a]Source: Yin et al. 1989.[52]

In his comprehensive review of leukemia cell types and benzene exposure (TABLE 13), Peter Infante (1995[39]) concluded:

> As a result of this review, it is apparent that all of the major types of leukemia and some of the more rare types of leukemia are associated with occupational benzene exposures, namely: acute myelogenous (AML), myelomonocytic (AMMoL), monocytic (AMoL), promyelocytic, erythroleukemia, (AEL), acute undifferentiated, acute lymphatic (ALL), chronic myelogenous (CML), chronic lymphatic (CLL), and hairy cell (HCL) as well as myelodysplastic syndrome and myelofibrosis have been reported in association with benzene exposure. In some case series, acute leukemia, predominately AML, was the most frequent form of leukemia reported, while in other case series, chronic forms of leukemia were common.

A case-control interview study of 125 adult patients with acute leukemia and 125 controls matched with respect to age (±four years), sex, and residence was carried out in central Sweden during the period from September 1980 to May 1983 (Lindquist et al., 1987[64]). Risk of developing acute leukemia was significantly higher in patients, found to have an increased solvent exposure rate, than in controls with an estimated odds ratio (OR) of 4.9, (95% CI 2.2–12.1). The most frequently exposed profession, painters, exhibits a relative risk of 13 (95% CI 2.0–554). These results suggested an etiological relationship between organic solvent exposure and the development of acute leukemia in man. Schwartz (1987[46]) found increased leukemias: PMR = 3.28 ($P < 0.05$).

CARCINOGENICITY OF ALKYLBENZENES

A number of earlier studies failed to detect benzene-induced carcinogenicity (IARC, 1982;[63] Huff, et al., 1989[30]), a rather puzzling phenomenon, but not unusual in scientific experimentation. In 1985, Maltoni, et al.[64] demonstrated that alkylbenzenes—namely, toluene, xylene, and ethylbenzene—cause cancers in laboratory animals. In 1989,[9] Maltoni called alkylbenzenes "slow carcinogens," which means that it takes a long time before cancers develop after exposure to alkylbenzenes. These findings are not surprising in that asbestos-caused mesothelioma, which takes 20 to 45 years to develop, is a slow carcinogen; yet there is no question that asbestos causes human cancers.

REGULATION

The threshold limit value–time weighted average (TLV-TWA) for benzene in 1946 was 100 ppm; in 1947, 50 ppm; 1948–1956, 35 ppm; 1957–1962, 25 ppm; 1977–1987, 10 ppm; it is currently 1 ppm. In July of 1990, the American Conference of Governmental Industrial Hygienists (ACGIH[65]) recommended that the TLV-TWA for benzene be reduced to 0.1 ppm. In 1993, the Collegium Ramazzini recommended that the occupational standard of exposure to benzene be limited to 0.1 ppm or less.

In September 1948, the American Petroleum Institute (API) issued a document entitled "API Toxicology Review: Benzene" prepared by P. Drinker; it was widely circulated to oil companies. This report states, "Inasmuch as the body develops no tolerance to benzene and there is a wide variation in individual susceptibility, it is generally considered that the only absolutely safe concentration for benzene is *zero.*" The API document further stated, "a limit of 50 ppm or less is strongly recommended, particularly where exposures are recurrent. Skin contact should be avoided." This API document–recommended level was 500-fold greater than that recently recommended by ACGIH. Thus, in spite of scientific evidence and general knowledge of the carcinogenicity of benzene, it is disconcerting that warning labels are not used to identify the carcinogenicity of benzene in products containing this chemical, especially gasoline. The current benzene content of gasoline is between 1.5% and 6%.

SUMMARY

Because we do not know of any safe benzene level above zero, the problems that have been plaguing the health protection process relative to benzene can perhaps be best resolved by setting current recommended maximum occupational levels of exposure to 0.004 to 0.1 ppm (Mehlman, 1991[66]) and, to the extent possible, avoiding any exposure at all to benzene and benzene-containing products. Studies of low levels of benzene at concentrations of 0.1–50 ppm that Maltoni proposed in the late 1970s have not yet been started.

REFERENCES

1. MALTONI, C. 1983. Myths and facts in the history of benzene carcinogenicity. *In* Advances in Modern Environmental Toxicology, Vol. IV. M.A. Mehlman, Ed.: 1–15. Princeton Scientific Publishers. Princeton, NJ.
2. LIGNAC, G.O.E. 1933. Die benzallekamie bei menschen und weissen mausen. III. Zweite Benzoiversuchareine-Von 54 Mausen Genen 8 An Leukamie Odar Lymphoblastoma Innkrans Aleucaemicum Zugrunde-Fruhere Teerbenzol-Versuche. Krankeitsforsch. **9:** 426–453.
3. KIRSCHBAUM, A. & L.D. STRONG. 1942. Influence of carcinogens on the age incidence of leukemia in the high leukemia F strain of mice. Cancer Res. **2:** 841–845.
4. AMIEL, J.L. 1960. Essai négatif d'induction de leucémies chez les souris par le benzéze. Rev. Fr. Etud. Clin. Biol. **5:** 198–199.
5. HIRAKI, K. *et al.* 1963. Development of subcutaneous sarcomas in Swiss mice given repeated injections of benzene in olive oil. Gann **54:** 427–431.

6. IARC. 1974. Some antithyroid and related substances, nitrofurans and industrial chemicals. IARC Monograph on the Carcinogenic Risk of Chemical to Man, Vol. 7. International Agency for Research on Cancer. Lyon, France.
7. MALTONI, C. & C. SCARNATO. 1977. Le prime prove experimentali del l'azione cancerogena del benzene. Gli. Ospedali della Vita **4:** 111–113.
8. MALTONI, C. & C. SCARNATO. 1979. First experimental demonstration of the carcinogenic effects of benzene. Long-term bioassays on Sprague-Dawley rats by oral administration. Med. Lav. **70:** 352–357.
9. MALTONI, C. *et al.* 1989. Benzene, an experimental multipotential carcinogen: results of the long-term bioassays performed at the Bologna Institute of Oncology. Environ. Health Perspect. **82:** 109–124.
10. SANTESON, C.C. 1897. Chronic poisoning with benzene. Arch. Fur. Hyg. **30/31:** 336–376.
11. LENOIRE, C. 1897. On a case of purpura attributed to benzene intoxication. (Trans.) Bull. Med. Soc. Hosp. Paris **14:** 1251–1261.
12. SELLING, L. 1916. Benzol as a leucotoxin. Johns Hopkins Hosp. Rep. **17:** 83–142.
13. GREENBURG, L. *et al.* 1939. Benzene (benzol) poisoning in the rotogravure printing industry in New York City. J. Ind. Hyg. Toxicol. **21:** 395–420.
14. HUNTER, F.T. 1939. Chronic exposure to benzene (benzol). II. the clinical effects. J. Ind. Hyg. Toxicol. **21:** 331–354.
15. MALLORY, T.B. *et al.* 1939. Chronic exposure to benzene (Benzol). III. the pathologic results. J. Ind. Hyg. Toxicol. **21:** 355–377.
16. VIGLIANI, E.C. & G. SAITA. 1964. Benzene and leukemia. N. Engl. J. Med. **271:** 872–876.
17. AKSOY, M. *et al.* 1972. Acute leukemia due to chronic exposure to benzene. Am. J. Med. **52:** 160–166.
18. AKSOY, M., *et al.* 1974. Leukemia in shoe-workers exposed chronically to benzene. Blood **44:** 837–841.
19. MEHLMAN, M.A. 1983. Advances in Modern Environmental Toxicology, Vol. IV. M.A. Mehlman, Ed. Princeton Scientific Publishers. Princeton, NJ.
20. GIRARD, R. *et al.* 1968. Les expositions benzéniques meconnues. Leur recherche systematique au cours des hémopaties graves. Enquetes chez 200 hemopathiques hospitalises. Arch. Mal. Prof. **29:** 723.
21. GIRARD, R. & L. ROVOL. 1970a. La fréquence e'une exposition benzénique au cours des hémopaties graves. Nou. Rev. Fr. Hematol. **10:** 477.
22. GIRARD, R. *et al.* 1970. Hydrocarbures benzéniques et hémopathies graves. Arch. Mal. Prof. **31:** 625.
23. GIRARD, R. *et al.* 1971. Comments on identification for benzene induced leukemia and aplasia. Arch. Mal. Prof. **32:** 581.
24. GIRARD, R. *et al.* 1971. Malignant hemopathies and benzene poisoning. Med. Lav. **62:** 71.
25. ISHIMARU, T. *et al.* 1971. Occupational factors in the epidemiology of leukemia in Hiroshima and Nagasaki. Am. J. Epidem. **93:** 157.
26. INFANTE, P. F. *et al.* 1977. Leukemia in benzene workers. Lancet **ii:** 76–79.
27. MALTONI, C. 1982. Benzene: a multipotential carcinogen. Acta Oncol. **3:** 1–4.
28. MALTONI, C. *et al.* 1983. Benzene: a multipotential carcinogen. Results of long-term bioassays performed at the Bologna Institute of Oncology. Am. J. Ind. Med. **4:** 589–630.
29. MALTONI, C. *et al.* 1987. Further evidence of benzene carcinogenicity: results on Wistar rats and Swiss mice treated by ingestion. Ann. N.Y. Acad. Sci. **534:** 412–426.
30. HUFF, J.E. *et al.* 1989. Multiple site carcinogenicity of benzene in Fisher 344 rats and B6CF1 mice. Envir. Health Persp. **82:** 125–163.
31. AKSOY, M. 1989. Hematotoxicity and carcinogenicity of benzene. Environ. Health Persp. **82:** 193–197.
32. BOND, G.G. *et al.* 1986. An update among chemical workers exposed to benzene. Br. J. Ind. Med. **43:** 685–691.
33. DECOUFLÉ, P. *et al.* 1983. Mortality among chemical workers exposed to benzene and other agents. Environ. Res. **30:** 16–25.
34. DELORE, P. & C. BORGOMANO. 1928. Leucémie algue au cours de l'intoxication benzénique. Sur l'origine toxique de certaines leucémies aigues et leurs rélations avec les anémies graves. J. Med. Lyon **9:** 227–233.

35. GOGUEL, A. et al. 1967. Les leucemies benzeniques de la region Parisienne entre, 1950 et 1965. (Etude de 50 observations). Nouv. Rev. Fr. Hematol. **7:** 465–480.
36. GOLDSTEIN, B.D. 1977. Hematotoxicity in humans. J. Toxicol. Environ. Health (Suppl.) **2:** 89–105.
37. HAYES, R.B. et al. 1997. Benzene and the dose-related incidence of hematologic neoplasms in China. J. Natl. Cancer. Inst. **89:** 1065.
38. INFANTE, P.F. & M. WHITE. 1985. Projections of leukemia risk associated with occupational exposure to benzene. Am. J. Med. **7:** 403–413.
39. INFANTE, P.F. 1995. Benzene and leukemia: cell types, latency and amount or exposure associated with leukemia. Adv. Occupat. Med. Rehabil. **1:** 107–119.
40. MCMICHAEL, A.J. et al. 1974. An epidemiologic study of mortality within a cohort of rubber workers, 1964–72. J. Occup. Med. **16:** 458–464.
41. MCMICHAEL, A.J. et al. 1975. Solvent exposure and leukemia among rubber workers: an epidemiologic study. J. Occup. Med. **17:** 234–239.
42. MCMICHAEL, A.J. et al. 1976. Mortality among rubber workers: relationship to specific jobs. J. Occup. Med. **18:** 178–185.
43. RINSKY, R.A. et al. 1981. Leukemia in benzene workers. Am. J. Ind. Med. **2:** 217.
44. RINSKY, R.A. et al. 1986. Benzene and leukemia: an epidemiologic risk assessment: 24. National Institute for Occupational Safety and Health. Cincinnati, OH.
45. SAVITZ, D.A. & K.W. ANDREWS. 1997. Review of epidemiologic evidence on benzene and lymphatic and hematopoietic cancers. Am. J. Ind. Med. **31:** 287–295.
46. SCHWARTZ, E. 1987. Proportionate mortality analysis of automobile mechanics and gasoline service station workers in New Hampshire. Am. J. Ind. Med. **12:** 91–99.
47. TRAVIS, L.B. et al. 1994. Hematopoietic malignancies and related disorders among benzene-exposed workers in China. Leuk. Lymphoma **14:** 91–102.
48. VIANNA, N.J. & A. POLAN. 1979. Lymphomas and occupational benzene exposure. Lancet. **30:** 1394–1395.
49. VIGLIANI, E.C. 1976. Leukemia associated with benzene exposure. *In* Occupational Carcinogenesis. Ann. N.Y. Acad. Sci. **271:** 143–151.
50. WONG, O. 1987. An industry wide mortality study of chemical workers occupationally exposed to benzene. I. General results. Br. J. Indust. Med. **44:** 365–381.
51. YIN, S.-N. et al. 1987. Leukemia in benzene workers: a retrospective cohort study. Br. J. Indust. Med. **44:** 124–128.
52. YIN, S.-N. et al. 1989. A retrospective cohort study of leukemia and other cancers in benzene workers. Environ. Health Persp. **82:** 207–213.
53. YIN, S.-N. et al. 1994. Cohort study among workers exposed to benzene in China. I. General methods and resources. Am. J. Med. **26:** 383–400.
54. YIN, S.-N. et al. 1996. A cohort study of cancer among benzene exposed workers in China: overall results. Am. J. Indust. Med. **29:** 227–235.
55. STEINECK, G. et al. 1990. Increased risk of urothelial cancer in Stockholm during 1985–87 after exposure to benzene and exhausts. Int. J. Cancer. **45:** 1012–1017.
56. SILVERMAN, D.T. et al. 1988. Occupation and cancer of the lower urinary tract in Detroit. J. Nat. Cancer Inst. **70:** 237–245.
57. SILVERMAN, D.T. et al. 1989. Occupational risks of bladder cancer in the United States: I. White men. J. Natl. Cancer Inst. **81:** 1472–1480.
58. SILVERMAN, D.T. et al. 1989. Occupational risks of bladder cancer in the United States: II. Nonwhite men. J. Natl. Cancer Inst. **81:** 1480–1483.
59. VINEIS, P. & C. MAGNANI. 1985. Occupation and bladder cancer in males: a case-control study. Int. J. Cancer. **35:** 599–606.
60. OTT, G.M. et al. 1978. Mortality among individuals exposed to benzene. Arch. Environ. Health **33:** 3–10.
61. RINSKY, R.A. et al. 1987. Benzene and leukemia: an epidemiologic risk assessment. N. Engl. J. Med. **316:** 1044–1050.
62. LINDQUIST, R. et al. 1987. Increased risk of developing acute leukemia after employment as a painter. Cancer **60:** 1378–1384.
63. IARC. 1982. IARC Monographs on the evaluation of the carcinogenic risk of chemicals to humans, Vol. 29: Some industrial chemical and dyestuffs. Benzene: 93–148. World Health Organization, International Agency for Research on Cancer. Lyon, France.

64. MALTONI, C. *et al.* 1985. Experimental studies on benzene carcinogenicity at the Bologna Institute of Oncology: current results and ongoing research. Am. J. Ind. Med. **7:** 415–446.
65. ACGIH. 1990. Notice of intended changes—benzene. Appl. Occup. Hyg. **5:** 453–463.
66. MEHLMAN, M.A. 1991. Benzene health effects: unanswered questions still not addressed. Am. J. Ind. Med. **20:** 707–711.

Carcinogenicity of Methyl-Tertiary Butyl Ether in Gasoline

MYRON A. MEHLMAN

Department of Environmental Medicine, The Mount Sinai Medical Center, New York, New York 10029, USA

ABSTRACT: Methyl tertiary butyl ether (MTBE) was added to gasoline on a nationwide scale in 1992 without prior testing of adverse, toxic, or carcinogenic effects. Since that time, numerous reports have appeared describing adverse health effects of individuals exposed to MTBE, both from inhalation of fumes in the workplace and while pumping gasoline. Leakage of MTBE, a highly water-soluble compound, from underground storage tanks has led to contamination of the water supply in many areas of the United States. Legislation has been passed by many states to prohibit the addition of MTBE to gasoline. The addition of MTBE to gasoline has not accomplished its stated goal of decreasing air pollution, and it has posed serious health risks to a large portion of the population, particularly the elderly and those with respiratory problems, asthma, and skin sensitivity. Reports of animal studies of carcinogenicity of MTBE began to appear in the 1990s, prior to the widespread introduction of MTBE into gasoline. These reports were largely ignored. In ensuing years, further studies have shown that MTBE causes various types of malignant tumors in mice and rats. The National Toxicology Program (NTP) Board of Scientific Counselors' Report on Carcinogens Subcommittee met in December 1998 to consider listing MTBE as "reasonably anticipated to be a human carcinogen." In spite of recommendations from Dr. Bailer, the primary reviewer, and other scientists on the committee, the motion to list MTBE in the report was defeated by a six to five vote, with one abstention. On the basis of animal studies, it is widely accepted that if a chemical is carcinogenic in appropriate laboratory animal test systems, it must be treated as though it were carcinogenic in humans. In the face of compelling evidence, NTP Committee members who voted not to list MTBE as "reasonably anticipated to be a human carcinogen" did a disservice to the general public; this action may cause needless exposure of many to health risks and possibly cancers.

KEYWORDS: MTBE; gasoline oxygenates; carcinogenicity

CARCINOGENICITY OF METHYL TERTIARY BUTYL ETHER IN GASOLINE

Gasoline is produced from crude petroleum by a variety of refining and manufacturing processes that modify the character of gasoline. These processes include catalytic cracking, coking, alkylation, and catalytic reforming.[1–3] Gasoline, which

---

Address for correspondence: Myron A. Mehlman, 7 Bouvant Dr., Princeton, NJ 08540. Voice: 609-683-4750; fax: 609-683-0838.
mehlman@rcn.com

contains more than 1,000 possible substances, constitutes one of the more complex mixtures to which humans are exposed. The composition of gasoline varies, depending on the geographic region, the season, performance requirements (octane rating), blending of stock, and the source of the crude oil. The hydrocarbons in liquid gasoline consist of 60–70% alkanes (paraffins), 25–30% aromatics, 5–10% alkenes (olefins), plus octane enhancers and lead scavenger agents.[4]

Of the thousands of chemicals and additives (such as MTBE at 11–19%) in gasoline, some are known to be carcinogenic. Chemicals found in gasoline include benzene, 1,3-butadiene, toluene, xylenes, ethylbenzene, MTBE, n-hexane, and many more.

Health Effects of Gasoline

In humans, exposure to gasoline vapors can cause symptoms such as facial flushing, staggering gait, slurred speech, and mental confusion. At higher concentrations (20,000 ppm), gasoline vapors act as central nervous system depressants and may cause unconsciousness, coma, and death from respiratory failure. Other signs of exposure include early acute hemorrhage of the pancreas, cloudy swelling and fatty degeneration of the liver, fatty degeneration of the proximal convoluted tubules of the kidney, and passive congestion of the spleen.[5]

Use of Animal Bioassays to Predict Cancer Risks to Humans

The "gold standard" for identification of chemical carcinogens is a long-term, chronic animal bioassay. If there is strong evidence that a chemical is carcinogenic in appropriate laboratory animal test systems, it must be treated as though it were carcinogenic in humans.[6] The usual animals used in such testing are rats and mice. The rationales for use of rodents in chronic oncogenic testing are:

- rodents have the same organs as humans, such as liver, lungs, kidneys, and heart, and these organs perform the same functions as in humans;
- rodents are physiologically similar to humans in that they have the same biochemical reactions, although rates may differ;
- rodents react in the same ways to outside chemicals, and differences between their reactions and human reactions are generally quantitative rather than qualitative;
- rodents are available, of convenient size, easy to handle, and relatively inexpensive.

The closeness in basic biochemistry and metabolism to humans is the basis for the widespread use of rodents in testing substances for adverse health effects in humans. Dr. David Rall, former Director of the National Toxicology Program of the National Institute of Environmental Health Sciences, put it this way[6]:

> Both theoretical consideration and experience indicate that it is possible to test in laboratory animals chemicals to which humans are or will be exposed and to use these test results to predict in general terms what is likely to occur in the human population. Essential to this premise is the knowledge, derived from considerable basic research, that biological processes of molecular, cellular, tissue, and organ functions that control life are strikingly similar from one mammalian species to another. Processes such as sodium and potassium transport and ion regulation, energy metabolism, and DNA

replication vary little in the aggregate as one moves along the phylogenetic ladder [i.e., from species to species]. The classic work on the transmission of neural impulses in the squid axon is directly relevant to humans. Extensive renal function studies in fish, rodents, and dogs set the basis for our current understanding of renal function and the treatment of hypertension in humans. ***Also, the processes of cell replication and development of cancer are analogous in all mammalian species.*** [Emphasis added.]

The bioassay is a classical experiment, in Mill's sense. One or more groups of animals are dosed with the chemical, often at various concentrations, and the control group is not. At various intervals, the animals are examined, and the presence of tumors is determined. Results of such bioassays are an important part of the basis for regulatory decisions about the carcinogenicity of chemicals. Belpoggi *et al.*[7] reported that natural mortality and longevity in rodents and the relative age at which spontaneous tumors develop correlate very well with those of humans, clearly demonstrating that animal studies can be used to predict epidemiological studies in humans without waiting decades to show causation of cancers by chemical agents.

There are now 119 chemicals, evaluated in the Monographs by the International Agency for Research on Cancer (IARC) of the World Health Organization, for which experimental data provide sufficient evidence of carcinogenicity but for which there are absolutely no human data. The IARC Directory of Ongoing Research in Cancer Epidemiology for 1985 reports that only nine of these 119 chemicals are subjects of epidemiological surveys.[8,9] The IARC of the World Health Organization, in its Supplement 7 of the Monograph[10] states:

> Information compiled from the first 41 volumes of IARC monographs shows that, of the 44 agents for which there is sufficient or limited evidence of carcinogenicity to humans, all 37 that have been tested adequately experimentally produce cancer in at least one animal species.... Thus, in the absence of adequate data on humans, it is biologically plausible and prudent to regard agents for which there is sufficient evidence of carcinogenicity in experimental animals as if they presented a carcinogenic risk to humans.

In an extensive animal study by Staffa and Mehlman entitled *Innovation in Cancer Risk Assessment* (ED_{01} study),[11] which was sponsored by the U.S. Environmental Protection Agency (EPA) and the U.S. Food and Drug Administration, we attempted to determine relevance of using various doses of exposure to evaluate a particular chemical for carcinogenicity. Conclusions of that study were that

> [t]here should be no debate over a key principle that has shaped both our investigations and the regulatory posture of the FDA and EPA namely that no level of exposure to a toxic substance greater than zero can be assumed to be without potentially harmful effects....The evidence from the ED_{01} study on 24,192 mice...has provided massive and overwhelming experimental profiles, and the database lends support to regulatory policies.

SYNERGISM BETWEEN CHEMICALS

Exposure can result in synergistic effects. TABLE 1 demonstrates that when workers are exposed to asbestos and smoke, their risk of dying from lung cancer is significantly increased. The risk of lung cancer in asbestos workers who do not smoke is 58 deaths/100,000 as compared to 601.6 deaths/100,000 in asbestos workers who smoke, a great increase in the risk of lung cancer in case-control studies of nonsmokers of 11.3/100,000 and smokers of 122.6/100,000. The same type of increased risk is seen when comparing smoking and nonsmoking workers exposed

TABLE 1. Synergism between asbestos work and cigarette smoking in the production of lung cancer[a]

Subjects	Lung cancer deaths[b]		Ratio Smokers:nonsmokers
	Smokers	Nonsmokers	
Asbestos workers	601.6	58.4	
Controls	122.6	11.3	10.8
Ratio, asbestos workers:controls	5.2	53.2	

[a]Data of Hammond, et al., 1976.[24]
[b]Rate/100,000 man-years, standardized for age on the distribution of man-years of all asbestos workers.

TABLE 2. Synergism between exposure to cadmium and cigarette smoking in the production of renal cancer[a]

Subjects	Renal cancer deaths[b]		Ratio Smokers:nonsmokers
	Smokers	Nonsmokers	
Exposed to cadmium	4.4	0.8	
Controls	1.0	1.0	1.0
Ratio, cadmium exposed:controls		.8	4.4

[a]Data of Kolonel, 1976.[25]
[b]Based on age-specific data.

TABLE 3. Synergism between uranium mining and cigarette smoking in the production of lung cancer[a]

Subjects	Lung cancer deaths[b]		Ratio Smokers:nonsmokers
	Smokers	Nonsmokers	
Uranium miners[b]	60	6	10
Controls[c]	15.5	1.5	
Ratio, uranium miners:controls	3.9		40

[a]Data of Lundin et al., 1969,[26] who observed smokers and nonsmokers for different numbers of person-years, namely, 26.392 and 9.047, respectively. If the person-years had been identical, the authors estimated that they would have found 10 times as many lung cancer deaths in smokers as in nonsmokers (17/10,000 person-years as compared to 1.7/10,000 person-years). In the above table, their results for nonsmokers have been transposed from 9.047 to 26.392 person-years of observation so that the numbers of person-years are identical for both subgroups of uranium miners.
[b]Cancer incidence observed in miners.
[c]Cancer incidence expected in controls.

TABLE 4. Selected types of interactions between toxic compounds

Pair of toxic exposures	Kind of interaction	Result
Cigarette smoking + asbestos	synergistic	lung cancer*[a]
Cigarette smoking + uranium (radon)	synergistic	lung cancer*
Cigarette smoking + carbon monoxide	synergistic	cardiac damage*
Cigarette smoking + -naphthylamine	additive or synergistic	bladder cancer
Carbon monoxide + methylene chloride	synergistic	cardiac damage*
Sulfur oxides + air particulate	synergistic	chronic obstructive pulmonary disease*
Benzene + toluene	antagonistic	chromosomal damage
Benzene + radiation	additive or synergistic	leukemia
Carbon tetrachloride + ethyl or isopropyl alcohol	synergistic	hepatic and renal damage*
Noise + solvent exposure	additive or synergistic	hearing loss[b]

[a](*) Indicates interaction is well established (Trieff & Corrigan, 1975[27]).
[b]Morata, 1989.[28]

to cadmium and the development of renal cancer (TABLE 2), and uranium miners and the production of lung cancer (TABLE 3). TABLE 4 shows a variety of synergistic effects.

CARCINOGENICITY STUDIES

Evidence described below from three separate animal bioassay studies in species of rats and in mice demonstrates that chronic exposure to MTBE by either oral or inhalation routes causes tumors in these animals.[12–15]

Carcinogenicity Bioassay by Inhalation in F344 Rats

Chun et al.[12] reported increased incidence of uncommon renal tumors in male rats exposed to MTBE compared with controls. Three of the eight renal tumors in the 3000-ppm group were carcinomas. Two male rats in the low-exposure group (400 ppm) had preneoplastic adenomatous hyperplasia in renal tubules. In female rats, there was one renal tubular adenoma in the middle-exposure group. This bioassay was significantly shortened for the high-dose group due to morbidity, and thus decreased the total number of animals at risk for developing tumors. An exposure-related increase in testicular tumors was also observed in male rats. TABLE 5 shows renal cancer incidence in male F344 rats.

TABLE 5. Kidney cancers in male F344 rats in inhalation study with MTBE

	Cancer incidence		
Dose (ppm)	Adenomas	Carcinomas	Combined adenoma and carcinoma
0	1/44 (2%)	1/44 (0%)	1/44 (2%)
400	0/46 (0%)	0/46 (0%)	0/46 (0%)
3000[a]	5/44 (11%)	3/44 (7%)	8/44 (18%)
8000[a]	3/39 (8%)	0/39 (0%)	3/39 (8%)

NOTE: 24-month study (Chun et al., 1992[12]).
[a]3000 and 8000 ppm rats were sacrificed at 97 and 82 weeks, respectively, points at which the appearance of cancer would not be detected.

TABLE 6. Incidence of cancers in Sprague-Dawley rats exposed to MTBE for two years

	Tumor incidence			
	Male		Female	
Dose (mg/kg)	Testicular Leydig cell (%)	Corrected no. (%)	Lymphomas & leukemia (%)	Corrected no. (%)
0 mg/kg	3.3	7.7	4.0	3.4
250 mg/kg	3.3	8.0	10.0	11.8
1000 mg/kg	18.3	34.4	20.0	25.5

Carcinogenicity Bioassay by Inhalation in CD-1 Mice

Burleigh-Flayer et al.[13] reported an increased incidence of hepatocellular carcinomas in male mice exposed to the highest dose of MTBE and an increased incidence of liver tumors (adenomas and carcinomas combined) in female mice exposed to the high dose. Because of increased mortality in the exposed group, the duration of the experiment was 18 months instead of the usual 24 months routinely used in cancer bioassays. Therefore, it is likely that additional tumors would have been detected had this study been continued for the usual 24-month exposure period.

Carcinogenicity Bioassay by Gavage in Sprague-Dawley Rats

Belpoggi et al.[14,15] reported that oral administration of MTBE causes a significant dose-related increase in interstitial cell adenomas (Leydig cell tumors) in male rats and significant increases in hemolymphoreticular cancers of lymphocytic origin (lymphoblastic lymphomas, lymphoblastic leukemias, and lymphoimmunoblastic lymphomas) in female rats. In addition, hemolymphoreticular dysplasias were increased in exposed female rats (TABLE 6).

TABLE 7. Weight of evidence for carcinogenicity for MTBE

Animal	Organ	Statistically significant
Male rat	Kidney tumor	Yes
Male rat	Testes tumor	Yes
Male rat	Hemolymphoreticular tumors	Yes
Male mouse	Liver	Yes
Female mouse	Liver	Yes

SOURCE: Mehlman, 1994.[29]

These carcinogenicity studies show that MTBE induces benign and malignant tumors at multiple sites, in multiple species, by multiple routes of exposure. Thus, the weight of evidence (TABLE 7) demonstrates that MTBE is a probable human carcinogen.[1,2,16–18] This means that some humans exposed to MTBE are likely to develop cancer.

NTP Criteria for Listing Chemicals as "Reasonably Anticipated to be a Human Carcinogen"

The National Toxicology Program (NTP) Board of Scientific Counselors' Report on Carcinogens Subcommittee held its fourth meeting in December 1998 at the National Institute of Environmental Health Sciences (NIEHS), Research Triangle Park, North Carolina. Dr. Ronald Melnick recommended that methyl *tert*-butyl ether (MTBE) be nominated for listing in the report as "reasonably anticipated to be a human carcinogen" based on evidence of benign and malignant tumor induction at multiple organ sites in long-term studies in two animal species. In addition, a recent review by the National Science and Technology Council (NSTC) concluded that MTBE is an animal carcinogen and has carcinogenic potential for humans. Dr. Melnick reported that U.S. production of MTBE increased enormously in the past 5 to 10 years and is currently about 24 billion pounds. MTBE is used primarily as a gasoline oxygenate. Human exposure is through the air, primarily from various types of emissions (industrial, vehicles, and gasoline stations), and from drinking water contaminated by leakage of underground storage tanks. Dr. Melnick stated that there are no identified studies of potential carcinogenicity in humans. He listed other information relating to carcinogens: MTBE is metabolized to formaldehyde and *t*-butyl alcohol (a renal carcinogen in male rats) in humans and rats.

Dr. Bailer, the primary reviewer, agreed with the proposed listing. He said that MTBE induces dose-related increases in renal tumors in male F344 rats and that the lower response at the higher exposure level was likely due to treatment-related toxicity yielding poor survival in the high-dose group. MTBE was also associated with dose-related increases in Leydig cell tumors of the testes in F344 rats, hepatocellular carcinomas in CD-1 male mice, significant increases in Leydig cell tumors in males, and leukemia/lymphomas combined in females in gavage studies in Sprague-Dawley rats. Dr. Bailer maintained that this supports the proposed listing.

Dr. Frederick, the secondary reviewer, said that he had evaluated the reports of other groups, such as the EPA and the State of California, on MTBE. He also thought a key point in the inhalation study in CD-1 mice and F344 rats was that the top dose, 8000 ppm, was clearly neurotoxic, and animals were clearly not normal for evaluating cancer effects. The animals at 3000 ppm appeared to be normal. He thought the Leydig cell tumors demonstrated a hormonal action of MTBE. With regard to kidney tumors in male rats, there appeared to be a mixed mechanism, in part due to alpha-2-globulin. Dr. Frederick said that, in his opinion, the very limited carcinogenicity data available at appropriate dose levels and the available mechanistic information were not sufficient to warrant listing in the report.

Public Comments

Dr. Susan Borghoff, CIIT, said her remarks would be confined to the kidney tumors in male rats. She noted that a number of studies looking at the ability of MTBE to induce alpha-2 nephropathy were not cited in the background document. In her work at CIIT and that of others, there is an increase in numbers of protein droplets in male rat kidney, and some aspects of the pathological sequence of lesions associated with alpha-2 nephropathy are seen. The response is mild compared with potent inducers of alpha 2, such as *d*-limonene. She said there is also increased cell proliferation in male but not female rats, which correlates with increases in alpha-2.

Dr. Robert Tardiff, the Sapphire Group, representing the Oxygenated Fuel Association, stated that MTBE is clearly an animal carcinogen, but the studies described by Dr. Borghoff lend support to the notion that male rat kidney tumors are not relevant to humans. With regard to female mouse liver tumors, he stated that there are differences of opinion on the relevance of these tumors as a model and suggested that they are not predictive of tumorigenic responses in humans.

Dr. Bailer moved that nomination of methyl *tert*-butyl ether for listing in the report as "reasonably anticipated to be a human carcinogen" be accepted. Dr. Bingham seconded the motion. Dr. Mirer noted that the criteria say that we need mechanistic reasons to downgrade the tumors observed. Dr. Hooper suggested deferral pending publication of data from the University of California study. Dr. Henry abstained from voting because of a perception that the petroleum industry has an interest. She agreed with Dr. Hooper's proposal to defer. Dr. Bailer's motion to list MTBE in the report was defeated by six *No* votes (Belinsky, Frederick, Hecht, Kelsey, Medinsky, Zahm) to five *Yes* votes (Bailer, Bingham, Hooper, Mirer, Russo), with one abstention (Henry) of API. Thus, MTBE was not recommended for listing in the ninth report.

Failure to List MTBE as a Probable Carcinogen

The NTP's vote against listing MTBE as "reasonably anticipated to be a carcinogen" was made despite strong and sufficient evidence in laboratory animals that it is a carcinogen. Some of the board members voting against it did acknowledge that MTBE is clearly an animal carcinogen. Their position contradicts the criteria established by NTP for listing agents in the *Report on Carcinogenesis*. These criteria specify that an agent be listed in the report as "reasonably anticipated to be a human carcinogen" if

> [t]here is sufficient evidence of carcinogenicity from studies in experimental animals which indicates that there is an increased incidence of malignant and/or a combination of malignant and benign tumors: (1) in multiple species or at multiple tissue sites; (2) by multiple routes of exposure; or (3) to an unusual degree with regard to incidence, site or type of tumor or age of onset.

The criteria also allow consideration of other relevant information: "For example, there may be substances for which there is evidence of carcinogenicity in laboratory animals but for which there are compelling data indicating that the agent acts through mechanisms which do not operate in humans and would therefore not reasonably be anticipated to cause cancer in humans." Board members who voted against listing MTBE noted that there was no compelling mechanistic data on MTBE showing that it acts through mechanisms that operate in humans. The board acknowledged that the mechanisms of MTBE-induced carcinogenicity are unknown.

Immunological and DNA Studies in Humans

Because long-term human consumption of water containing gasoline and MTBE has been associated with abnormal apoptosis and cell cycle progression, Vojdani and Brautbar[19] studied two population groups. Group 1 consisted of 20 subjects who were exposed to MTBE through ingestion and bathing in contaminated water for a period of 8 months. Group 2 consisted of 20 healthy controls with no known significant exposure to gasoline other than pumping gasoline once a week into vehicles; they were otherwise comparable to the exposed group.

DNA single strand breaks, which lead to DNA adducts and apoptosis, were measured in each group. Lymphocyte DNA adducts were significantly increased in the exposed group compared to healthy controls. These changes could not be attributed to other known exposures or factors, such as immunological disease, chemotherapy, or heavy smoking. These findings suggest that long-term ingestion of MTBE, that is, in contaminated drinking water, is associated with DNA damage. Similarly natural killer cell activity in 65% of the exposed individuals was significantly abnormal, further suggesting an effect on an arm of the immune system. They suggested the following chain of events: MTBE in gasoline and/or its metabolites bind to DNA, increase DNA adducts, and induce the DNA strand break which ends with changes in cell cycle and cell function that may lead to enhanced programmed cell death (apoptosis).

According to the authors, MTBE and gasoline are associated with changes in DNA adducts (Vojdani *et al.*[20]). The DNA adducts represent genotoxic damage that may signal initiation of carcinogenicity. Further, DNA binding metabolites from chemicals have been demonstrated *in vitro* for at least 13 chemicals that are known human carcinogens. In target tissue, DNA binding has been shown to correlate with carcinogenic potency.

In 1998, Vojdani and Brautbar[21] reported that subjects exposed to MTBE and benzene concentrations above the permissible levels through drinking, bathing, and showering with contaminated water had severe headaches, fatigue, exhaustion, cognitive dysfunction, abdominal pain, indigestion, and depression. These symptoms were correlated with the exposure period to MTBE and benzene and are very similar to symptoms reported in patients being treated with MTBE for dissolution of gallstones.[22]

Previous Evaluations of MTBE

Keller, in a report, dated November 1998, to the governor and the legislature of the State of California concluded that MTBE is an animal carcinogen with potential to cause cancers in humans. This is further supported by a report of the White House National Science and Technology Council (1997) that "there is sufficient evidence that MTBE is an animal carcinogen" and that "the weight of evidence supports MTBE as having carcinogenic hazard potential for humans." In 1997, the U.S. EPA[23] concluded that the "weight of the evidence indicates that MTBE is an animal carcinogen and MTBE poses a carcinogenic potential to humans."

CONCLUSIONS

There is no question that the data presented demonstrated that MTBE is a probable human carcinogen, which means that MTBE will cause cancers in some humans. Dr. Marvin Schneiderman (1975), the Director of the Biostatistical Branch of the National Cancer Institute, wrote in *Persons at High Risk for Cancer* that

> [i]t is "not" necessary to know what individual or complex of chemicals in cigarette smoke causes cancer [i.e., the mechanism] in order to take action now to reduce the risks of cigarette smoking by reducing cigarette smoking. Clean water and good sewers nearly eliminated cholera in Europe well before anyone knew what in dirty water or in sewage "caused" cholera. Controlling cancer must not get in the way of basic research, and the need for basic research must not delay controlling cancer.

REFERENCES

1. MEHLMAN, M.A. 1990. Dangerous properties of petroleum refining products: carcinogenicity of motor fuels (gasoline). Teratog. Carcinog. Mutagen. **10:** 399–408.
2. MEHLMAN, M.A. 1991. Response to API's questions on "Dangerous properties of petroleum refining products: Carcinogenicity of motor fuels (gasoline)." Teratog. Carcinog. Mutagen. **11:** 220–226.
3. MEHLMAN, M.A. 1992. Dangerous and cancer-causing properties of products and chemicals in the oil refining and petrochemical industry. VII. Health effects of motor fuels: Carcinogenicity of gasoline—scientific update. Environ. Res. **59:** 238–249.
4. PAGE, N. & M.A. MEHLMAN. 1989. Health effects of gasoline refueling vapors and measured exposures at service stations. Toxicol. Ind. Health **5:** 869–890.
5. SITTING, M. 1981. Handbook of Toxic and Hazardous Chemicals. Noyer Data Corp. Park Ridge, NJ. p. 348.
6. RALL, D. 1986. Relevance of results from laboratory animal toxicology studies. *In* Toxicology of Environmental Disease. J. Last, Ed. Public Health and Preventive Medicine, 12th edition. Appleton–Century–Crofts. New York.
7. BELPOGGI, F. *et al.* 2002. Carcinogenicity bioassays on gasoline-oxygenated additives and on trimethyl pentane (TMP). In press.
8. TOMATIS, L. 1985. Directory for Ongoing Research in Cancer Epidemiology, 10th Anniversary Issue, S. Muir & G. Wagner, Eds. International Agency for Research on Cancer. Lyon, France.
9. TOMATIS, L. 1988. The contribution of the IARC monographs program to the identification for cancer risk factors. Ann. N.Y. Acad. Sci. **534:** 31–38.
10. INTERNATIONAL AGENCY FOR RESEARCH ON CANCER (IARC). 1987. Supplement 7. International Agency for Research on Cancer. Lyon, France.
11. STAFFA, J.A. & M.A. MEHLMAN. 1979. Innovations in Cancer Risk Assessment (ED_{01} Study). Pathotox Publishers, Inc. Park Forrest, IL.
12. CHUN, J.S. *et al.* 1992. Methyl Tertiary Butyl Ether: Vapor Inhalation Oncogenicity Study in Fischer 344 Rats. Bushy Run Research Center; BRRC report 91N0013B. November 13. Union Carbide Chemicals and Plastics Company, Inc. Submitted to the U.S. EPA under TSCA Section 4 Testing Consent Order 40 CFR 799.5000 with cover letter dated November 1992. EPA/OPTS#42098. Export, PA: Bushy Run Research Center.

13. BURLEIGH-FLAYER, H.D. et al. 1992. Methyl Tertiary Butyl Ether: Vapor Inhalation Oncogenicity Study in CD-1 Mice. Bushy Run Research Center: BRRC report 91N0013A. October 15. Union Carbide Chemicals and Plastics Company, Inc. Submitted to the U.S. EPA under TSCA Section 4 Testing Consent Order 40 CFR 799.5000 with cover letter dated October 29, 1992. EPA/OPTS#42098. Export, PA: Bushy Run Research Center.
14. BELPOGGI, F. et al. 1995. Methyl-tertiary butyl ether [MTBE]—a gasoline additive— causes testicular and lympho-haematopoietic cancers in rats. Toxicol. Ind. Health **11:** 119–149.
15. BELPOGGI, F. et al. 1998. Pathological characterization of testicular tumors and lymphomas-leukemias, and of their precursors observed in Sprague-Dawley rats exposed to methyl-tertiary butyl ether (MTBE). Eur. J. Oncol. **3:** 201–206.
16. MEHLMAN, M.A. 1998. Human health effects from exposure to gasoline containing methyl-tert-butyl ether. Eur. J. Oncol. **3:** 171–189.
17. MEHLMAN, M.A. 2001. Methyl-tertiary-butyl ether (MTBE) misclassified. Am. J. Int. Med. **39:** 505–508.
18. RUDO, K. 1995. Review of the Wisconsin DHSS report on health concerns attributed to reformulated gasoline use in Southeastern Wisconsin. Toxicol. Ind. Health **11:** 463–466.
19. VOJDANI, A. & N. BRAUTBAR. 1999. White blood cell DNA adducts and immunological abnormalities in humans exposed to contaminated drinking water containing gasoline and MTBE. Eur. J. Oncol. **4:** 573–578.
20. VOJDANI, A. et al. 1997. Abnormal apoptosis and cell cycle progression in humans exposed to methyl-tertiary-butyl ether and benzene contaminating water. Hum. Environ. Toxicol. **16:** 485–494.
21. VOJDANI, A. et al. 1998. Contaminated drinking water with MTBE and gasoline: Immunological and cellular effects. Eur. J. Oncol. **3:** 191–199.
22. MEHLMAN, K.N. & M.A. MEHLMAN. 2001. Methyl tertiary butyl ether (MTBE) in the treatment of gallstone disease: Toxicity and potential carcinogenicity in humans. Eur. J. Oncol. **6:** 239–242.
23. U.S. ENVIRONMENTAL PROTECTION AGENCY. 1997. Drinking Water Advisory: Consumer Acceptability Advice and Health Effects Analysis on Methyl Tertiary-Butyl Ether (MtBE). December. Face Sheet 4 pp. EPA-822-F-97-009. ODW 4304. Health and Ecological Criteria Division, Office of Science and Technology, Office of Water, U.S. Environmental Protection Agency. Washington, DC: US EPA.
24. HAMMOND, E.C. et al. 1979. Asbestos exposure, cigarette smoking and death rates. Ann. N.Y. Acad. Sci. **330:** 473–490.
25. KOLONEL, L.N. 1976. Association of cadmium with renal cancer. Cancer **37:** 1782–1787.
26. LUNDIN, F.E., JR. 1969. Mortality of uranium miners in relation to radiation exposure, hard-rock mining and cigarette smoking—1950 through September 1967. Health Phys. **16:** 571–578.
27. TRIEFF, N.M. & G.E. CORRIGAN. 1975. Contemporary work-related environmental diseases. Tex. Rep. Biol. Med. **33:** 107–144.
28. MORATA, T.C. 1989. Study of the effects of simultaneous exposure to noise and carbon disulfide on workers' hearing. Scand. Audiol. **18:** 53–58.
29. MEHLMAN, M.A. 1994. Dangerous and cancer-causing properties of products and chemicals in the oil refining and petrochemical industry. Part XX: Health dangers of petroleum hydrocarbons: Gasoline methyl-tertiary butyl ether, benzene, 1,3-butadiene and alkylbenzenes. J. Clean Technol. Environ. Sci. **4:** 37–57.

Asbestos Fibers Contributing to the Induction of Human Malignant Mesothelioma

YASUNOSUKE SUZUKI AND STEVEN R. YUEN

Department of Community and Preventive Medicine, Mount Sinai School of Medicine, New York, New York 10029, USA

ABSTRACT: To elucidate the features of the asbestos fibers contributing to the induction of human malignant mesothelioma, we used high-resolution analytical electron microscopy to determine the type, number, and dimensions of asbestos fibers in lung and mesothelial tissues in 168 cases of mesothelioma. Results: 1. Asbestos fibers were present in almost all of the lung and mesothelial tissues from the mesothelioma cases. 2. The most common types of asbestos fibers in lung were either an admixture of chrysotile with amphiboles, amphibole alone, and occasionally chrysotile alone. In mesothelial tissues, most asbestos fibers were chrysotile. 3. In lung, amosite fibers were greatest in number followed by chrysotile, crocidolite, tremolite/actinolite, and anthophyllite. In mesothelial tissues, chrysotile fibers were 30.3 times more common than amphiboles. 4. In some mesothelioma cases, the only asbestos fibers detected in either lung or mesothelial tissue were chrysotile fibers. 5. The average number of asbestos fibers in both lung and mesothelial tissues was two orders of magnitude greater than the number found in the general population. 6. The majority of asbestos fibers in lung and mesothelial tissues were shorter than 5 µm in length. Conclusions: 1) Fiber analysis of both lung and mesothelial tissues must be done to determine the types of asbestos fibers associated with the induction of human malignant mesothelioma; 2) short, thin asbestos fibers should be included in the list of fiber types contributing to the induction of human malignant mesothelioma; 3) Results support the induction of human malignant mesothelioma by chrysotile.

KEYWORDS: asbestos, fibers, mesothelioma, chrysotile, amphiboles

INTRODUCTION

It is well accepted that asbestos fibers are the cause of virtually all cases of human malignant mesothelioma.[1–4] It is also known that all asbestos types, including chrysotile and amphiboles, have been shown in epidemiological and toxicological studies to be fully capable of inducing the tumor.[5–9] In addition to heavy (occupational) asbestos exposure, milder asbestos exposure (bystanders and family contact) can also induce the tumor.[10–12] Presently, no data are available to support a threshold limit for exposure to asbestos below which there is no risk of malignant mesothelioma.[4]

Address for correspondence: Yasunosuke Suzuki, M.D., Department of Community and Preventive Medicine, Mount Sinai School of Medicine, 1 Gustave L. Levy Place, New York, New York 10029. Voice: 212-241-4777; fax: 212-996-0407.
yasunosuke.suzuki@mssm.edu

Human malignant mesothelioma develops after a long latency period: it takes fifteen years or more from the first asbestos exposure to death from malignant mesothelioma. Latency periods greater than 40 years have been reported.[2–5] There is no cure for human malignant mesothelioma.

Asbestos fibers are durable in nature and are not easily digested or dissolved by either phagocytic cells or the tissue fluid. Asbestos fibers are identified as asbestos bodies by light microscope or as naked asbestos fibers by an electron microscope.

Some inhaled asbestos fibers translocate from lung into regional lymph nodes,[13–15] pleural and peritoneal mesothelial tissues,[16–24] and other organs.[13,25] Fibers may pass from lung to other organs by direct migration[26,27] via lymphatic capillary system,[13–15,19] and by hematogenous spread.[28,29]

Up to now, most investigators have focused exclusively on asbestos fibers in the lung for identification of asbestos fibers that contribute to induction of human malignant mesothelioma.[30–34] We questioned the adequacy of such an approach because the primary site of malignant mesothelioma is not the lung but the mesothelial tissue and because the type and number of asbestos fibers in lung may not be identical to those in mesothelial tissue owing to possible translocation of lung fibers to other organs including mesothelial tissues.

Short, thin asbestos fibers, i.e., 0.06 µm long and 0.02–0.03 µm wide, can easily be identified by the high resolution analytical electron microscope, a transmission electron microscope with an energy dispersive X-ray spectrometer, but not by scanning electron microscope with resolution of 0.3 to 0.4 µm. It is noteworthy that the number and dimensions of asbestos fibers obtained from a high resolution analytical electron microscope will be different from those obtained by a scanning electron microscope.

It has been proposed on the basis of animal studies that long (greater than 8 µm in length) and thin (less than 0.25 µm in width) mineral fibers were strongly carcinogenic for the induction of malignant mesothelioma in rats (Stanton's hypothesis)[35] and that shorter fibers pose less risk. Stanton's hypothetical dimensions were derived from his experimental studies using direct administration of heavy doses of various mineral fibers of different dimensions into rats pleural cavities. Stanton stated that direct application of his results to the problem in man would be unwise.[36] However his hypothetical model of asbestos fibers' relative carcinogenicity has been directly applied to the counting of the asbestos fibers in man.

To evaluate airborne fibrous dusts in industrial atmospheres, the current Occupational Safety and Health Administration (OSHA) method by light microscopy (phase microscopy) counts only those asbestos fibers that are longer than 5 µm with an aspect ratio larger than 3 to 1, assuming that fibers shorter than 5 µm are not carcinogenic. Even using the electron microscopic level, using the same assumption, some investigators have neglected to count short asbestos fibers (≤ 5 µm).[37–40]

Our previous tissue burden studies[22] using high resolution analytical electron microscopy revealed that the majority of asbestos fibers from human lung and mesothelial tissues of human mesothelioma patients were less than 5 µm long (81.4%) and less than 0.25 µm in diameter. The narrow width of asbestos fibers (≤ 0.25 µm in diameter, another parameter in Stanton's hypothesis) has been emphasized as an important parameter not only for the fibers' carcinogenicity, but also for the fibers' ability to penetrate into the peripheral part of the lung by an aerodynamic mechanism.[35,36,41–46] Both OSHA's method using light microscopy and some asbestos tissue burden studies using electron microscopy did not pay serious attention to fiber diameter when counting of asbestos fibers. It was noteworthy that our study[22] revealed that only 4% of all asbestos fibers detected in the lung and mesothelial tissues from mesothelioma patients fit Stanton's criteria.

An asbestos tissue burden study is an effective approach to clarify whether chrysotile fibers are capable of inducing human malignant mesothelioma. If the asbestos fiber type seen in both the lung and mesothelial tissues is solely chrysotile, such mesothelioma cases can be considered to have been caused by chrysotile exposure. Indeed, such cases have been reported elsewhere.[21,22,47]

Our objective in this study was to characterize the features of the asbestos fibers contributing to the induction of human malignant mesothelioma. To achieve this goal, the type, number, and dimensions of the asbestos fibers in both lung and mesothelial tissues taken from human malignant mesothelioma cases were investigated.

MATERIALS AND METHODS

Both lung and mesothelial tissues (the mesotheliomatous and/or fibroplastic serosal tissues) from 168 cases of human malignant mesothelioma (164 males and 4 females; 156 pleural and 12 peritoneal; definite or probable diagnostic certainty) were used. The mesotheliomatous tissue was selected from the primary serosal tumor where the tumor was intimately associated with fibrosis and/or hyaline plaque. Asbestos fibers were studied in both the lung and mesothelial tissues in 74 of the 168 cases, exclusively in lung in 45 of the 168 cases, and exclusively in mesothelial tissue in the remaining 49 cases.

Patients' occupational history was diverse and included asbestos insulators, pipe fitters, electricians, shipyard workers, U.S. Navy servicemen, sheet metal workers, power plant workers, boiler men, brake lining mechanics, fire fighters, family members of asbestos workers, etc.

To prepare electron microscopic specimens, bulk tissues were digested using bleach or KOH solution, or a low temperature ashing technique of 25 μm thick section, or both were used. Details of these techniques have been reported elsewhere.[18,47–49]

A high-resolution analytical electron microscope (JEOL 100CX equipped with an EDX spectrometer) was used for the identification and characterization of asbestos fibers in these tissues; ultrastructure, energy dispersive X-ray spectrophotometry and, in a limited number of cases, selected area electron diffraction were utilized. Asbestos fibers were measured in printed electron micrographs, and those with an aspect ratio of 3:1 and greater were counted even if they were shorter than 1 μm in length.

RESULTS

1. Types of asbestos fibers in lung and mesothelial tissues in the 168 malignant mesothelioma cases studied are outlined below.

TABLE 1. Type of asbestos fibers in lung and mesothelial tissues in 168 malignant mesothelioma cases

A. Asbestos tissue burden study was performed in both the lung and mesothelial tissues: 74 of 168 cases

Lung tissue	Mesothelial tissue	No. of cases
C + A	C	19
C	C	18
A	C	16
C + A	C + A	9
A	C + A	4
A	–	3
–	C	2
C	C + A	2
A	A	1
	Total	74

B. Asbestos tissue burden study was performed in the lung tissues alone: 45 of 168 cases

Lung tissue	No. of cases
A	19
C + A	15
C	11
–	0
Total	45

C. Asbestos tissue burden study was performed in mesothelial tissues alone: 49 of 168 cases

Mesothelial tissue	No. of cases
C	35
C + A	7
–	6
A	1
Total	49

C = chrysotile; A = amphibole(s); C + A = chrysotile and amphibole(s); – = not detected.

A. In 74 of the 168 cases, asbestos fiber analysis was performed in both the lung and mesothelial tissues, using digested bulk samples, ashed sections or both. Results are summarized in TABLE 1A.

1) Types of asbestos fibers detected in the lung were quite often different from those in the mesothelial tissue. The combination of asbestos type between the lung and mesothelial tissues was as follow:

  (i) chrysotile plus amphibole(s) in lung, and chrysotile alone in mesothelial tissues: 19/74 (25.7%);
  (ii) chrysotile in lung, and chrysotile in mesothelial tissues: 18/74 (24.3%);
  (iii) amphibole(s) in lung, and chrysotile in mesothelial tissues: 16/74 (21.6%);

(iv) chrysotile plus amphibole(s) in lung, and chrysotile plus amphibole(s) in mesothelial tissues: 9/74 (12.2%);
(v) amphibole(s) in lung, and chrysotile plus amphibole(s) in mesothelial tissues: 4/74 (5.4%);
(vi) amphibole in lung, and no asbestos fibers in mesothelial tissues: 3/74 (4.0%);
(vii) no asbestos fibers in lung, and chrysotile in mesothelial tissues: 2/74 (2.7%);
(viii) chrysotile in lung, and chrysotile plus amphibole(s) in mesothelial tissues: 2/74 (2.7%); and
(ix) amphibole in lung, and amphibole in mesothelial tissues: 1/74 (1.4%).

In summary, a disproportion of the type of asbestos fibers between the two tissues was seen in the majority (46 of 74; 62.2%) of cases.

2) Asbestos types identified in lung were chrysotile (49/74; 66.2%) and amosite (49/74; 66.2%), followed by tremolite (15/74; 20.3%), crocidolite (13/74; 17.6%) and anthophyllite (12/74; 16.4%).

3) Chrysotile was the most common asbestos type detected in mesothelial tissues. It was present in 70 of the 74 cases (94.6%); chrysotile was exclusively detected in 55 of the 74 cases (74.3%).

4) When chrysotile was exclusively seen in the lung, the asbestos type detected in mesothelial tissues was also exclusively chrysotile in 18 of 20 cases (90%).

5) When amphibole(s) was exclusively found in lung, the asbestos type(s) in mesothelial tissues was exclusively amphibole(s), although this occurred rarely (1/24; 4.2%). Other asbestos types found in mesothelial tissues were chrysotile alone (16/24; 66.7%), chrysotile plus amphibole(s) (4/24; 16.7%), and none (3/24; 12.5%).

B. In 45 of the 168 cases, an asbestos tissue burden study was carried out exclusively in the lung using digested bulk samples, ashed tissue section, or both. Results are summarized in TABLE 1B.

1) Asbestos types detected in lung of the 45 cases varied. They were amphibole(s) alone (19/45; 42.2%) followed by chrysotile plus amphibole(s) (15/45; 33.3%), and chrysotile only (11/45; 24.5%).

2) The subtype of amphiboles in the lung of 34 of the 45 cases was amosite alone (18/34; 52.9%), followed by amosite plus tremolite/actinolite (5/34; 14.7%), crocidolite alone (4/34; 11.8%), tremolite alone (2/34; 5.9%), amosite plus crocidolite (2/34; 5.9%), amosite plus crocidolite plus anthophyllite (2/34; 5.9%), and amosite plus anthophyllite (1/34; 2.9%).

C. In 49 of the 168 cases, an asbestos tissue burden study was done on mesothelial tissues only, using digested bulk samples, ashed sections, or both. Results are summarized in TABLE 1C.

1) Again, chrysotile fibers were the major asbestos type detected in mesothelial tissues.

2) Asbestos types seen in mesothelial tissues were chrysotile alone (35/49; 71.4%) followed by chrysotile with amphibole (7/49; 14.3%), no asbestos fibers detected (6/49; 12.3%), and amphibole alone (1/49; 2.0%).

Findings presented in 1, A, B and C are summarized as follows.

(1) Asbestos fibers were present in almost all of the lung tissue (117/119; 98.3%) as well as in the mesothelial tissue (114/123; 92.7%).

TABLE 2. Type and number of asbestos fibers in lung parenchyma, pleural plaque, and mesotheliomatous tissues among 22 cases of mesothelioma

Case #	Occupation	Site	Disease	Chry	Amos	Croc	Anth	Tr/Ac	DL	Total #
1	insulation	L	pl meso	28.3	125	<DL	2.83	<DL	2.83	156.1
		P		12.1	1.29	<DL	<DL	<DL	0.16	13.4
2	insulation	L	pl meso	28.6	194	<DL	3	3	1.5	228.6
		P		39.2	0.6	<DL	<DL	<DL	0.6	39.8
		T		62.1	<DL	<DL	<DL	<DL	1.27	62.1
3	insulation	L	pe meso	24	139	7.37	<DL	11.4	1.26	181.8
		P		36.3	6.34	<DL	<DL	<DL	0.58	42.6
		T		14.8	<DL	<DL	<DL	<DL	0.76	14.8
4	insulation	L	pe meso	111	282	25.6	4.3	<DL	2.13	422.9
		P		31.8	6.81	<DL	<DL	<DL	0.76	38.6
		T		16.5	0.52	<DL	<DL	<DL	0.17	17
5	insulation	L	pe meso	25.5	120	<DL	<DL	<DL	0.77	145.5
		P		29.4	1.8	<DL	<DL	<DL	0.6	31.2
		T		12.6	1.76	<DL	<DL	<DL	0.44	14.4
6	insulation	L	pe meso	91.9	213	86.4	<DL	3.68	1.84	395
		T		50.1	1.79	<DL	<DL	<DL	0.6	51.9
		T		43.7	<DL	<DL	<DL	<DL	0.48	43.7
7	insulation	L	pe meso	18.8	415	11.3	<DL	11.3	3.75	456.44
		T		90	14	<DL	<DL	<DL	1.42	104
8	insulation	L	pe meso	1.5	7.1	<DL	<DL	<DL	0.29	8.5
		T/P		17	<DL	<DL	<DL	<DL	0.26	17
9	engineer	L	pl meso	<DL	2.5	0.53	<DL	<DL	0.18	3
		T		22.5	<DL	0.22	0.22	<DL	0.22	22.94
10	aircraft inspector	L	pl meso	61	<DL	<DL	<DL	0.7	0.35	61.7
		T		120	<DL	<DL	<DL	<DL	0.35	120
11	power plant	L	pl meso	<DL	47	<DL	<DL	<DL	2.9	47
		T		240	<DL	<DL	<DL	<DL	2.9	240
12	shipyard/ power plant	L	pl meso	<DL	2.6	<DL	<DL	<DL	0.22	2.6
		T		51.3	<DL	<DL	<DL	<DL	0.27	51.3
13	power plant	L	pl meso	<DL	1.3	<DL	0.15	0.15	0.15	1.6
		T		2.6	0.3	<DL	<DL	<DL	0.15	2.9
14	welder	L	pl meso	0.62	<DL	<DL	0.26	<DL	0.26	0.88

TABLE 2. Type and number of asbestos fibers in lung parenchyma, pleural plaque, and mesotheliomatous tissues among 22 cases of mesothelioma *(Continued)*

Case #	Occupation	Site	Disease	Chry	Amos	Croc	Anth	Tr/Ac	DL	Total #
		T		0.7	0.4	<DL	0.3	<DL	0.09	1.4
15	US Navy	L	pl meso	27	<DL	<DL	<DL	<DL	4.4	27
		T		22	<DL	<DL	<DL	<DL	0.88	22
16	electrician	L	pl meso	<DL	19.2	2.9	<DL	<DL	1.45	22.1
		T/P		228.2	1.8	<DL	<DL	<DL	2.9	230
17	firefighter	L	pl meso	32.5	1.4	<DL	<DL	<DL	1.77	33.9
		T/P		16.6	<DL	<DL	<DL	<DL	0.22	16.6
18	US Navy/ railroad	L	pl meso	<DL	0.08	<DL	<DL	<DL	0.02	0.08
		T		0.06	<DL	<DL	<DL	<DL	0.03	0.06
19	US Navy	L	pl meso	<DL	0.52	<DL	<DL	<DL	0.03	0.53
		T		2.6	<DL	<DL	<DL	<DL	0.11	2.6
20	sheetmetal	L	pl meso	0.49	<DL	<DL	0.04	<DL	0.04	0.53
		T		0.19	<DL	<DL	0.04	<DL	0.04	0.23
21	roofer	L	pl meso	1.5	0.03	<DL	<DL	<DL	0.03	1.53
		T		0.3	<DL	<DL	<DL	<DL	0.03	0.3
22	boiler mechanic	L	pl meso	<DL	0.54	0.07	<DL	<DL	0.02	0.6
		T		<DL	0.15	<DL	<DL	<DL	0.03	0.15

NOTE: Figures represent asbestos fibers $\times 10^6$/gram (dry tissue).

ABBREVIATIONS: L, lung; P, plaque; T/P, tumor/plaque; DL, detection limit; <DL, under detection limit (no detection); Chry, chrysotile; Amos, amosite; Croc, crocidolite; Anth, anthophyllite; Tr/Ac, tremolite/actinolite; pl, pleura; pe, peritoneum; meso, mesothelioma.

(2) A disproportion in the types of asbestos fibers between lung and mesothelial tissues was common and was present in 49 of 74 cases (66.2%).

(3) The most common asbestos types in lung were an admixture of chrysotile with amphiboles (43/119 36.1%) or amphiboles alone (43/119; 36.1%). Chrysotile alone was seen occasionally (31/119; 26.1%). Rarely, no asbestos fibers were seen (2/119; 1.7%).

(4) In mesothelial tissues, the major asbestos type was chrysotile alone (90/123; 73.2%) followed by chrysotile plus amphibole (22/123; 17.9%), no asbestos fibers detected (9/123; 7.3%), and amphibole alone (2/123; 1.6%). The amphiboles included anthophyllite mixed with chrysotile in 15 cases and amphiboles alone in 1 case, followed by tremolite mixed with chrysotile in 4 cases, amosite mixed with chrysotile in 3, and amosite alone in 1 case.

2. Quantitative analysis of asbestos fibers in the tissues (number of fibers/dry gram) was performed in both digested lung and digested mesothelial tissues taken from 22

TABLE 3. Type and number of asbestos fibers in lung parenchyma in 27 additional cases of mesothelioma

Case #	Occupation	Site	Chry	Amos	Croc	Anth	Tr/Ac	DL	Total #
1	electrician	L (L)	0.04	0.08	0.02	0.02	<DL	0.02	0.16
		L (R)	12.1	1.29	<DL	<DL	<DL	0.03	13.4
2	US Navy	L	<DL	3.3	<DL	<DL	<DL	0.17	3.3
3	insulation	L	<DL	0.6	<DL	<DL	0.9	0.26	1.5
4	family contact	L-1	<DL	0.11	<DL	0.11	0.33	0.17	0.55
		L-2	<DL	<DL	<DL	0.31	0.31	0.29	0.62
5	jet plane mechanic	L	260	<DL	<DL	<DL	<DL	0.22	260
6	mechanic	L	76	0.98	<DL	0.16	<DL	0.12	77.1
7	construction	L	<DL	9.9	<DL	<DL	<DL	0.13	9.9
8	US Navy	L	<DL	2.78	<DL	0.22	<DL	0.11	3.0
9	insulation	L	<DL	7.06	<DL	<DL	<DL	0.11	7.0
10	insulation	L	<DL	26	<DL	<DL	<DL	0.22	26.0
11	construction	L	36	7.5	<DL	<DL	0.75	0.75	44.3
12	electrician	L	1.5	1	0.5	<DL	<DL	0.25	3.0
13	pipe fitter	L	1.26	0.63	2.8	0.63	<DL	0.33	5.32
14	US Navy	L	16	<DL	0.22	<DL	<DL	0.22	16.2
15	insulation	L	88	<DL	<DL	<DL	<DL	0.44	88.0
16	US Navy	L	<DL	1.64	0.12	0.5	<DL	0.12	2.26
17	shipyard	L	0.66	1.32	<DL	<DL	<DL	n/a	1.98
18	US Navy	L	0.94	0.38	<DL	<DL	<DL	n/a	1.32
19	boiler repair	L	<DL	0.35	<DL	0.07	0.97	0.02	1.39
20	pipe fitter	L (R)	3	<DL	<DL	<DL	<DL	0.05	3.0
		L (L)	0.03	<DL	<DL	<DL	<DL	0.03	0.03
21	boiler repair	L	2.9	<DL	<DL	<DL	0.07	0.04	3.6
22	shipyard	L	<DL	0.08	<DL	<DL	<DL	0.03	0.08
23	shipyard	L	<DL	0.11	0.1	<DL	0.08	0.02	0.29
24	machinist	L	<DL	0.06	<DL	<DL	<DL	0.02	0.06
25	boiler worker	L	<DL	0.1	<DL	0.04	0.04	0.02	0.18
26	shipfitter	L	<DL	1.99	<DL	<DL	0.07	0.07	2.06
27	pipe coverer	L	<DL	<DL	0.58	<DL	<DL	0.05	0.58

NOTE: Figures represent asbestos fibers $\times 10^6$/gram (dry tissue).
ABBREVIATIONS: L, lung; DL, detection limit; <DL, under detection limit (no detection); Chry, chrysotile; Amos, amosite; Croc, crocidolite; Anth, anthophyllite; Tr/Ac, tremolite/actinolite; (L), left; (R), right; n/a, not available.

mesothelioma cases (TABLE 2) and from the digested lung alone in an additional 27 mesothelioma cases (TABLE 3).

A. As shown in TABLE 2, the total number of asbestos fibers detected in lung tissue was 456.4×10^6 fibers/dry gram maximum, and 0.08×10^6 fibers/dry gram minimum, and 99.9×10^6 fibers/dry gram on average. The average number of intrapulmonary asbestos fibers seen in the 22 cases was greatest for amosite (71.4×10^6 fibers/dry gram) followed by chrysotile (20.6×10^6 fibers/dry gram), and crocidolite (6.1×10^6 fibers/dry gram), tremolite/actinolite (1.37×10^6 fibers/dry gram and 0.48×10^6 fibers/dry gram).

B. In the mesothelial tissues, the total number of asbestos fibers was 240×10^6 fibers/dry gram maximum, 0.06×10^6 fibers/dry gram minimum, and 46.5×10^6 fibers/dry gram on average. The average number of intramesothelial asbestos fibers was greatest for chrysotile (45.2×10^6 fibers/dry gram) followed by amosite (1.3×10^6 fibers/dry gram), anthophyllite (0.03×10^6 fibers/dry gram), crocidolite (0.01×10^6 fibers/dry gram), and tremolite (0.0×10^6 fibers/dry gram). Total number of chrysotile fibers was compared with that of amphiboles in the mesothelial tissues from 22 cases. The total number of chrysotile fibers was 30.3 times greater than the total number of amphiboles fibers in mesothelial tissues

C. Asbestos fiber analysis was done exclusively in the lung in an additional 27 mesothelioma cases (TABLE 3). The total number of asbestos fibers detected in lung was 260×10^6 fibers/dry gram maximum, 0.08×10^6 fibers/dry gram minimum, and 21.0×10^6 fibers/dry gram on average. The average number of intrapulmonary asbestos fibers among asbestos types seen in these 27 cases was greatest for chrysotile (18.2×10^6 fibers/dry gram) followed by tremolite/actinolite (3.18×10^6 fibers/dry gram), amosite (2.46×10^6 fibers/dry gram), crocidolite (0.16×10^6 fibers/dry gram), and anthophyllite (0.06×10^6 fibers/dry gram).

Combined data for both the type and number of intrapulmonary asbestos fibers in the 49 mesothelioma cases (22 from TABLE 2 and 27 from TABLE 3) was as follows:

(i) Total number of asbestos fibers detected in lung of the 49 cases was 456.4×10^6 fibers/dry gram maximum, 0.08×10^6 fibers/dry gram minimum and 56.4×10^6 fibers/dry gram on average.

(ii) The most common asbestos type(s) seen in lung in the 49 cases was amphibole alone (23/49; 47.0%), followed by amphibole(s) plus chrysotile (22/49; 44.9%) and chrysotile alone (4/49; 8.1%).

(iii) Among the various asbestos type seen in lung of the 49 cases, amosite fibers were greatest in number (36.9×10^6 fibers/dry gram on average) followed by chrysotile (19.4×10^6 fibers/dry gram on average), crocidolite (3.13×10^6 fibers/dry gram on average), tremolite/actinolite (0.69×10^6 fibers/dry gram on average) and anthophyllite (0.27×10^6 fibers/dry gram on average).

Findings obtained from 2A, B and C (based on TABLES 2 and 3) are summarized as follows:

(1) Except for 3 cases, the number of asbestos fibers in lung tissue of 49 mesothelioma cases (22 from TABLE 2 group and 27 from TABLE 3 group) was greater than the average number in lung in the general population (0.44×10^6 fibers/dry gram).[22]

(2) The number of asbestos fibers in mesothelial tissues taken from 22 mesothelioma cases (TABLE 2 group) was also greater in the majority of

TABLE 4. Dimensions of 10,575 asbestos fibers detected in lung and mesothelial tissues: totals for the 168 cases

Tissue	Length					Width				
	N.	G.M.	G.S.D.	Min.	Max.	N.	G.M.	G.S.D.	Min.	Max.
					Amosite					
Lung	1577	5.08	3.12	0.20	82.4	1577	0.19	2.47	0.02	6.50
Plaque	48	2.38	3.62	0.15	28.0	48	0.14	2.61	0.02	1.10
Tumor	66	4.55	3.50	0.30	62.0	66	0.21	2.19	0.03	1.10
					Chrysotile					
Lung	2921	0.42	2.26	0.08	18.5	2921	0.04	1.46	0.01	3.00
Plaque	1208	0.39	2.37	0.08	38.0	1208	0.04	1.40	0.01	0.20
Tumor	4412	0.35	1.97	0.07	15.0	4412	0.04	1.36	0.01	0.70
					Crocidolite					
Lung	230	4.63	2.34	0.40	31.5	230	0.14	1.96	0.03	1.50
Plaque*	–	–	–	–	–	–	–	–	–	–
Tumor	2	3.53	1.09	3.33	3.75	2	0.40	1.37	0.32	0.50
					Tremolite					
Lung	54	5.80	2.75	0.60	34.5	54	0.33	2.60	0.05	1.80
Plaque*	–	–	–	–	–	–	–	–	–	–
Tumor	11	1.61	2.18	0.40	3.72	11	0.19	2.37	0.03	0.80
					Anthophyllite					
Lung	38	5.57	3.11	0.25	49.6	38	0.54	2.37	0.10	2.90
Plaque*	3	2.30	2.26	1.00	5.10	3	0.23	2.08	0.10	0.40
Tumor	5	4.40	3.01	1.60	26.3	5	0.39	1.78	0.24	1.00

N = number; G.M. = geometric mean; G.S.D. = geometric standard deviation.

cases (18/22) than the average number of the general population (0.41 × 10^6 fibers/dry gram).[22]

(3) The average number of each type of asbestos fibers in lung (49 cases) was greatest for amosite (36.9 × 10^6 fibers/dry gram), followed by chrysotile (19.4 × 10^6 fibers/dry gram), crocidolite (3.13 × 10^6 fibers/dry gram, tremolite/actinolite (1.69 × 10^6 fibers/dry gram) and anthophyllite (0.27 × 10^6 fibers/dry gram). In contrast, in mesothelial tissues (22 cases), the average number of asbestos fibers was greatest for chrysotile (45.2 × 10^6 fibers/dry gram) followed by amosite (1.3 × 10^6 fibers/dry gram), anthophyllite (1.03 × 10^6 fibers/dry gram), crocidolite (0.01 × 10^6 fibers/dry gram) and tremolite/actinolite (0 × 10^6 fibers/dry gram). It was obvious that a disproportion of the average number of asbestos types was present between lung and mesothelial tissues in the 22 cases of malignant mesothelioma.

3. From the 168 cases of human malignant mesothelioma, dimensions (length and diameter) of a total of 10,575 asbestos fibers detected in lung and mesothelial tissues (mesotheliomatous tissue and fibrotic serosa including hyaline plaque) were measured. Findings are summarized in TABLE 4 and TABLE 5.

A. 1) As shown in TABLE 4, the 10,575 asbestos fibers consisted of 8,536 chrysotile fibers (2,921 in lung, 1,203 in plaque, 4,412 in tumor), 1,691 amosite fibers (1,577 in lung, 48 in plaque, 66 in tumor), 232 crocidolite fibers (230 in lung, 0 in plaque, 2 in tumor), 65 tremolite/actinolite fibers (54 in lung, 0 in plaque, 11 in tumor), and 46 anthophyllite fibers (38 in lung, 3 in plaque, 5 in tumor).

2) The chrysotile fibers obtained were generally short in length (geometric mean [G.M.]: 0.42 µm in lung, 0.39 µm in hyaline plaque, 0.35 µm in tumor) and thin in diameter (G.M.: 0.04 µm in lung, 0.04 µm in both plaque and tumor). Amosite fibers were greater in length (G.M.: 5.08 µm in lung, 2.38 µm in plaque, 4.55 µm in tumor) and thicker in diameter (G.M.: 0.19 µm in lung, 0.14 µm in plaque, 0.21 µm in tumor). Although other amphibole fibers, such as crocidolite, tremolite/actinolite and anthophyllite fibers were much less common: crocidolite fiber length was 4.63 µm (G.M.) in lung, not available in plaque, and 3.53 µm (G.M.) in tumor; their diameter was 0.14 µm (G.M.) in lung, not available in plaque, and 0.40 µm (G.M.) in tumor. Tremolite/actinolite fiber length was 5.80 µm (G.M.) in lung, not available in plaque, and 1.61 µm (G.M.) in tumor; their diameter was 0.33 µm (G.M.) in lung, not available in plaque, and 0.19 µm (G.M.) in tumor. Anthophyllite fiber length was 5.57 µm (G.M.) in lung, 2.30 µm in plaque, and 4.40 µm (G.M.) in tumor; diameter was 0.54 µm (G.M.) in lung, 0.23 µm (G.M.) in plaque, and 0.39 µm (G.M.) in tumor.

3) The numerical distribution of each type of asbestos fiber was compared between the lung and mesothelial tissues (hyaline plaque plus mesotheliomatous tissue). Distributions were quite different for chrysotile fibers and amphibole fibers. 34.2% (2,921/8,536) of chrysotile fibers were detected in lung, and 65.8% (5,616/8,536) were present in mesothelial tissues. The majority of amphibole fibers were seen in lung (amosite: 93.3%; 1,577/1,691; crocidolite: 99.1%; 230/232; tremolite/actinolite: 83.0%; 54/65; and anthophyllite: 82.6%; 38/46). Only a small proportion of these amphibole fibers were present in mesothelial tissues. This finding supported that compared with amphibole(s) fibers, chrysotile fibers had a much stronger capacity to translocate from the lung to the mesothelial tissue.

TABLE 5A. Total number of asbestos fibers in lung, plaque, and mesotheliomatous tissues greater than or equal to 5 µm in length from the 168 cases

Amosite	873/1691	(51.6%)
Crocidolite	112/232	(48.3%)
Tremolite	30/65	(45.5%)
Anthophyllite	23/46	(50.0%)
Chrysotile	83/8541	(1.0%)
Fibers ≥ 5µm	1121/10,575	(10.6%)
Fibers ≤ 5µm	9454/10,575	(89.4%)

TABLE 5B. Number of fibers found with length greater than or equal to 8 μm and diameter less than or equal to 0.25 μm from the 168 cases

	Lung	Plaque	Tumor
Amosite	171/1577 (10.8%)	2/48 (4.2%)	7/66 (10.6%)
Crocidolite	48/230 (20.9%)	0/0 (0.0%)	0/2 (0.0%)
Tremolite	4/54 (7.4%)	0/0 (0.0%)	0/11 (0.0%)
Anthophyllite	0/38 (0.0%)	0/3 (0.0%)	0/5 (0.0%)
Chrysotile	4/2921 (0.1%)	7/1208 (0.6%)	4/4412 (0.1%)
		Totals:	247/10,575 (2.3%)

B. (1) Asbestos fibers greater than 5 μm in length were counted in the 10,575 fibers. As shown in TABLE 5A, only 10.6% (1,121/10,575) of fibers were longer than 5 μm, and 89.4% were shorter than 5 μm. The proportion of long fibers, i.e., those > 5 μm, was greatest for amosite (873/1,691; 51.6%) followed by anthophyllite (23/46; 50.0%), crocidolite (112/232; 48.3%), tremolite/actinolite (30/65; 45.5%), and chrysotile (83/8,541; 1.0%).

(2) TABLE 5B shows that of the 10,575 fibers, only 247 fibers (2.3%) fit Stanton's hypothetical dimensions (> 8 μm in length and < 0.25 μm in diameter). The proportion of asbestos fibers that fit the hypothetical dimensions among asbestos types in these tissues was greatest for crocidolite (48/232; 20.7%), followed by amosite (180/1,691; 10.6%), tremolite /actinolite (4/65; 6.2%), chrysotile (15/8,541; 0.2%) and anthophyllite (0/46; 0%). None of crocidolite (2), tremolite/actinolite (11) and anthophyllite (5) fibers seen in mesothelial tissues (hyaline plaque and mesotheliomatous tissue) fit Stanton's hypothetical dimensions.

On the basis of these findings, it was concluded that asbestos fibers detected in lung and mesothelial tissues from mesothelioma patients were predominantly short and thin; 89.4% of asbestos fibers in these tissues were shorter than 5 μm, and the percentage of asbestos fibers that confirms Stanton's hypothetical dimensions (longer than 8 μm in length and smaller than 0.25 μm in diameter) was only 2.3%.

DISCUSSION

Translocation of asbestos fibers from the lung to other organs, including the pleura and peritoneum, has been well documented.[16–24] On the light microscopic level, asbestos bodies translocated from lung of deceased asbestos factory workers have been found in various organs such as kidney, heart, liver, spleen, adrenal, pancreas, brain, prostate, and thyroid.[25] Asbestos bodies have also been documented in hilar, mediastinal and abdominal lymph nodes, peritoneal mesotheliomatous tissue, and intestinal wall taken from mesothelioma cases.[13]

On the level of electron microscopy, in 1973, LeBouffant et al.[16] revealed the presence of numerous uncoated short, thin chrysotile fibers in pleural hyaline plaques taken from asbestos workers using a transmission electron microscope. This was an important finding at that time because pathologists could not obviously identify coated or uncoated asbestos fibers in the hyaline plaque in routine histopathological slides under light microscopy although they knew that this unique pleural change was intimately related to exposure to asbestos. Several years later, Sébastian et al.[17] found an obvious disproportion in the type and number of asbestos fibers between lung and parietal pleura of 29 asbestos workers: most of asbestos fibers seen in the parietal pleura were short chrysotile fibers. Dodson et al.[19] also found asbestos fibers (predominantly chrysotile) in pleural hyaline plaque taken from tissues of eight shipyard workers. Boutin et al.[20] also found highly concentrated asbestos fibers in black spots (glomerate lymphatic capillaries stained dark from anthracitic pigmentation) in the parietal pleura and stated that amphiboles outnumbered chrysotile in the black spots. Dodson et al.[24] detected asbestos fibers (predominantly amosite fibers) in omentum and mesentery taken from 20 (17 pleural and 3 peritoneal) mesothelioma cases; they concluded that asbestos fibers could translocate from the lung to the peritoneal cavity.

Our previous studies[18,21,22] revealed that the types of asbestos fibers were quite often different between the lung and mesothelial tissues in mesothelioma cases and that the major asbestos type seen in the pleural and peritoneal mesothelial tissues was short, thin chrysotile fibers. The capacity of such short, thin chrysotile fibers to translocate from alveoli to the pulmonary interstitium and finally to the pleura has been documented in experimental animal studies.[26,27]

Our present study based on a larger number of mesothelioma cases confirmed the same disproportion in fiber types (as shown in TABLE 1A) and number (as shown in TABLE 2) of asbestos fibers between lung and mesothelial tissues. We have already suggested that such a disproportion was caused by the strong ability of chrysotile fibers to translocate from lung to the pleura and peritoneum.[18,21,22]

We proposed that, to clarify the features of the asbestos fibers contributing to the induction of malignant mesothelioma, asbestos fiber analysis should be done of both lung and mesothelial tissues obtained from deceased mesothelioma patients. This approach is essential to grasp the total picture of asbestos exposure in malignant mesothelioma cases.

The number of asbestos fibers (per dry gram) counted in both the digested lung (49 cases) and mesothelial tissues (22 cases) varied among the mesothelioma cases. However, it was greater than the general population average number in lung (0.44×10^6 fibers/dry gram) in 45/49 (91.8%). It was also greater than the general population average number in the mesothelial tissues (0.41×10^6 fibers/dry gram) in 18/22 (81.8%). The numerical ratio between chrysotile fibers and amphibole(s) fibers in the mesothelial tissues was examined in 13 of the 22 mesothelioma cases and was found to be 30.3 (chrysotile):1 (amphibole[s]) in the mesothelial tissues.

Our present study revealed that the majority of asbestos fibers detected in the lung and mesothelial tissues were shorter than 5 µm; only 10.5% (1,115/10,575) of the fibers exceeded 5 µm in length.

Thinness of asbestos fibers has been emphasized as an important factor in their penetration from the proximal area to the peripheral part in the lung and for their translocation from the lung to the pleura.[44–46] It was also suggested that the thinness was related to the carcinogenicity of asbestos fibers.[35,36,42,43] The present study supports this concept in that the vast majority of asbestos fibers that translocated into the mesothelial tissues, the original site from which malignant mesothelioma develops, were very thin (0.04 µm in G.M.).

Our present study also revealed that asbestos fibers fitting Stanton's hypothetical dimensions (≥ 8 μm in length and ≤ 0.25 μm in width) comprised only 2.3% (247/10,575) of the fibers detected in both the lung and mesothelial tissues. From these findings, it is obvious that if we exclusively count only asbestos fibers longer than 5 μm or if we select only asbestos fibers fitting Stanton's hypothetical dimensions, a large proportion of asbestos fibers in these tissues will be omitted.

We conclude that short, thin asbestos fibers should be considered carcinogenic because they were the principal type of asbestos fiber encountered in the lung and mesothelial tissues taken from human mesothelioma cases.

It has been generally accepted, that like other asbestos types, chrysotile fibers are capable of inducing human malignant mesothelioma. This conclusion has been obtained from various sources including molecular biological studies,[50–54] animal experimental,[23,35,36,41,42,55–57] epidemiological studies,[58–70] case reports,[71–73] and asbestos tissue burden studies.[21,22,47]

The present study of asbestos tissue burden further supports the notion that chrysotile fibers are capable of inducing human malignant mesothelioma, because a) chrysotile was seen exclusively in both the lung and the mesothelial tissues in 18/74 (24.3%) cases, in the lung alone in 11/45 (24.4%), and in the mesothelial tissues alone in 35/49 (71.4%) cases; and b) chrysotile was the most common asbestos types in mesothelial tissue (112/123; 91.1%), which is the original site of the induction of malignant mesothelioma.

REFERENCES

1. WAGNER, J.C., C.A. SLEGGS, & P. MARCHAND. 1960. Diffuse pleural mesothelioma and asbestos exposure in the North Western Cape Province. Br. J. Ind. Med. **17:** 260–271.
2. SELIKOFF, I.J. & H. SEIDMAN. 1991. Asbestos associated deaths among insulation workers in the United States and Canada, 1967–1987. Ann. N.Y. Acad. Sci. **643:** 1–14.
3. COCHRANE, J.C. & I. WEBSTER. 1978. Mesothelioma in relation to asbestos fibre exposure. S. Afr. Med. J. **54:** 279–281.
4. HILLERDAL, G. 1999. Mesothelioma: Cases associated with non-occupational and low dose exposures. Occup. Environ. Med. **56:** 505–513.
5. SELIKOFF, I.J., C. HAMMOND & J. CHURG. 1972. Carcinogenicity of amosite asbestos. Arch. Environ. Health **25:** 183–186.
6. MCDONALD, J.C., F.D.K. LIDDELL, A. DUFRESNE, et al. 1993. 1891–1920 birth cohort of Quebec chrysotile miners and millers: mortality 1976–88. Br. J. Ind. Med. **50:** 1073–1081.
7. KARJALAINEN, A., L. MEURMAN & E. EUKKALA. 1994. Four cases of mesothelioma among Finnish anthophyllite miners. Occup. Environ. Med. **51:** 212–215.
8. LUCE, D., I. BUGEL, P. GOLDBERG, et al. 2000. Environmental exposure to tremolite and respiratory cancer in New Caledonia: a case-control study. Am. J. Epidemiol. **151:** 259–265.
9. WORLD-HEALTH-ORGANIZATION. 1986. Asbestos and other natural mineral fibers. Environ. Hlth. Criteria, pp. 1–194.
10. NEWHOUSE, M.L. & H. THOMPSON. 1965. Mesothelioma of pleura and peritoneum following exposure to asbestos in the London area. Br. J. Ind. Med. **22:** 261–269.
11. HARRIES, P.G. 1968. Asbestos hazards in naval deckyard. Ann. Occup. Hyg. **11:** 136.
12. ANDERSON, H.A., R. LILIS, S.M. DAUM, et al. 1970. Household-contact asbestos neoplastic risk. Ann. N.Y. Acad. Sci. **271:** 311–323.
13. GODWIN, M.C. & J. JAGATIC. 1970. Asbestos and mesothelioma. Environ. Res. **3:** 391–416.

14. LAUWERYS, J.M. & J.H. BARET. 1977. Alveolar clearance and the role of the pulmonary lymphatics. Am. Rev. Respir. Dis. **115:** 625–683.
15. DODSON, R.F., J. HUANG & J.R. BRUCE. 2000. Asbestos content in the lymph nodes of non-occupationally exposed individuals. Am. J. Ind. Med. **37:** 169–174.
16. LEBOUFFANT, L., J.C. MARTIN, S. DURIF, *et al.* 1973. Structure and composition of pleural plaque. *In* Biological Effects of Asbestos. P. Bogovski, J.C. Gilson, V. Timbrell & J.C. Wagner, Eds.: **8:** 249–257. I.A.R.C. Scientific Publication, WHO/International Agency for Research on Cancer, Lyon, France.
17. SÉBASTIEN, P., X. JANSON, A. GAUDICHET, *et al.* 1980. Asbestos retention in human respiratory tissues: comparative measurements in lung parenchyma and in parietal pleura. *In* Biological Effects of Mineral Fibres. J.C. Wagner, Ed.: **30:** 237–246. I.A.R.C. Scientific Publications, WHO/International Agency for Research on Cancer, Lyon, France.
18. KOHYAMA, N. & Y. SUZUKI. 1991. Analysis of asbestos fibers in lung parenchyma, pleura and mesothelioma tissues of North American insulation workers. Ann. N.Y. Acad. Sci. **643:** 27–52.
19. DODSON, R.F., M.G. WILLIAMS, C.J. CORN, *et al.* 1991. A comparison of asbestos burden in lung parenchyma, lymph nodes and plaques. Ann. N.Y. Acad. Sci. **643:** 53–60.
20. BOUTIN, C., P. DUMORTIER, F. REY, *et al.* 1996. Black spots concentrate oncogenic asbestos fibers in the parietal pleura—thoracoscopic and mineralogic study. Am. J. Respir. Crit. Care Med. **153:** 444–449.
21. SUZUKI, Y., S. YUEN, R. ASHLEY, *et al.* 1998. Asbestos fibers and human malignant mesothelioma. *In* Proceeding of the 9th International Conference on Occupational Respiratory Diseases. K. Chiyotani, Y. Hosoda & Y. Aizawa, Eds.: 709–713. Kyoto, Japan 13–16 October, 1997.
22. SUZUKI, Y. & S.R. YUEN. 2001 Asbestos tissue burden study on human malignant mesothelioma. Industrial Health **39:** 150–160.
23. FASSKE, E. 1988. Experimental lung tumors following specific intrabronchial application of chrysotile asbestos. Respiration **33:** 111–127.
24. DODSON, R.F., M.F. O'SULLIVAN, J. HUANG, *et al.* 2000. Asbestos in extrapulmonary sites—omentum and mesentery. Chest **117:** 486–493.
25. AUERBACH, O., A.S. CONSTON, L. GARFINKEL, *et al.* 1980. Presence of asbestos bodies in organs other than the lung. Chest **77:** 133–137.
26. BRODY, A.R. & L.H. HILL. 1982. Interstitial accumulation of inhaled chrysotile asbestos fibers and consequent formation of microcalcifications. Am. J. Pathol. **109:** 107–114.
27. VIALLAT. J.R., F. RAYBUAD, M. PASSAREL, *et al.* 1986. Pleural migration of chrysotile fibers after intratracheal injection in rats. Arch. Environ. Hlth. **41:** 282–286.
28. CUNNINGHAM, H.M. & R.D. PONTEFRACT. 1974. Placental transfer of asbestos. Nature. **249:** 177–178.
29. HAQUE, A.K., D.M. VRAZEL & K.D. BURAU. 1996. Is there transplacental transfer of asbestos? A study of 40 stillborn infants. Pediatr. Pathol. **16:** 877–892.
30. CHURG, A., B. WIGGS, L. DEPAOLI, *et al.* 1984. TB done exclusively in lung. Lung asbestos content in chrysotile workers with mesothelioma. Am. Rev. Resp. Dis. **130:** 1042–1045.
31. MCDONALD, J.C., B. ARMSTRONG, B.W. CASE, *et al.* 1989. Mesothelioma and asbestos fiber type - evidence from lung tissue analyses. Cancer **63:** 1544–1547.
32. ROGGLI, V., P.C. PRATT & A.R. BRODY. 1993. Asbestos fiber type in malignant mesothelioma: an analytical scanning electron microscopic study of 94 cases. **23:** 605–614.
33. DUFFRESNE, A., R. BEGIN, A. CHURG, *et al.* 1996. Mineral fiber content of lungs in patients with mesothelioma seeking compensation in Quebec. Am. J. Respir. Crit. Med. **153:** 711–718.
34. DODSON, R.F., M. O'SULLIVAN, C.J. CORN, *et al.* 1997 Analysis of asbestos fiber burden in lung tissue from mesothelioma patients. Ultrast. Pathol. **21:** 321–336.
35. STANTON, M.F., M. LAYARD, A. TEGERIS, *et al.* 1981 Relation of particles dimension to carcinogenicity in amphibole asbestoses and fibrous minerals. JNCI **67:** 965–975.
36. STANTON, M.F. & C. WRENCH. 1972. Mechanisms of mesothelioma induction with asbestos and fibrous glass. JNCI **48:** 797–821.

37. CASE, B.W. & P. SÉBASTIAN. 1987. Environmental and occupational exposure to chrysotile asbestos: A comparative micro analytical study. Arch. Environ. Hlth. **42:** 185–191.
38. SÉBASTIAN. P., J.C. MCDONALD, A.D. MCDONALD, et al. 1989. Respiratory cancer in chrysotile textile and mining industries: Exposure inferences from lung analysis. Br. J. Ind. Med. **46:** 180–187.
39. CASE, M.W. & P. SÉBASTIAN. 1989. In Fibre levels in lung and correlation to mineral fibers. J. Bignon, J. Peto & R. Saracci Eds.: **90:** 207–218. I.A.R.C. Scientific Publication, International Agency for Research on Cancer, Lyon, France.
40. CASE, B.W. 1991. Health effects of tremolite. Now and in the future. Ann. N.Y. Acad. Sci. **643:** 491–504.
41. WAGNER, J.C., G. BERRY, & V. TIMBRELL. 1973. Mesothelioma in rats after inoculation with asbestos and other materials. Br. J. Cancer. **28:** 173–185.
42. POTT, F. & K.H. FRIEDRICHS. 1973. Tumoren der Ratte nach i.p.-Injection faserformiger Staube. Naturwissenschaften **59:** 318–324.
43. WYLIE, A.G., K.F. BAILEY, J. KELSE, et al. 1993. The importance of width in asbestos fiber carcinogenicity and its implications for public policy. Am. Ind. Hyg. Assoc. J. **54:** 239–252.
44. TIMBRELL, V., D.M. GRIFFITHS & F.D. POOLEY. 1971. Possible biological importance of fibre diameters of South African amphiboles. Nature **232:** 55–56.
45. TIMBRELL, V. 1973. Physical factors as etiological mechanisms. In Biological Effects of Asbestos. P. Bogovski, V. Timbrell, J.C. Gilson & J.C. Wagner, Eds.: 295–303. Lyon, France. I.A.R.C.
46. HARINGTON, J.S. 1981. Fiber carcinogenesis: epidemiological observations and the Stanton hypothesis. JNCI **67:** 977–989.
47. MORINAGA, K., N. KOHYAMA, N. YOKOYAMA, et al. 1989. Asbestos fibre content of lungs with mesotheliomas in Osaka, Japan: a preliminary report. In Non-Occupational Exposure to Mineral Fibres. J. Bignon, J. Peto & R. Saracci, Eds.: **90:** 438–443. I.A.R.C. Scientific Publication, International Agency for Research on Cancer, Lyon, France.
48. HIROSHIMA, K. & Y. SUZUKI. 1993. Characterization of asbestos bodies and uncoated fibers in lung of hamster. J. Electron Microsc. **42:** 41–47.
49. KOHYAMA, N., H. KYONO, K. YOKOYAMA, et al. 1993. Evaluation of low-level asbestos exposure by transbronchial lung biopsy with analytical electron microscopy. J. Electron Microsc. **42:** 315–327.
50. APPEL, J.D., D.S. FASY, D.S. KOHTZ, et al. 1988. Asbestos fibers mediate transformation of monkey cells by exogeneous plasmid DNA. Proc. Natl. Acad. Sci. USA **85:** 7670–7674.
51. HEI, T.K., C.Q. PIAO, Z.Y. HE, et al. 1992. Chrysotile fiber is a strong mutagen in mammalian cells. Cancer Res. **52:** 6305–6309.
52. GAN, L., F.F. SAVRANSKY, T.M. FASY, et al. 1993. Transfection of human mesothelial cells mediated by different asbestos fiber types. Environ. Res. **62:** 28–42.
53. LEZON-GEYDA, K., C.M. JAIME, H. GODBOLD, et al. 1996. Chrysotile asbestos fibers mediate homologous recombination in rat fibroblast: implication for carcinogenesis. Mutat. Res. **361:** 113–120.60.
54. OKAYASU, R., S. TAKAHASHI, S. YAMADA, et al. 1999. Asbestos and DNA double strand break. Cancer Res. **59:** 298–300.
55. WAGNER, J.C., G. BERRY, J.W. SKIDMORE, et al. 1974. The effects of the inhalation of asbestos in rats. Br. J. Cancer. **29:** 252–269.
56. REEVES, A.L., H.E. PULO & R.G. SMITH. 1974. Inhalation carcinogenesis from various forms of asbestos. Environ. Res. **8:** 178–202.
57. SUZUKI, Y. & N. KOHYAMA. 1984. Malignant mesothelioma following intraperitoneal administration of asbestos and zeolite. Environ. Res. **35:** 277–292.
58. SMITH, A.H. & C.C. WRIGHT. 1996. Chrysotile asbestos is the main cause of pleural mesothelioma. Am. J. Ind. Med. **30:** 252–266.
59. STAYNER, L.T., D.A. DANKOV & R.A. LEMEN. 1996. Occupational exposure to chrysotile asbestos and cancer risk: a review of the amphibole hypothesis. Am. J. Public Health. **86:** 179–186.

60. Asbestos, asbestosis, and cancer: the Helsinki criteria for diagnosis and attribution. 1997. (Consensus report). Scan. J. Work Environ. Health. **23:** 311–316.
61. LANDRIGAN, P.J., W.J. NICHOLSON & Y. SUZUKI, *et al.* 1999. The hazards of chrysotile asbestos: a critical review. Ind. Health. **37:** 271–280.
62. CULLEN, M.R. & R.S. BALOYI. Chrysotile asbestos and health in Zimbabwe. I: Analysis of miners and millers compensated for asbestos-related diseases since independence. Am. J. Ind. Med. **19:** 161–169.
63. FINKELSTEIN, M.M. 1989. Mortality among employees of an Ontario factory that manufacture construction materials using chrysotile asbestos and coal tar pitch. Am. J. Ind. Med. **16:** 281–287.
64. PIOLATTO, G., E. NEGRI, C. LAVECCHIA, *et al.* 1990. An update of cancer mortality among chrysotile miners in Balangero, Northern Italy. **47:** 810–814.
65. SHIQU, Z., W. YONGXIAN, M. FUSHENG, *et al.* 1990. Retrospective mortality study of asbestos workers in Laiyuan. *In* Proceedings of the VII International Pneumoconioses Conference, Part II: August 23–26, 1988; Pittsburgh, PA. National Institute for Occupational Safety and Health 1242–1244. DHHS publication 90–109 Part II.
66. BEGIN, R., J. GAUTHIER, M. DESMEULES, *et al.* 1992. Work-related mesothelioma in Quebec, 1967–1990. Am. J. Ind. Med. **22:** 531–541.
67. MCDONALD, J.C., E.D.K. LIDDELL, A. DUFRESNE, *et al.* 1993. The 1891–1920 birth cohort of Quebec chrysotile miners and millers: mortality 1976–1988. Br. J. Ind. Med. **50:** 1073–1081.
68. DEMENT, J.M., D.P. BROWN & A. OKUN. 1994. A mortality among chrysotile asbestos textile workers: cohort mortality and case-control analyses. Ann. Occup. Hyg. **38:** 525–532.
69. YANO, E., Z.M. WANG, Z.R.WANG, *et al.* 2001. Cancer mortality among workers exposed to amphibole-free chrysotile asbestos. Am. J. Epidemiol. **154:** 538–543.
70. GOODWIN, M.C. & G. JAGARIC. 1968. Asbestos and mesothelioma. JAMA Letters **204:** 1009.
71. LANGER, A.M. & E.T.E. MCCAUGHEY. 1982. Mesothelioma in a brake repair worker. The Lancet **ii** (Nov. 13); (**8307**) 1101–1103.
72. HUNCHAREK, M. 1987. Chrysotile asbestos exposure and mesothelioma. Br. J. Ind. Med. **44:** 287–266 (Correspondence).
73. HUNCHAREK, M., J. MUSCAT & V. CAPOTORTO. 1989. Pleural mesothelioma in a brake mechanic. Br. J. Ind. Med. **46:** 69–71.

Carcinogenicity and Mechanistic Insights on the Behavior of Epoxides and Epoxide-Forming Chemicals

RONALD L. MELNICK

National Institute of Environmental Health Sciences, Research Triangle Park, North Carolina, USA

ABSTRACT: Many epoxides and their precursors are high production volume chemicals that have major uses in the polymer industry and as intermediates in the manufacture of other chemicals. Several of these chemicals were demonstrated to be carcinogenic in laboratory animal studies conducted by the Ramazzini Foundation (e.g., vinyl chloride, acrylonitrile, styrene, styrene oxide, and benzene) and by the National Toxicology Program (e.g., ethylene oxide, 1,3-butadiene, isoprene, chloroprene, acrylonitrile, glycidol, and benzene). The most common sites of tumor induction were lung, liver, harderian gland, and circulatory system in mice; Zymbal's gland and brain in rats; and mammary gland and forestomach in both species. Differences in cancer outcome among studies of epoxide chemicals may be related to differences in study design (e.g., dose, duration, and route of exposure; observation period; animal strains), as well as biological factors affecting target organ dosimetry of the DNA-reactive epoxide (toxicokinetics) and tissue response (toxicodynamics). $N7$-Alkylguanine, $N1$-alkyladenine, and cyclic etheno adducts, as well as K-*ras* and *p53* mutations, have been detected in animals and/or workers exposed to several of these chemicals. The classifications of these chemical carcinogens by IARC and NTP are based on animal and human data and results of mechanistic studies. Reducing occupational and environmental exposures to these chemicals will certainly reduce human cancer risks.

KEYWORDS: vinyl chloride; vinyl halides; butadiene; styrene; ethylene oxide; glycidol; epoxides; carcinogenicity; toxicokinetic modeling; DNA adducts; etheno adducts; K-*ras* mutations; cancer risk; human carcinogens

INTRODUCTION

Because of their reactivity, several epoxides are important intermediates in the chemical industry. In addition, epoxides may also be formed *in vivo* during the metabolic elimination of their respective alkene or aromatic precursors. Human exposure to epoxides is a concern because these chemicals are alkylating agents that can react *in vivo* with nucleophilic sites on proteins and DNA.

Address for correspondence: Ronald L. Melnick, Ph.D., National Institute of Environmental Health Sciences, P. O. Box 12233, 111 T. W. Alexander Drive, Research Triangle Park, NC 27709. Voice: 919-541-4142; fax: 919-541-3647.
melnickr@niehs.nih.gov

Ann. N.Y. Acad. Sci. 982: 177–189 (2002). © 2002 New York Academy of Sciences.

Epoxide and epoxide-forming chemicals have been used extensively in the manufacture of synthetic polymers (e.g., polyvinyl chloride plastic, polyethylene, styrene-butadiene rubber, acrylonitrile-butadiene-styrene, polystyrene, and acrylic fibers), resins, and other chemicals (e.g., ethylene oxide from ethylene; ethylene glycol from ethylene oxide).[1] Epoxide chemicals are some of the highest production volume chemicals in the United States[2] (TABLE 1) and worldwide. Ethylene oxide is also used as a sterilant, while ethylene and isoprene are also natural products of plants and animals. In addition to occupational exposure, several of these agents are environmental pollutants present in tobacco smoke, automobile exhaust, and industrial releases.

Epoxides are oxygen-containing heterocyclic compounds. Because of the large strain associated with the three-membered ring structure,

$$CH_2 \underset{O}{-\!\!\triangle\!\!-} CH - R$$

epoxides are reactive molecules. Several epoxides and epoxide-forming chemicals were demonstrated to be carcinogenic in experimental studies conducted by the Ramazzini Foundation (e.g., vinyl chloride, acrylonitrile, styrene, styrene oxide, and benzene) and by the National Toxicology Program (e.g., ethylene oxide, 1,3-buta-

TABLE 1. U.S. production volume of epoxides and epoxide-forming chemicals

Chemical	Annual US Production x10^9 lbs	Structure
Ethylene	55.4	$CH_2=CH_2$
Ethylene oxide	8.5	CH_2-CH_2 with O bridge
Vinyl chloride	14.4	$CH_2=CH-Cl$
Vinyl bromide	0.05	$CH_2=CH-Br$
Vinyl fluoride	0.007	$CH_2=CH-F$
Acrylonitrile	3.4	$CH_2=CH-CN$
Butadiene	4.4	$CH_2=CH-CH=CH_2$
Isoprene	0.6	$CH_2=CH-C(CH_3)=CH_2$
Chloroprene	0.3	$CH_2=CH-C(Cl)=CH_2$
Glycidol	—	$CH_2-CH_2-CH_2OH$ with O bridge
Benzene	17.0	(phenyl ring)
Styrene	11.9	(phenyl)-$CH_2=CH_2$
Styrene oxide	—	(phenyl)-$CH-CH_2$ with O bridge

diene, isoprene, chloroprene, glycidol, acrylonitrile, and benzene). These laboratory animal studies prompted additional investigations of the potential carcinogenicity of other epoxide-forming chemicals (e.g., vinyl bromide, vinyl fluoride), as well as extensive mechanistic studies on epoxide carcinogenesis. This paper reviews laboratory studies on the carcinogenicity of this family of chemicals, toxicokinetic issues that impact on target organ doses, and toxicodynamic issues concerning the effects of epoxides at target sites. In addition to data from epidemiological studies that have been performed on several of these chemicals, animal data and mechanistic information are important for evaluating human cancer risk.

CARCINOGENICITY STUDIES IN LABORATORY ANIMALS

Several epoxides and epoxide-forming chemicals were demonstrated to be carcinogenic in laboratory animal studies conducted by the Ramazzini Foundation (RF) or by the National Toxicology Program (NTP). Other research groups have also contributed to the growing cancer database of epoxide chemicals. Although there are experimental design differences among these studies, some general observations are worth noting (TABLE 2). First, many of these chemicals induce tumors at multiple organ sites in rats and mice.[3-25] Second, for this family of chemicals, it is not unusual to observe differences in sites of induced neoplasms between rats and mice. The most common sites of tumor induction among these chemicals are the lung, liver, harderian gland, and circulatory system (hemangiosarcomas) in mice; the brain and Zymbal's gland in rats; and the mammary gland and forestomach in both species. Benzene is included in this table because its first metabolite is benzene oxide; however, it is recognized that the mechanism of benzene-induced carcinogenicity is not known and that other metabolites including muconaldehyde, hydroquinone, benzoquinone, and/or catechol may be equal or more important contributors to the toxicity and carcinogenicity of benzene. In spite of their structural and metabolic differences, it is remarkable how similar are the patterns of tumor induction by benzene[8] and 1,3-butadiene[11] in mice (8 of 9 corresponding sites).

Among the three vinyl halides that have been tested (vinyl chloride, vinyl fluoride, and vinyl bromide), hemangiosarcomas of the liver were observed in all species studied, lung and mammary gland tumors were induced in mice, and liver and Zymbal's gland tumors were induced in rats.[3-5] The similar tumor patterns among these chemicals, especially the induction of the rare hemangiosarcomas of the liver, strongly suggest a common mechanism of carcinogenesis for these chemicals.

Differences in cancer outcome across species and among the various studies of epoxides and epoxide-forming chemicals are likely due to differences in experimental design, as well as to biological factors that affect the behavior of these chemicals in exposed animals. Experimental designs differed with respect to the dose or exposure concentrations, the duration of exposure (hours per day, as well as number of weeks of exposure), the time at which studies were terminated (1 year, 2 years, or lifetime/natural death), the route of exposure (inhalation, gavage), and the use of different strains of animals (e.g., mice were B6C3F1, Swiss, or CD-1; rats were Sprague-Dawley, Wistar, or F344).

TABLE 2. Sites of tumor induction by epoxides and epoxide-forming chemicals in mice and rats

Site	VCl[1] mice	VCl[1] rats	VF[4] mice	VF[4] rats	VBr[5] mice	VBr[5] rats	ACN[6,7] mice	ACN[6,7] rats	Benzene[8,9] mice	Benzene[8,9] rats	Butadiene[10-12] mice	Butadiene[10-12] rats	Chloroprene[13] mice	Chloroprene[13] rats	Isoprene[14-16] mice	Isoprene[14-16] rats	Ethy[17] rats	EtO[18-20] mice	EtO[18-20] rats	Styr[21-23] mice	Styr[21-23] rats	StyO[21,24] mice	StyO[21,24] rats	Glycidol[25] mice	Glycidol[25] rats
Circulatory	+	☆	+	☆							+		+		+									+	
Liver	+	☆	+	☆	☆	☆			+	+	+		+		+							+		+	
Lung	+		+				+?		+	+	+		+	☆	+			+	+	+				+	☆
Harderian gl	+		+				+		+		+		+		+			+						+	☆
Mammary gl	+	☆	+					☆	+	+	+	☆	+	☆		☆		+			☆			+	☆
Forestomach		☆					+		+	☆	+		+		+							+	☆	+	
Lymphatic/hematopoietic																									☆
Zymbal's gl		☆		☆		☆		☆	+			☆													☆
Brain		☆						☆		☆		☆				☆?			☆						
Ovary											+														
Thyroid gl												☆		☆											☆
Testis										☆		☆				☆									
Uterus												☆						+							
Oral cavity														☆											
Skin																								+	☆
Kidney		☆									+?		+	☆		☆								+	☆
Small intest																									☆
Mesothelium																									☆

NOTE: VCl = vinyl chloride; VF = vinyl fluoride; VBr = vinyl bromide; ACN = acrylonitrile; Ethy = ethylene; EtO = ethylene oxide; Styr = styrene; StyO = styrene oxide. **+** represents the reported site of tumor induction in mice; ☆ represents the reported site of tumor induction in rats; ? represents an equivocal response. Superscripts after the listing of each chemical identify the references for the cited carcinogenicity studies.

EVALUATIONS OF TARGET ORGAN DOSIMETRY

In conjunction with the rodent toxicity and carcinogenicity studies, the NTP conducts toxicokinetic studies to characterize the behavior of test chemicals in the rodent models used in the bioassay. Toxicokinetic data are used to describe relationships between exposure and tissue concentrations of parent compound and metabolites, to relate target organ dosimetry to adverse effects resulting from exposure, and to improve low-dose and interspecies extrapolations.

Alkenes are metabolically activated by cytochrome P450–dependent monooxygenases to alkylating epoxides that can react with nucleophilic sites on proteins and DNA. The major P450 isozyme that catalyzes alkene oxidation is CYP2E1; however, other isozymes such as CYP2A6, CYP2B6, and CYP1A2 can also catalyze this reaction. Epoxide detoxification involves conjugation with glutathione (catalyzed by glutathione-S-transferase) to form precursors of mercapturic acids, or hydrolysis to diols catalyzed by epoxide hydrolases (FIG. 1). Halogenated epoxides may undergo rearrangement to form haloacetaldehydes, which can also react with DNA and proteins. The primary intermediate of 1,3-butadiene oxidation, epoxybutene, can undergo further oxidation to form diepoxybutane. Hydrolysis of epoxybutene prior to the second oxidation, or hydrolysis of diepoxybutane, leads to the production of epoxybutanediol. All three of these DNA-reactive epoxide intermediates may participate in the mutagenicity and carcinogenicity of 1,3-butadiene.

Physiologically based pharmacokinetic (PBPK) models have been developed to provide time- and dose-dependent estimates of tissue concentrations of parent compound and metabolites in animals exposed to toxic chemicals. A PBPK model consists of a series of mass-balance differential equations that are formulated to represent in quantitative terms the complex physiological and biochemical processes that affect the behavior of the chemical in the intact animal. The animal is represented as being divided into separate organ compartments, including the site where the chemical enters the body and the sites where it is subsequently stored or metabo-

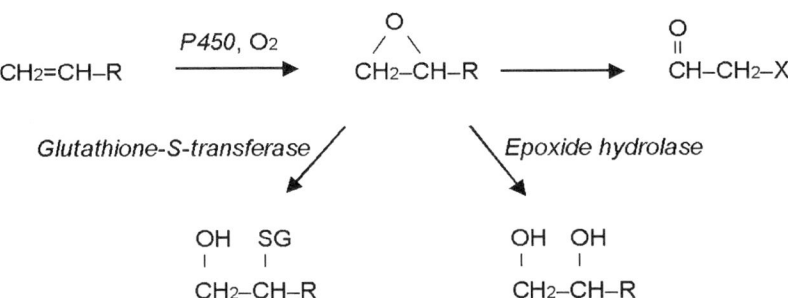

FIGURE 1. Activation and detoxication of epoxide-forming chemicals. R: H (ethylene), Cl (vinyl chloride), F (vinyl fluoride), Br (vinyl bromide), C≡N (acrylonitrile), CH=CH$_2$ (1,3-butadiene), or ⬡ (styrene). Haloethylene oxides (R: Cl, Br, or F) also undergo rearrangement to form haloacetaldehydes.

lized. The organ compartments are connected by arterial and venous blood flow. These models characterize the absorption (inhalation, oral, dermal) of the agent, distribution via the blood to all body organs, metabolism (activation and detoxication), and elimination of the parent compound and its metabolites. Validated PBPK models can provide a biologically based approach for using a tissue dose metric of the putative carcinogenic intermediate(s), rather than exposure concentrations, to characterize tumor dose-response relationships at occupational or environmental exposure levels.

A PBPK model of butadiene disposition based on evaluations of the metabolic elimination of butadiene and of epoxybutene in rats and mice, as well as measurements of epoxybutene in blood and urinary metabolites, indicated that epoxides formed *in vivo* have a privileged access to epoxide hydrolase compared to epoxides delivered to metabolizing organs from external exposures.[26] Metabolic elimination data obtained from animals exposed directly to the epoxide intermediates result in overprediction of circulating epoxides produced by oxidative metabolism of the precursor vinyl compounds. To be useful for dose-response analyses, PBPK models must accurately represent the physiological and biochemical processes that regulate tissue dosimetry. Substitution of animal parameters with human values measured in diverse populations can provide a sound scientific basis for extrapolating tissue dosimetry across species (laboratory animals to humans), while accounting for the range of interindividual variability.

Pharmacodynamic models provide estimates of the rate of tissue response per unit of delivered dose. Pharmacodynamic factors involved in chemical carcinogenesis may include mutagenic effects, altered gene expression, altered enzymatic activities, induced oxidative DNA damage, etc. A combination of validated pharmacokinetic and pharmacodynamic models (FIG. 2) can provide a biological approach for relating exposure to human cancer risk.

DNA ADDUCTS

Measurements of specific DNA adducts in animals and humans can serve as valuable biomarkers of exposure and effect because they represent the integrated consequences of exposure, absorption, distribution, metabolism, reaction with DNA, and the kinetics of DNA repair. The major site of DNA alkylation by simple epoxides is the $N7$ position of guanine, the most nucleophilic center in the DNA molecule. The adduct products include $N7$-(2-oxoethyl)guanine or $N7$-(2-hydroxyethyl)guanine derivatives (FIG. 3). $N7$-Alkylguanine adducts are generally not promutagenic because this position is not a base-pairing site. However, alkylation of guanine at the $N7$ position and subsequent spontaneous or glycosylase-mediated depurination of this product yield apurinic sites that may lead to point mutations due to miscoding during replication and/or to frameshift mutations due to polymerase slippage. The $N7$-guanine adduct of ethylene oxide may serve as a biomarker of *in vivo* conversion of ethylene to ethylene oxide and as an indicator of ethylene cancer risk.[27]

Exocyclic etheno adducts are also produced *in vitro* and *in vivo* by reaction of certain bifunctional electrophiles (e.g., monosubstituted epoxides) with DNA. For example, the halogenated epoxide intermediates of vinyl chloride, vinyl bromide, or vinyl bromide or cyanoethylene oxide from acrylonitrile may react with adenine to

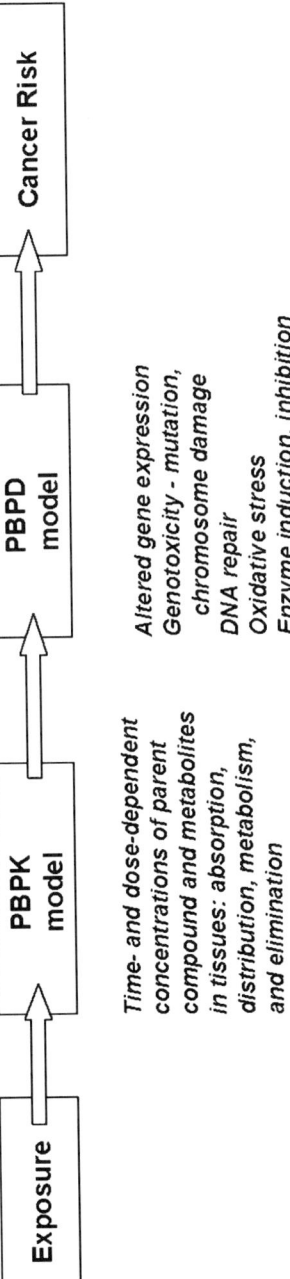

FIGURE 2. Linking exposure to cancer risk.

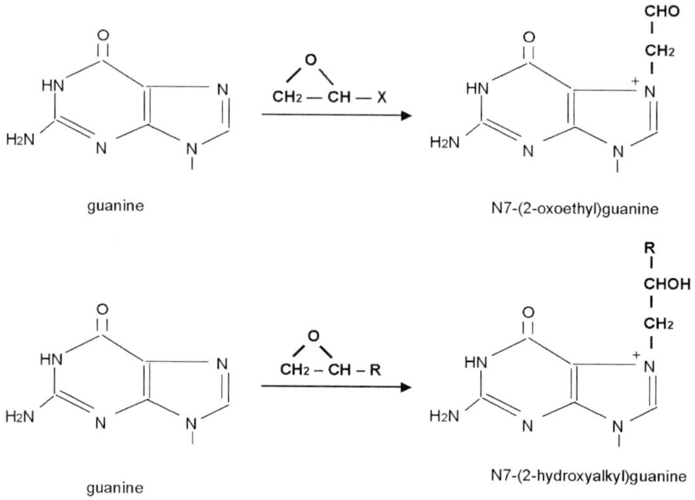

FIGURE 3. N7-Guanine adducts of (*top*) mono-substituted epoxides [X: Cl (vinyl chloride), Br (vinyl bromide), F (vinyl fluoride), or C≡N (acrylonitrile)] or (*bottom*) simple epoxides [R: H (ethylene oxide), CH_2OH (glycidol), $CH=CH_2$ (epoxybutene), $CHOH$-CH_2OH (epoxybutane diol), or ⌬ (styrene oxide)].

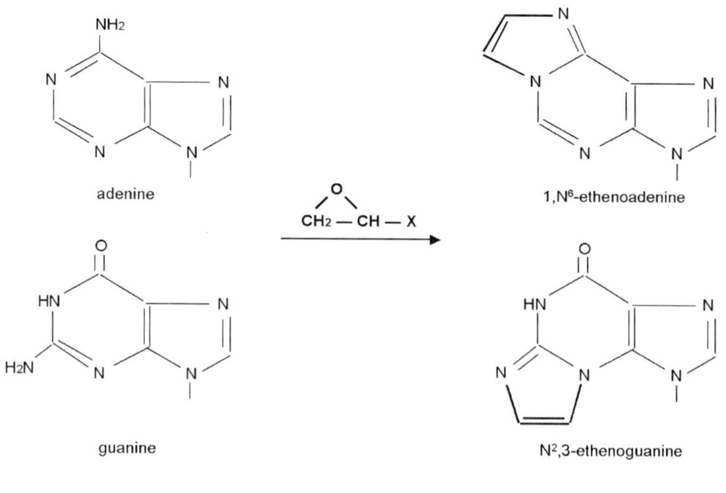

FIGURE 4. Cyclic etheno adducts formed by reaction of mono-substituted epoxides with DNA.

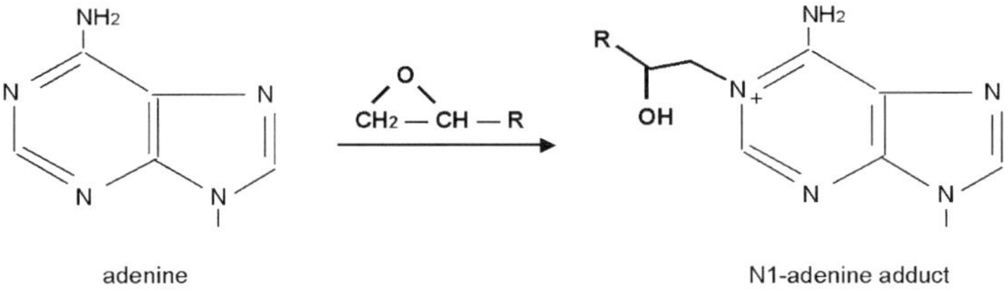

FIGURE 5. $N1$-Adenine adducts of styrene oxide and epoxybutanediol identified in workers exposed to 1,3-butadiene or styrene. R: CHOH-CH$_2$OH (epoxybutanediol) or ⬡ (styrene oxide).

produce 1,N^6-ethenoadenine, while reaction with guanine produces N^2,3-ethenoguanine (FIG. 4). These adducts, which are minor products compared to the $N7$-alkylguanine adducts, are formed by reaction of an epoxide carbon with the nucleophilic ring nitrogen ($N1$ of adenine or $N3$ of guanine), followed by opening of the epoxide ring, loss of the hydrogen halide or HCN, and cyclization with the exocyclic nitrogen (N^6 of adenine or N^2 of guanine). Etheno adducts are relatively resistant to repair; how they cause mutations and possibly cancer is not clear, but may be due to a blocking effect on DNA polymerases causing misincorporations.[28]

$N1$-Adenine adducts of styrene oxide and of epoxybutanediol have been identified in lymphocyte DNA samples obtained from workers occupationally exposed to styrene[29] or 1,3-butadiene,[30] respectively (FIG. 5). These stable $N1$-adenine adducts, which have also been detected in tissues of rodents exposed to styrene[31] or 1,3-butadiene,[32] may be important premutagenic lesions because they reflect the reaction of xenobiotic electrophiles at base-pairing sites in the DNA. $N1$-Adenine adducts may be responsible for the A→G transition mutations that have been detected in butadiene-induced tumors.

MUTATIONS IN CANCER GENES

The NTP collects tumor tissue from animals exposed to environmental agents in the bioassay program in order to identify and characterize mutations in proto-oncogenes or in tumor suppressor genes as indicators of molecular events that occur in the pathogenesis of the induced tumor response. This information is very relevant to human carcinogenesis since activated K-*ras* proto-oncogenes and inactivating mutations in the *p53* tumor suppressor gene have been frequently detected in a wide variety of human cancers.

The types of mutations detected in epoxide-induced tumors are consistent with the types of DNA adducts formed by these agents, suggesting that the DNA lesions described above may be important in epoxide-induced carcinogenicity. K-*ras* mutations (G→A transitions) and *p53* mutations (A→T transversions) occurred in a high proportion of liver hemangiosarcomas taken from vinyl chloride–exposed workers.[33] Similar *p53* mutations were observed in liver hemangiosarcomas induced by vinyl chloride in rats.[34] Analyses of hepatocellular carcinomas showed K-*ras* and *p53* mutations (G→A transitions) in workers exposed to vinyl chloride[35,36] and H-*ras* mutations (A→T transversions) in vinyl chloride–exposed rats.[37] K-*ras* and *p53* mutations have also been detected in hemangiosarcomas induced by 1,3-butadiene in mice.[38] In addition, K-*ras* mutations were observed in lung carcinomas and in harderian gland carcinomas in mice exposed to 1,3-butadiene, chloroprene, or isoprene.[39] These findings indicate that *ras* and *p53* mutations likely play an important role in the molecular pathogenesis of tumors induced by vinyl chloride, 1,3-butadiene, chloroprene, or isoprene. For vinyl chloride, there is compelling evidence for the involvement of these genes in tumors induced in both animals and humans.

CONCLUSIONS

Rodent models have been extremely valuable in identifying the carcinogenic potential of epoxides and epoxide-forming chemicals. Laboratory studies by the RF and the NTP have shown that most of these compounds induce neoplasms at multiple organ sites in rats and mice. Observed differences in response among the various studies of these chemicals are likely due to a combination of differences in experimental design as well as the molecular mechanisms by which these chemicals induce their carcinogenic effects. Animal and human carcinogenicity data on epoxide chemicals have stimulated extensive research aimed at understanding and characterizing the mechanistic basis for the carcinogenic effects of these agents. Laboratory animals have served as valuable models to elucidate toxicokinetic factors that affect target tissue dosimetry of the various epoxides and toxicodynamic factors that determine tissue response to these DNA-reactive chemicals.

Several epoxide and epoxide-forming chemicals are classified by the International Agency for Research on Cancer (IARC) and by the NTP as known human carcinogens (e.g., vinyl chloride, ethylene oxide, and benzene), while others are classified as probable, possible, or reasonably anticipated human carcinogens (e.g., vinyl bromide, vinyl fluoride, acrylonitrile, chloroprene, isoprene, glycidol, styrene, and styrene oxide). 1,3-Butadiene is listed in the *NTP Report on Carcinogens* as a known human carcinogen based on sufficient evidence of carcinogenicity from studies in humans, including epidemiological and mechanistic information, which indicates a causal relationship between exposure to butadiene and excess mortality from lymphatic and/or hematopoietic cancers. In contrast, a recent IARC review retained the listing of 1,3-butadiene as a probable human carcinogen.

Based on comparative animal carcinogenicity data and mechanistic information on metabolism and DNA adduct formation, it is reasonable to regard vinyl chloride, vinyl bromide, and vinyl fluoride as human carcinogens that act by a common mechanism(s). Further, the finding that human liver microsomes and purified human CYP2E1 metabolize vinyl chloride and vinyl bromide at similar rates to intermediates that react with adenosine to form 1,N^6-ethenoadenosine[40] shows that a common metabolic pathway is involved in the bioactivation of these chemicals in humans. The same product was formed, but at a slower rate, when acrylonitrile was used as the substrate.[40] Thus, human risk among these chemicals may differ quantitatively, but not qualitatively.

Although a carcinogenicity study of ethylene has not been performed in mice and there are no regulatory standards for this chemical, it would be reasonable to use the major DNA adduct formed by ethylene oxide (N7-hydroxyethylguanine) as a dose metric for evaluating ethylene cancer risk. Likewise, DNA adducts of styrene oxide may serve as a reasonable dose metric for evaluating styrene cancer risk. The detection of N1-adenine adducts of styrene oxide in workers exposed to styrene and of N1-adenine adducts of epoxybutanediol in workers exposed to 1,3-butadiene raises a high level of concern of cancer risk associated with occupational exposure to these chemicals.

To date, DNA adduct studies following exposure to 1,3-butadiene implicate epoxybutene and epoxybutanediol in the carcinogenicity of this chemical in laboratory animals and humans; no adducts specific for diepoxybutane have been detected in animals exposed to 1,3-butadiene. The findings that N7-guanine adducts of epoxybutanediol were formed in the highest amounts in tissues of rats and mice exposed to 1,3-butadiene and that the N7-guanine adducts of epoxybutene were present at about 10-fold lower levels[32] suggest that epoxybutanediol may be the major epoxide metabolite involved in the mutagenesis and carcinogenesis of 1,3-butadiene.

The fact that humans metabolize alkenes to DNA-reactive epoxide intermediates similar to rodents, the fact that several epoxide and epoxide-forming chemicals are classified as known or probable human carcinogens, and the fact that DNA adducts as well as K-*ras* and *p53* mutations have been detected in animals and/or workers exposed to several of these chemicals raise serious concerns of occupational and environmental cancer risk from exposure to epoxide chemicals.

Future studies to better characterize cancer risk would benefit from information on the range and distribution of human metabolic parameters for these compounds in order to develop PBPK models that reflect interindividual differences in activation and clearance of the reactive epoxides, as well as additional information on the rates of DNA adduct formation and repair. With a greater understanding of the qualitative and quantitative relationships between DNA adducts, mutagenicity, and carcinogenicity, it should be possible to provide more accurate estimates of the distribution and range of cancer risks in heterogeneous populations exposed to epoxide chemicals. Even without this additional information, one thing is certain: reducing occupational and environmental exposures to these chemicals will reduce human cancer risks.

REFERENCES

1. McCoy, M., M.S. Reisch, *et al.* 2001. Facts and figures for the chemical industry. Chem. Eng. News **79**: 42–51.
2. Budavari, S. 1996. The Merck Index: An Encyclopedia of Chemicals, Drugs, and Biologicals. Twelfth edition. Merck. Whitehouse Station, NJ.
3. Maltoni, C., G. Lefemine, A. Ciliberti, *et al.* 1981. Carcinogenicity bioassays of vinyl chloride monomer: a model of risk assessment on an experimental basis. Environ. Health Perspect. **41**: 3–29.
4. Bogdanffy, M.S., G.T. Makovec & S.R. Frame. 1995. Inhalation oncogenicity bioassay in rats and mice with vinyl fluoride. Fundam. Appl. Toxicol. **26**: 223–238.
5. Benya, T.J., W.M. Busey, *et al.* 1982. Inhalation carcinogenicity bioassay of vinyl bromide in rats. Toxicol. Appl. Pharmacol. **64**: 367–379.
6. National Toxicology Program. 2002. Toxicology and Carcinogenesis Studies of Acrylonitrile (CAS No. 107-13-1) in B6C3F$_1$ Mice (Gavage Studies). Technical Report Series No. 506. NIH Publ. No. 02-4440. U.S. Department of Health and Human Services, NIH. Research Triangle Park, NC.

7. MALTONI, C., A. CILIBERTI, et al. 1988. Long-term carcinogenicity bioassays on acrylonitrile administered by inhalation and by ingestion to Sprague-Dawley rats. Ann. N.Y. Acad. Sci. **534:** 179–202.
8. HUFF, J.E., J.K. HASEMAN, et al. 1989. Multiple-site carcinogenicity of benzene in Fischer 344 rats and B6C3F$_1$ mice. Environ. Health Perspect. **82:** 125–163.
9. MALTONI, C., A. CILIBERTI, G. COTTI, et al. 1989. Benzene, an experimental multipotential carcinogen: results of the long-term bioassays performed at the Bologna Institute of Oncology. Environ. Health Perspect. **82:** 109–124.
10. HUFF, J.E., R.L. MELNICK, et al. 1985. Multiple organ carcinogenicity of 1,3-butadiene in B6C3F$_1$ mice after 60 weeks of inhalation exposure. Science **227:** 548–549.
11. MELNICK, R.L., J. HUFF, et al. 1990. Carcinogenicity of 1,3-butadiene in C57BL/6 × C3H F$_1$ mice at low exposure concentrations. Cancer Res. **50:** 6592–6599.
12. OWEN, P.E., J.R. GLAISTER, et al. 1987. Inhalation toxicity studies with 1,3-butadiene: 3. Two-year toxicity/carcinogenicity studies in rats. Am. Ind. Hyg. Assoc. J. **48:** 407–413.
13. MELNICK, R.L., R.C. SILLS, et al. 1999. Multiple organ carcinogenicity of inhaled chloroprene (2-chloro-1,3-butadiene) in F344/N rats and B6C3F$_1$ mice and comparison of dose-response with 1,3-butadiene in mice. Carcinogenesis **20:** 867–878.
14. MELNICK, R.L., R.C. SILLS, et al. 1994. Isoprene, an endogenous hydrocarbon and industrial chemical, induces multiple organ neoplasia in rodents after 26 weeks of inhalation exposure. Cancer Res. **54:** 5333–5339.
15. PLACKE, M.E., L. GRIFFIS, et al. 1996. Chronic inhalation oncogenicity study of isoprene in B6C3F$_1$ mice. Toxicology **110:** 253–262.
16. MELNICK, R.L. & R.C. SILLS. 2001. Comparative carcinogenicity of 1,3-butadiene, isoprene, and chloroprene in rats and mice. Chem. Biol. Interact. **135:** 27–42.
17. HAMM, T.E., D. GUEST & J.G. DENT. 1984. Chronic toxicity and oncogenicity bioassay of inhaled ethylene in Fischer 344 rats. Fundam. Appl. Toxicol. **4:** 473–478.
18. NATIONAL TOXICOLOGY PROGRAM. 1987. Toxicology and Carcinogenesis Studies of Ethylene Oxide (CAS No. 75-21-8) in B6C3F$_1$ Mice (Inhalation Studies). Technical Report Series No. 326. NIH Publ. No. 88-2582. U.S. Department of Health and Human Services, NIH. Research Triangle Park, NC.
19. SNELLINGS, W.M., C. WEIL & R.R. MARONPOT. 1984. A two-year inhalation study of the carcinogenic potential of ethylene oxide in Fischer 344 rats. Toxicol. Appl. Pharmacol. **75:** 105–117.
20. LYNCH, D.W., T.R. LEWIS, et al. 1984. Carcinogenic and toxicologic effects of inhaled ethylene oxide and propylene oxide in F344 rats. Toxicol. Appl. Pharmacol. **79:** 69–84.
21. CONTI, B., C. MALTONI, et al. 1988. Long-term carcinogenicity bioassays on styrene administered by inhalation, ingestion, and injection and styrene oxide administered by ingestion in Sprague-Dawley rats, and para-methylstyrene administered by ingestion in Sprague-Dawley rats and Swiss mice. Ann. N.Y. Acad. Sci. **534:** 203–234.
22. CRUZAN, G., J.R. CUSHMAN, et al. 2001. Chronic toxicity/oncogenicity study of styrene in CD-1 mice by inhalation exposure for 104 weeks. J. Appl. Toxicol. **21:** 185–198.
23. CRUZAN, G., J.R. CUSHMAN, et al. 2001. Chronic toxicity/oncogenicity study of styrene in CD rats by inhalation exposure for 104 weeks. J. Toxicol. Sci. **46:** 266–281.
24. LIJINSKY, W. 1986. Rat and mouse forestomach tumors induced by chronic oral administration of styrene oxide. J. Natl. Cancer Inst. **77:** 471–476.
25. IRWIN, R.D., S.L. EUSTIS, et al. 1996. Carcinogenicity of glycidol in F344 rats and B6C3F1 mice. J. Appl. Toxicol. **16:** 201–209.
26. KOHN, M.C. & R.L. MELNICK. 2001. Physiological modeling of butadiene disposition in mice and rats. Chem. Biol. Interact. **135:** 285–301.
27. WALKER, V.E., K.Y. WU, et al. 2000. Biomarkers of exposure and effect as indicators of potential carcinogenic risk arising from *in vivo* metabolism of ethylene to ethylene oxide. Carcinogenesis **21:** 1661–1669.
28. GUENGERICH, F.P., S. LANGOUET, et al. 1999. Formation of etheno adducts and their effects on DNA polymerases. *In* Exocyclic DNA Adducts in Mutagenesis and Carcinogenesis, pp. 137–145. IARC Sci. Publ. No. 150. IARC. Lyon.
29. KOSKINEN, M., P. VODICKA & K. HEMMINKI. 2001. Identification of 1-adenine DNA adducts in workers occupationally exposed to styrene. J. Occup. Environ. Med. **43:** 694–700.

30. ZHAO, C., P. VODICKA, et al. 2000. Human DNA adducts of 1,3-butadiene, an important environmental carcinogen. Carcinogenesis **21:** 107–111.
31. KOSKINEN, M. & K. PLNA. 2000. Specific DNA adducts induced by some monosubstituted epoxides *in vitro* and *in vivo*. Chem. Biol. Interact. **129:** 209–229.
32. KOC, H., N.Y. TRETYAKOVA, et al. 1999. Molecular dosimetry of *N*-7 guanine adduct formation in mice and rats exposed to 1,3-butadiene. Chem. Res. Toxicol. **12:** 566–574.
33. MARION, M.J. & S. BOIVIN-ANGELE. 1999. Vinyl chloride–specific mutations in humans and animals. IARC Sci. Publ. **150:** 315–324.
34. BARBIN, A., O. FROMENT, et al. 1997. *p53* gene mutation pattern in rat liver tumors induced by vinyl chloride. Cancer Res. **57:** 1695–1698.
35. WEIHRAUCH, M., M. BENICK, et al. 2001. High prevalence of K-*ras*-2 mutations in hepatocellular carcinomas in workers exposed to vinyl chloride. Int. Arch. Occup. Environ. Health **74:** 405–410.
36. WEIHRAUCH, M., G. LEHNERT, et al. 2000. *p53* mutation pattern in hepatocellular carcinoma in workers exposed to vinyl chloride. Cancer **88:** 1030–1036.
37. BOIVIN-ANGELE, S., L. LEFRANCOIS, et al. 2000. *Ras* gene mutations in vinyl chloride–induced liver tumors are carcinogen-specific, but vary with cell type and species. Int. J. Cancer **85:** 223–227.
38. HONG, H-H.L., T.R. DEVEREUX, et al. 2000. Mutations of *ras* protooncogenes and *p53* tumor suppressor gene in cardiac hemangiosarcomas from B6C3F1 mice exposed to 1,3-butadiene for 2-years. Toxicol. Pathol. **28:** 529–534.
39. SILLS, R.C., H.L. HONG, et al. 1999. High frequency of codon 61 K-*ras* A→T transversions in lung and harderian gland neoplasms of B6C3F1 mice exposed to chloroprene (2-chloro-1,3-butadiene) for 2 years and comparisons with the structurally related chemicals isoprene and 1,3-butadiene. Carcinogenesis **20:** 657–662.
40. GUENGERICH, F.P., D.H. KIM & M. IWASAKI. 1991. Role of human cytochrome P-450 IIE1 in the oxidation of many low molecular weight cancer suspects. Chem. Res. Toxicol. **4:** 169–179.

Primary Prevention Protects Public Health

LORENZO TOMATIS

Cave 25/r, 34011 Aurisina (Trieste), Italy

ABSTRACT: It is widely accepted that epidemiological data provide the only reliable evidence of a carcinogenic effect in humans, but epidemiology is unable to provide early warning of a cancer risk. The experimental approach to carcinogenicity can ascertain and predict potential cancer risks to humans in time for primary prevention to be successful. Unfortunately, only in rare instances were experimental data considered sufficiently convincing per se to stimulate the adoption of preventive measures. The experimental testing of environmental agents is the second line of defense against potential human carcinogens. The first line is the testing of synthesized agents, be these pesticides, medical drugs, or industrial chemical/physical agents, at the time of their development. We do not know, however, how many substances have been prevented from entering the environment because most tests are carried out by commercial or private laboratories and results are rarely released. A better understanding of the mechanisms underlying the sequence of events of the carcinogenesis process will eventually lead to a more accurate characterization and quantification of risks. However, the ways that mechanistic data have been used lately for evaluating evidence of carcinogenicity have not necessarily meant that the evaluations were more closely oriented toward public health. A tendency has surfaced to dismiss the relevance of long-term carcinogenicity studies. In the absence of absolute certainty, rarely if ever reached in biology, it is essential to adopt an attitude of responsible caution, in line with the principles of primary prevention, the only one that may prevent unlimited experimentation on the entire human species.

KEYWORDS: cancer; carcinogenicity; health; human; primary prevention; risk

THE DECLINE OF EXPERIMENTAL CHEMICAL CARCINOGENESIS

Implementation of primary prevention by avoiding exposure to agents and mixtures recognized as human carcinogens and agents for which experimental evidence of carcinogenicity is available has always encountered difficulty; lately, its very principles appear under attack. There have been occasional peaks of absurdity, for instance, that "the only demonstrable usefulness" of the regulation of potential carcinogenic risks "is to the fortunes of assorted bureaucracies," that "science has been abused," and that "conjectural cancer risks" have cost the United States over 700 billion dollars in 1999.[1] The undermining of primary prevention, however, has ancient roots.

Address for correspondence: Lorenzo Tomatis, M.D., Cave 25/r, 34011 Aurisina (Trieste), Italy. Fax: 39 040 202549.
 ltomatis@hotmail.com

Experimental chemical carcinogenesis prevailed in cancer research from 1922, when Passey provided experimental confirmation of the carcinogenicity of soot extracts,[2] until the late 1960s.[3] Cancer research during this long period was oriented towards etiology and primary prevention. In spite of delays, there has been positive achievement of the banning of occupational carcinogens[4] from production and use. One of the main reasons that primary prevention has not obtained greater success is the blockage by industry concerned with identification of their products as potentially carcinogenic to humans, implying costly investments and jeopardizing the profitability of production. Much of the success of primary prevention remains unknown because potentially carcinogenic and mutagenic agents were prevented from entering the marketplace. We also do not have a clear idea of how risk-benefit criteria are used to allow introduction of certain chemicals in spite of evidence of their carcinogenicity and/or mutagenicity.

Epidemiology began to prevail in the assessment of cancer risk to humans after scientists were unable to confirm in experimental animals the evidence of carcinogenicity provided by epidemiological studies. In addition, after having formulated the hypothesis that carcinogenesis is the result of a multistep, multifactorial process,[5,6] experimentalists failed to develop adequate methods for identifying and evaluating roles of various factors in carcinogenesis. The hypothesis was widely accepted, but its validation, following several unconvincing attempts to extrapolate results obtained in the mouse skin model to other organs and species, and the difficulties in identifying promoters other than croton oil and TPA, relied perhaps more on epidemiological than experimental data. In the late 1960s, statisticians and epidemiologists developed criteria for assessing the causation of chronic diseases that took into consideration biological plausibility, but which were based primarily, if not entirely, on epidemiological evidence.[7]

In the early 1970s, after the premise that epidemiological data alone could provide proof of causation had become accepted, a further concept was introduced that only the epidemiological approach could provide convincing evidence of a causal association between exposure and human cancer.[3] The corollary, in spite of abundant evidence to the contrary,[8,9] was that experimental results are not valid predictors of similar effects in humans and cannot substitute for observations in humans. As epidemiological studies cannot give early warning of a cancer risk, acceptance of the corollary is equivalent to accepting that a potential hazardous effect of an environmental agent can be assessed only *a posteriori*, after the agent has had time to cause its harmful effects. Even in the face of this situation, the greatest fear of many epidemiologists appears to be that they may be accused of producing false-positive results. The production of false-negative results does not seem to be as troublesome, but public health is certainly not well served if people are allowed to be exposed to carcinogens based on allegedly false negatives.

BEHIND THE SHINING SHIELD

Experimental results, having been obscured by the overwhelming predominance of epidemiological findings, have regained importance and significance in recent years. Nevertheless, behind the shining shield of a basic research that has produced spectacular results in molecular biology and genetics, a negative attitude towards primary prevention is surging. Primary prevention, the argument goes, may become useless in view of the continuous progress of diagnostic capacity and therapeutic efficiency, even if, in spite of such advances, it would still seem preferable not to develop a cancer in the first place. Moreover, because measures of primary prevention may impose restrictions on the expansion of industrial production and restraints on consumption, they may seem to be having a negative impact on the economy. Presently, the growing consumption of costly drugs is already resulting in a steep rise in the cost of health.[10,11]

PRETEXTS FOR IGNORING SUFFICIENT WARNINGS

A rare example in which experimental results were considered, at least in one country, as sufficient *a priori* warning of similar effects in humans is that of 4-aminobiphenyl, a chemical that was produced and used in the United States from 1935 to 1955. The results of a study on the carcinogenicity of 4-aminobiphenyl in rats published in 1952[12] and confirmed two years later in dogs[13] were the main reason why the compound was not produced or used in the United Kingdom. In the United States, the first report of cases of urinary bladder cancer attributed to 4-aminobiphenyl was published in 1955[14] and, subsequently, the production of the chemical was stopped.

In many more situations, experimental evidence of carcinogenicity was ignored. Diethylstilbestrol was shown to produce tumors in mice in the 1930s and in several other animal species in the 1950s.[15] This notwithstanding, it became a popular drug for administration to women of reproductive age, with known consequences to their offspring.[16] Another example was an unusual clustering of lung cancer cases noted among workers involved in the production of bis(chloromethyl) ether (BCME) in the early 1960s. No attention was paid to the evidence of its carcinogenicity, first reported in 1968 and confirmed in 1969,[17] after dermal application or subcutaneous injection in mice. No preventive measures were taken until the 1970s, when rats were reported to have developed lung and nasal cavity tumors after receiving BCME by inhalation,[18] the main route by which humans were presumed to be exposed.

The pretext for ignoring the evidence of the initial long-term tests was that the route of exposure and the tumor type induced were different from those in humans. Maltoni's demonstration of the multipotential carcinogenicity of benzene given to mice and rats by the oral route[19] has provided clear evidence of the absurdity of this pretext. In other instances, the pretext was the dose, described as being unrealistically high or given for too long a time. It is not clear whether results obtained in genetically modified mouse models, an area presently of intense study and in rapid evolution,[20,21] will be considered more trustworthy than those obtained in traditional long-term bioassays.

THE USE OF MECHANISTIC DATA

In the face of uncertainty in evaluating the extent and sometimes even the existence of increased risk, it is reasonable and almost unavoidable to turn to mechanistic studies in the hope that they might provide useful clues or elements that would refute or strengthen the biological plausibility of an association between exposure and increased risk for cancer. Placing all the emphasis on elucidating the pathogenesis of a disease, however, can lead to forgetting the primary goal: understanding the etiology of the disease.

After a long debate, the International Agency for Research on Cancer (IARC) in 1992 included relevant data on mechanisms among the criteria used to evaluate carcinogenicity.[22] Subsequently, an agent could be included in Group 1, "agents definitely carcinogenic to humans," when the "evidence of carcinogenicity in humans is less than sufficient, but there is sufficient evidence of carcinogenicity in experimental animals and strong evidence in exposed humans that the agent or mixture acts through a relevant mechanism of carcinogenicity". This additional criterion was used to include ethylene oxide[23] and dioxins[24] among the recognized human carcinogens, but it was not used similarly for 1,3-butadiene, in spite of the fact that the experimental and human evidence, as well as the data on the mechanism of action, pointed overwhelmingly to its carcinogenicity in humans,[25] as was acknowledged fully in the *Ninth Report on Carcinogens*.[26]

A better understanding of the mechanisms underlying the sequence of events of the carcinogenesis process will eventually lead to a more accurate characterization and quantification of risks, but it is not at all evident that, at present, mechanistic data can prove the nonexistence of a risk. Lately, they have been used to downgrade the classifications of several agents based on the claim that the experimental evidence, even if it is sufficient per se, is not relevant if "the mechanism of carcinogenicity in experimental animals does not operate in humans."[27] With the use of mechanistic data based largely on hypotheses that have not been adequately tested experimentally, atrazine, saccharin, di(2-ethylhexyl)phthalate, glass wool, and rock wool have been downgraded from IARC Group 2B, "possibly carcinogenic," to Group 3, "not classifiable as to their carcinogenicity to humans." In assigning these agents to the latter group, their production and use are unrestricted; however, it may be necessary to impose strict regulations to avoid possible ominous consequences on public health.

A LARGE PARKING LOT

The combination of the sometimes exaggerated caution of epidemiologists and the rising mistrust in the results of long-term tests for carcinogenicity is also at the origin of the large number of agents now included in Group 2B of the IARC classification of carcinogens. This group contains agents classified as "possibly carcinogenic to humans" on the basis of either limited evidence of carcinogenicity in humans and less than sufficient evidence in experimental animals, or of inadequate evidence in humans and sufficient evidence in experimental animals. Originally, Group 2B was conceived as a temporary grouping of agents for which further investigation was urgently needed. It has now become a huge parking lot, with over 200 agents or mixtures of agents, in which the available data are of variable quality and quantity. The possibility that new data will become available on these agents in the near future now appears remote.[28] Apart from the initiatives promoted or coordinated by the National Toxicology Program and the bioassays carried out at the Ramazzini Institute, few short- and long-term carcinogenicity tests have been carried out by independent laboratories or institutes. Agents and mixtures in IARC Group

2B have not raised the interest of epidemiologists, given the difficulties of designing adequate studies in several instances. For many of these agents and mixtures, epidemiological data are absent or inadequate, but they cannot be held equivalent to negative findings, nor can they be considered less relevant to public health than positive experimental results.

HESITANCY AND CONTRASTING RESULTS

The inherent limitations of the epidemiological approach may lead to hesitancy in reaching definitive conclusions. In a WHO publication prepared by a group of experts chaired by Richard Doll in 1964, air pollution was considered a relevant cause of lung cancer in humans.[29] A few years later, however, Richard Doll, after pointing to the preponderant role of tobacco smoke in the causation of lung cancer, attributed a minimal role to environmental factors, including air pollution.[30] The seesaw attitude to the evidence for a cancer risk from air pollution has persisted for 20 years. As soon as an epidemiological survey is published showing that exposure to air pollutants is associated with increased mortality from cardiovascular disease and lung cancer, a number of papers reporting ambiguous or allegedly negative results suddenly appear and cast doubt on the association. Several tactics have been used to undermine sound scientific results or to delay their official acceptance: research may be sponsored to counter adverse scientific evidence, research results may be directly or indirectly altered, or scientists may be openly or surreptitiously hired to contradict sound scientific results.[31] Most recently, convincing evidence for a determining role of inhaled small particles in increasing mortality from cardiovascular disease and lung cancer has received considerable attention,[32,33] and additional evidence of adverse health effects due to exposure to multiple pollutants has become available.[34] It is to be hoped that proper measures to decrease air pollution will finally be implemented, but one wonders why it took so long.

The saga of the possible role of organochlorine compounds and of environmental agents in general in increasing the risk for breast cancer[35] has been running for decades. As for air pollution, publication of results indicating a causal association has generally been followed by publication of doubtful or allegedly negative results. It was reported recently that the concentrations of certain polychlorinated biphenyls in plasma lipids were significantly associated with an increased risk for breast cancer,[36] and more attention is being given to the role of environmental chemicals.[37] It remains to be seen whether these findings will initiate a new phase in research into the etiology of breast cancer.

Electromagnetic fields (EMFs) provide another example of controversy. There is apparent consensus on the epidemiological evidence for an increased risk of childhood leukemia, with a far from negligible relative risk of 2.0 for postnatal exposure to 0.4 μT.[38,39] Although it has been stated authoritatively that the energy of the radiation emitted by EMFs is insufficient to damage DNA directly and that EMFs cannot therefore have tumor initiation activity, at the same time the need to identify possible mechanisms of action has been emphasized. A severe limitation of the experimental approach for studying the possible carcinogenic effect of EMFs is use of the same criteria traditionally applied to study the carcinogenicity of chemical agents. In the sequence of events that lead to malignant transformation, mechanisms other than direct interaction with or damage to DNA may be involved. Thus, experimental approaches should be used that can support or refute alternative mechanisms that could modify the risk of cancer. In the absence of a mechanism that can satisfactorily justify the increased risk, EMFs have joined the hundreds of agents in IARC Group 2B. However, in contrast to its companions in this large parking lot, EMF has actually stimulated the interest of epidemiologists and, importantly, the release of funds for additional research.

Another relevant example of uncertainty in etiological research is non-Hodgkin's lymphoma (NHL), whose incidence has continued to increase throughout the past quarter of a century. While there has been remarkable progress in understanding what is going wrong within the cells, not much progress has been made in identification of what induces the cells to malfunction. Several risk factors have been incriminated, including nitrites in drinking water, EMFs, hair colorants, dietary habits, and (above all) occupational exposure to organic solvents, pesticides, and herbicides. Whatever the role of each of these factors, none alone can clearly explain the continuous, widespread increase in the incidence of NHL. In the absence of certainty, one hypothesis for the increase in incidence that is attracting attention is that of mild immunodepression combined with occasional stimulation of the immune system. This hypothesis implies acceptance that something is happening to the entire population that is difficult to identify because the epidemiological tools at our disposal are incapable of measuring the effects of risks of very low potency.

PRIMARY PREVENTION IN THE ABSENCE OF ABSOLUTE CERTAINTIES

We may find ourselves facing a choice between an active attitude, expressed as the adoption of measures of primary prevention in the absence of certainties, and a passive attitude that finds in the etiological uncertainties justification to disregard prudent primary prevention. In the case of NHL, the first attitude would involve drastic measures to reduce the use of pesticides, herbicides, and organic solvents; evacuation, at least temporarily, of residences with high levels of EMFs; and orientation towards a utopic, but essential, reduction in consumption. Clearly, this first choice reflects the widely discussed precautionary principle,[40] which indicates that urgent intervention is justified in the face of a potentially serious risk even in the absence of incontestable scientific evidence of a cause–effect relationship.

A cautious, prudent attitude is sometimes interpreted as antitechnological and antiscientific. In fact, those who champion an attitude of caution are simply recognizing that predictive knowledge in most instances is of lesser quality and remains at a lower level than technological knowledge. Recognition of our limited capacity to predict the long-term consequences of our knowledge can only lead to learning more and, thus, it represents a stimulus, and certainly not an impediment, to research.[41]

By adopting an attitude of responsible caution, we also accept that we have a duty to provide accurate information on possible or potential risks and to prevent relevant data from being ignored or concealed. Only with such an attitude can we avoid the entire human species being exposed to everything that technological progress can invent.

ACKNOWLEDGMENTS

I thank Ron Melnick and James Huff for their thoughtful comments and Elisabeth Heseltine for editing the manuscript.

REFERENCES

1. GORI, G.B. 2001. The costly illusion of regulating unknowable risks. Regul. Toxicol. Pharmacol. **34:** 205–212.
2. PASSEY, R.D. 1922. Experimental soot cancer. Br. Med. J. **ii:** 1112–1113.
3. TOMATIS, L. & J. HUFF. 2002. Evolution of research in cancer etiology. *In* The Molecular Basis of Human Cancer, pp. 189–201. Humana Press. Totowa, NJ.
4. MONTESANO, R. & L. TOMATIS. 1977. Legislation concerning chemical carcinogens in several industrialized countries. Cancer Res. **37:** 310–316.
5. BERENBLUM, I. 1941. The mechanism of carcinogenesis: a study of the significance of carcinogenic action and related phenomena. Cancer Res. **1:** 897–914.
6. BERENBLUM, I. & P. SHUBIK. 1947. A new quantitative approach to the study of the stages of chemical carcinogenesis in the mouse's skin. Br. J. Cancer **1:** 383–391.
7. BRADFORD-HILL, A. 1965. The environment and disease: association or causation? Proc. R. Soc. Med. **56:** 295–300.
8. TOMATIS, L. 1979. The predictive value of rodent carcinogenicity tests in the evaluation of human risk. Annu. Rev. Pharmacol. Toxicol. **19:** 511–530.
9. HUFF, J. 1993. Chemicals and cancer in humans: first evidence in experimental animals. Environ. Health Perspect. **100:** 201–210.
10. CHARATAN, F. 2001. US healthcare spending to rise sharply. Br. Med. J. **322:** 692.
11. KRUGMAN, P. 2002. Wealth versus health. New York Times, 19 April 2002.
12. WALPOLE, A.L., M.H.C. WILLIAMS & D.C. ROBERTS. 1952. The carcinogenic action of 4-aminodiphenyl and 3:2′-dimethyl-4-aminodiphenyl. Br. J. Ind. Med. **9:** 255.
13. WALPOLE, A.L., M.H.C. WILLIAMS & D.C. ROBERTS. 1954. Tumours of the urinary bladder in dogs after ingestion of 4-aminodiphenyl. Br. J. Ind. Med. **11:** 105.
14. MELICK, M.R., L.G. KOSS, A. RICCI, *et al.* 1955. The first reported cases of human bladder tumours due to a new carcinogen—xenylamine. J. Urol. **74:** 760.
15. LACASSAGNE, A. 1938. Apparition d'adenocarcinomes mammaires chez des souris males traitees par une substance oestrogene synthetique. C. R. Soc. Biol. (Paris) **129:** 641–643.
16. IARC. 1979. Monographs on the Evaluation of the Carcinogenic Risk of Chemicals to Humans. Vol. 21. Sex Hormones. IARC. Lyon.
17. VAN DUUREN, B.L., A. SIVAK, B.M. GOLDSCHMIDT, *et al.* 1969. Carcinogenicity of halo-ethers. J. Nat. Cancer Inst. **48:** 1431.
18. LASKIN, S., M. KUSHNER, R.T. DREW, *et al.* 1971. Tumors of the respiratory tract induced by inhalation of bis(chloromethyl) ether. Arch. Environ. Health **23:** 135.
19. MALTONI, C., B. CONTI & C. SCARNATO. 1983. Benzene: a multi-potential carcinogen—results of long-term bioassays performed at the Bologna Institute of Oncology. Am. J. Ind. Med. **4:** 441–445.
20. TENNANT, R., J. HASEMAN & R.E. STOLL. 2001. Transgenic assays and the identification of carcinogens. Environ. Mol. Mutagen. **37:** 86–91
21. JONKERS, J. & A. BERNS. 2002. Conditional mouse models of sporadic cancer. Nat. Rev. **2:** 251–265.
22. VAINIO, H., P. MAGEE, D. MCGREGOR & A. MCMICHAEL, Eds. 1992. Mechanisms of Carcinogenesis in Risk Identification. IARC Sci. Pub. No. 116. IARC. Lyon.
23. IARC. 1994. Monographs on the Evaluation of Carcinogenic Risks to Humans. Vol. 60. IARC. Lyon.
24. IARC. 1997. Monographs on the Evaluation of Carcinogenic Risks to Humans. Vol. 69. IARC. Lyon.
25. IARC. 1999. Monographs on the Evaluation of Carcinogenic Risks to Humans. Vol. 71. IARC. Lyon.

26. NATIONAL TOXICOLOGY PROGRAM. 2000. Ninth Report on Carcinogens. U.S. Dept. of Health and Human Services, NTP. Research Triangle Park, NC.
27. TOMATIS, L. 2002. The IARC Monographs Programme: changing attitudes towards public health. Int. J. Occup. Environ. Health **8:** 155–163.
28. KARSTADT, M. 1998. Availability of epidemiologic data for chemicals known to cause cancer in animals: an update. Am. J. Ind. Med. **34:** 519–525.
29. WORLD HEALTH ORGANIZATION. 1964. Prevention of cancer. WHO Tech. Rep. No. 276. Geneva.
30. DOLL, R. & R. PETO. 1981. The causes of cancer: quantitative estimates of avoidable risks of cancer in the United States today. J. Natl. Cancer Inst. **66:** 1191–1308.
31. ROSENSTOCK, L. & L. JACKSON LEE. 2002. Attacks on science: the risks to evidence-based policy. Am. J. Public Health **92:** 14–18.
32. POPE, C.A., R.T. BURNETT, N.J. THUN, et al. 2002. Lung cancer, cardiopulmonary mortality, and long-term exposure to fine particulate air pollution. JAMA **287:** 1132–1141.
33. PYNE, S. 2002. Small particles add up to big disease risk. Nature **295:** 1994.
34. KYLE, A.D., T.J. WOODRUFF, P.A. BUFFLER & D.L. DAVIS. 2002. Use of an index to reflect the aggregate burden of long-term exposure to criteria air pollutants in the United States. Environ. Health Perspect. **110:** 95–102.
35. HUFF, J., J.A. BOYD & J.C. BARRETT. 1996. Hormonal carcinogenesis and environmental influences. Prog. Clin. Biol. Res. **394:** 3–23.
36. DEMERS, A., P. AYOTTE, J. BRISSON, et al. 2002. Plasma concentrations of polychlorinated biphenyls and the risk of breast cancer: a congener-specific analysis. Am. J. Epidemiol. **155:** 629–635.
37. DEBRUIN, L. & P.D. JOSEPHY. 2002. Perspectives on the chemical etiology of breast cancer. Environ. Health Perspect. **110:** 119–128.
38. NIEHS. 1998. Report on Health Effects from Exposure to Power-line Frequency Electric and Magnetic Fields. NIH Pub. No. 1.
39. IARC. 2002. Monographs on the Evaluation of Carcinogenic Risks to Humans. IARC. Lyon. In press.
40. KRIEBEL, D., J. TICKNER, P. EPSTEIN, et al. 2001. The precautionary principle in environmental science. Environ. Health Perspect. **109:** 871–876.
41. JONAS, H. 1990. Il Principio Responsabilità [Italian text]. Einaudi. Turin.

The National Toxicology Program Rodent Bioassay

Designs, Interpretations, and Scientific Contributions

JOHN R. BUCHER

National Toxicology Program, National Institute of Environmental Health Sciences, Research Triangle Park, North Carolina 22709, USA

ABSTRACT: The National Toxicology Program rodent cancer bioassay program design evolved from that of the National Cancer Institute in the 1970s. Groups of 50 or more mice are assigned to control or treatment groups. Test substances are given at three dose levels by intubation, dietary or drinking water consumption, or dermal or inhalation exposure. Dosing starts at age 5–6 weeks and lasts for 2 years, when surviving animals receive a complete histopathologic examination. Statistical approaches accommodate survival differences and no longer require differentiation between fatal and incidental tumors. Photocarcinogenicity studies, employing SKH-1 hairless mice, evaluate onset of skin papillomas and incidences at 1 year. Top doses are chosen to expose animals to a minimally toxic challenge and lower doses to operate within the linear range of kinetics. This dosing allows comparison of results across studies. Bioassay and ancillary studies successfully identify tumor-causing agents in rodents, provide information on dose–response, and characterize other chemical-related toxicities. NTP and Ramazzini Foundation bioassay designs differ in several aspects, but bioassays at both institutions provide chemical-specific information for predicting human carcinogens, thus providing for protection of public health. Bioassays constitute an essential information reference set for new assay development and further investigations into mechanisms of action. The scientific community and the public owe a huge debt of gratitude to Dr. Cesare Maltoni of the European Foundation of Oncology and Environmental Sciences and to Dr. David P. Rall of the National Institute of Environmental Health Sciences for their foresight and wisdom in creating and nurturing these bioassay programs.

KEYWORDS: National Toxicology Program; rodent; bioassays; carcinogenesis; phototoxicology; photocarcinogenicity; dioxins

THE NTP RODENT BIOASSAY

As of April 2002, the National Cancer Institute and subsequently the National Toxicology Program (NTP) had published a combined total of 505 Technical Reports presenting the results of two-year rodent cancer studies on discrete substances,

---

Address for correspondence: John R. Bucher, Ph.D., National Toxicology Program, National Institute of Environmental Health Sciences, 79 T.W. Alexander Drive, Building 4401, P.O. Box 12233, Research Triangle Park, NC 27709. Voice: 919-541-4532; fax: 919-541-4255.
bucher@niehs.nih.gov

mixtures, or physical agents. An additional 70 studies are either completed or ongoing. The substances studied have included a variety of industrial chemicals, drugs, pesticides, common contaminants in air and water, metals, hormones, natural toxins, and selected physical agents such as electromagnetic fields. Recently, many unregulated herbal medicines and dietary supplements, a number of drinking-water disinfection byproducts, and some DNA-based therapeutic agents have been entered into study. Other substances are being evaluated in our new Phototoxicology Laboratory at the National Center for Toxicological Research (NCTR), in Jefferson, Arkansas (USA). The NTP strives to initiate and report the findings for 8 to 10 substances studied in traditional chronic rodent cancer assays per year.

Creation and Elements of Study Designs

The designs of the bioassays have evolved from the time of initiation of this program in the mid 1970s to the present, but retain a basic set of common elements. Groups of 50 to 100 males and females of selected strains of rats and mice, typically the F344/N inbred rat and the B6C3F1 hybrid mouse, are randomly assigned to receive the chemicals at one of three dose levels or to a control group that receives no added chemicals. Chemicals are administered by oral intubation, consumption in diet or drinking water, or dermal or inhalation exposure. The in-life portion of the studies starts at 5 to 6 weeks of age and lasts for 2 years (or 78 weeks for some early mouse bioassays), at which time, surviving animals are killed and evaluated by gross observation and complete histopathologic examination. Statistical approaches to data analysis have evolved to accommodate survival differences between groups, and no longer require different methods for tumors considered fatal vs. incidental ones. Phototoxicology studies usually employ the SKH hairless mouse and evaluate skin papillomas as the major end point in studies of 1 year's duration.

Although simple in concept, the "typical" NTP bioassay is in reality a highly complicated endeavor. After a multistep interagency government review and selection process, selected substances are assigned to a lead study scientist. The study scientist assembles a design team comprising experts in areas of carcinogenesis, toxicology, pathology, chemistry, etc., that are relevant to the substance at hand. Appropriate literature is searched, manufacturers and regulators consulted, and a complete evaluation of the existing toxicological and carcinogenic understanding of the substance is assembled. Study protocols addressing data needs are developed and reviewed. A subset of specialists proficient in metabolism and kinetics also reviews the literature on the substance, and designs any studies that might be needed to characterize quantitative dose–response relationships for any anticipated toxic effects. These scientific teams remain active for the duration of the studies and make periodic adjustments to the designs as warranted.

Initial chemistry activities involve development of analytical tools for identity and purity analyses, assessments of stability and homogeneity in various delivery vehicles, exposure generation and monitoring for inhalation studies, and methods for extraction and analysis of parent compounds and major metabolites from biological fluids and tissues. Work is also begun on studies to address absorption, distribution, metabolism, and elimination of the substance, as well as to establish basic kinetic parameters.

Pre-chronic toxicology studies typically involve short-term studies (14- or 28-day exposures), followed by studies of 3 months' duration. These studies and the 2-year rodent cancer bioassays are performed in independent contract laboratories or at the NCTR and follow a stringent statement of work for performance. The statement of work and accompanying standard operating procedures detail acceptable laboratory design and study performance standards for all aspects of the studies including such things as laboratory animal management and disease surveillance, animal source, diet, sanitation methods, pathology, and electronic data capture, to name a few items. All studies are done according to Good Laboratory Practices guidelines (http://www.fda.gov/ora/compliance_ref/bimo/7348_808/part_I.html).

The 3-month studies include clinical pathology measurements taken at several times, as well as assessments of gross behavioral changes, body and organ weights, food and water consumption, and sperm morphology and estrous cycling measures. These assessments serve as screens to identify the need for targeted in-depth study of neurobehavior, immunotoxicity, or reproductive and developmental effects under other protocols. Red blood cells and bone marrow are evaluated for evidence of micronuclei as an indication of genetic damage. Additional chemical-specific measures of such things as tissue glutathione depletion or CYP450 activities are frequently performed during the 3-month studies. Doses chosen typically range from no-adverse-effect levels to those that show some toxicity greater than that which would be tolerated in a 2-year study.

Dose Selection for the 2-Year Bioassay

Once a complete set of information is assembled from the pre-chronic toxicology and kinetic studies, designs are established for the 2-year rodent bioassay. The selection of doses for the assay is one of the most critical decisions in the success or failure of the assay. Because of the inherent limitations in statistical power of any bioassay employing from 50 to 100 animals per sex/species and dose group,[1] it is essential to expose the animals to top doses that provide a robust, but minimally toxic challenge (minimally toxic dose [MTD]), and to lower doses that operate within the linear range of kinetics for absorption, metabolism, and elimination, and that do not overly perturb the basic hormonal and physiological status of the animal. The issue of dose selection for chronic rodent cancer studies has been exhaustively reviewed. Interested readers should consult more complete references.[2–4]

Pathologists working in the NTP have become adept at understanding and predicting the typical progression of lesions in a variety of organs during a 2-year study. Examples of toxic lesions in various organs that, if observed in a 3-month study, would eliminate that dose from consideration for a longer study have been described[2,3] (TABLE 1). Consistent use of this working knowledge from bioassay to bioassay allows one to compare results with some assurance that the doses used were selected according to a common philosophy. This ability to compare one study to the next adds considerably to the value of the collective data produced by the NCI/NTP bioassay program.

Target-tissue pathology is the determining factor for dose selection for only about one-half of the bioassays performed by the NTP. Various other factors influence top dose selections in 2-year rodent studies. These can range from clinical signs such as agitation or narcosis, through acute mortality, to poor body weight gain from general

TABLE 1. Examples of "acceptable" and "unacceptable" pathology in prechronic studies for selection of the top dose for a 2-year bioassay

Organ	Pathology
Liver	*Acceptable*: minimal to mild hepatocellular hypertrophy; pigmentation of hepatocytes and/or pigmentation of Kupffer cells
	Unacceptable: hepatocellular necrosis, inflammation; cytoplasmic degeneration/vacuolization and/or bile duct or oval cell hyperplasia
Kidney	*Acceptable*: minimal to mild karyomegaly; pigmentation and papillary edema or a slight increase in nephropathy
	Unacceptable: increased nephropathy associated with distinct necrosis of the tubular epithelium or the presence of protein casts and/or papillary necrosis

toxicity or poor palatability of dosed diet or water. For nontoxic agents, there are a number of technical limitations based on practical considerations including the explosive limit concentrations in inhalation studies, the solubility of the chemical in dosing vehicles for oral or dermal administration studies, and the simple volumes of materials that can be given to rats and mice by gavage or in skin-painting procedures.[3] In phototoxicology and photocarcinogenicity studies, use is made of the experimentally determined MED, or minimal erythemal dose, as a way of assuring comparable light challenge and attendant skin pathology from study to study.

Selection and spacing of lower doses also depend on a number of factors including the kinetics or bioaccumulative properties of the substance, the certainty one has in the adequacy of the top dose, and the limited statistical power of the assay. Gaylor[5] has pointed out the statistical difficulty of detecting a tumor response at a dose of 1/4 or less of the MTD when the response is sublinear.

Pathology Reviews

A second crucial aspect of NTP studies is the attention paid to uniform and consistent study evaluation and interpretations. Use of the guidelines set out by the Pathology Working Group (PWG) has been central to the effort to standardize diagnoses, and to ensure consistency from study to study. This process, described by Ward *et al.*,[6] is the capstone of a three-tiered pathology evaluation procedure involving independent assessments by the laboratory study pathologist, a review of all tumors and all tissues with possible treatment-related effects by a quality-assurance pathologist, and finally a review of representative lesions and all diagnostic discrepancies by a panel of expert pathologists (PWG). Only after this process is completed are final histopathology tables prepared. In recent years, many of the tumors observed in 2-year studies have been probed for selected genetic lesions in oncogenes or tumor suppressor genes. In this way, hints have been obtained about the mechanisms of carcinogenesis involved.[7,8]

Study Interpretation and Reporting

The interpretation of tumor responses in 2-year assays is an art. Although statistical significance is a critical consideration, non-significant tumor responses have been determined to be chemically related in certain instances, and statistically significant responses have been discounted in others. All relevant information is considered, including evidence of biological response in the target organ, presence of preneoplastic changes, and the rarity of the tumor in relation to the historical database, as well as such things as tumor multiplicity.

The methods, results, and interpretations of the NTP toxicology and carcinogenesis studies are prepared by the study scientists, reviewed and revised extensively by staff, and finally made available on the World Wide Web and in print for public inspection in the form of draft technical reports. These draft reports are reviewed by the Technical Reports Subcommittee of the NTP's Board of Scientific Counselors in open public sessions once or twice per year. The subcommittee is composed of scientists knowledgeable about rodent cancer bioassays, toxicology, statistics, pathology, and other relevant disciplines. They review the reports for scientific adequacy and for assignment of the study to one of a number of evidence categories, ranging from "no evidence," through "equivocal," "some," or "clear" evidence for carcinogenic activity. The draft reports are then revised and published as part of the NTP Technical Report Series.

Novel Uses of the Rodent Bioassay

In recent years, the bioassay has been increasingly used to address fundamental issues in toxicology and carcinogenesis. Studies have been designed and are under way to examine two of the more important and far-reaching issues related to current regulatory policy for exposures to environmental contaminants. These are the questions concerning the use of the Toxic Equivalence Factor[9] (TEF) approach for regulating the cancer hazard for 2,3,7,8-tetrachlorodibenzo-*p*-dioxin (TCDD) and the hundreds of chemicals and mixtures in the environment that exhibit weak dioxin-like agonistic and antagonistic activity, and the uncertainty over the potential hazards of exposure to endocrine active agents during critical windows of development.

The dioxin TEF project comprises a series of seven bioassays using the female Sprague-Dawley rat.[9] Several hundred polyhalogenated aromatic compounds share biological response characteristics with dioxins. Such compounds include certain polychlorinated and polybrominated biphenyls, polychlorodibenzofurans, and other polychlorinated dibenzo-*p*-dioxins. TEFs have been generated for many of these chemicals, primarily through short-term studies of noncancer end points that compared the chemical's potency in producing pleiotropic responses to that of dioxin. Although studies to date have indicated a good correspondence of TEFs to cancer promotion in initiation–promotion studies, this correspondence has never been adequately tested using the traditional rodent bioassay.

FIGURE 1. Schematic design of exposures in a multigenerational study of endocrine-active agents under way at the NCTR, illustrating how the bioassay can be made part of a larger investigation of cancer and noncancer end points. Each generation is evaluated for a variety of developmental landmarks, focusing on the reproductive system and on neurobehavioral and other measures from birth through pnd 140. Subsets of the F1 generation are carried on through a chronic bioassay with and without continuous dosing. Animals in the F4 generation, who experienced exposure to the chemical only by virtue of gestational exposure to their parents, are also evaluated for the expression of tumors. Genistein and ethinyl estradiol are being evaluated using this design.

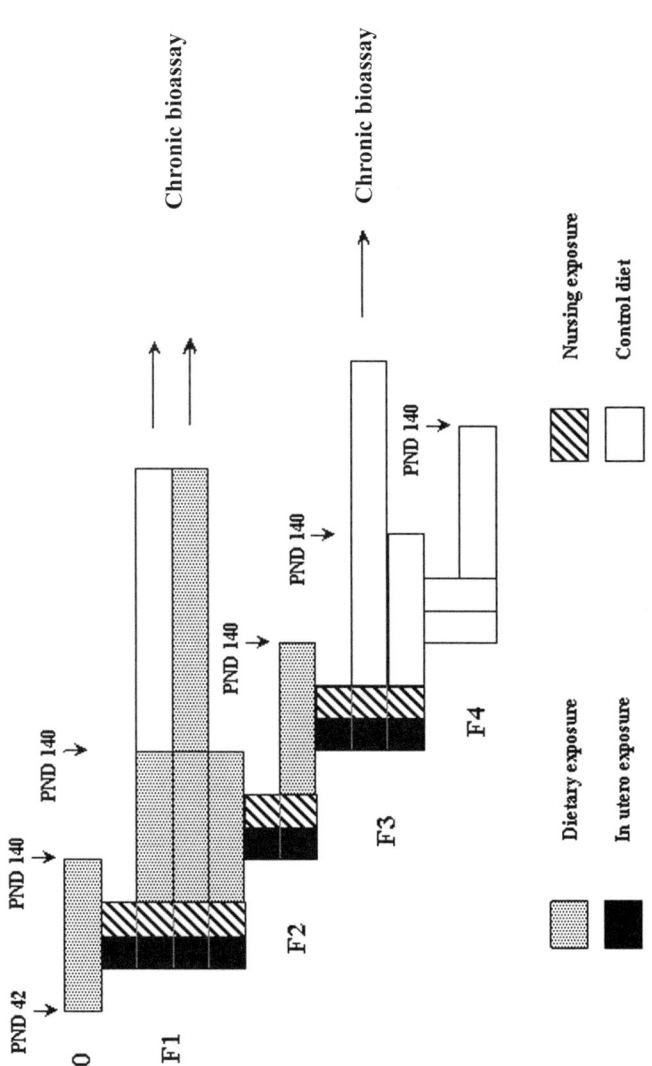

FIGURE 1. *See previous page for legend.*

Under study are the major dioxin-like chemicals found in human adipose tissue. These include 2,3,4,4′,5-pentachlorobiphenyl (PCB 118), 2,3,4,7,8-pentachlorodibenzofuran, and 3,3′,4,4′,5-pentachlorobiphenyl (PCB 126), as well as TCDD. A noncoplaner, nondioxin-like PCB, 2,2′,4,4′,5,5′-hexachlorobiphenyl (PCB 153), is also being studied alone and in a binary combination with TCDD to examine possible synergism or antagonism for the carcinogenic effects of TCDD. Finally, a combination of all chemicals other than PCB 153 is under test as a representative mixture of dioxin-like chemicals. Doses were selected to provide a rigorous test of the relationship between TEFs and carcinogenic potency. Selected tissue concentrations of the chemicals are being determined along with a subset of measures typically used to determine TEFs at various points during the studies such as CYP 1A1, 1A2, 1B1, thyroid hormone levels, and others.

Because of the potential importance of these studies in determining future regulatory approaches to dioxin-like chemicals, the NTP has made available tissues from these studies as well as a limited number of small research grants (RO3) through the NIEHS extramural grants program. This RO3 program is providing data to provide a firmer foundation for the TEF concept and its relationship to the carcinogenic activity of TCDD.

The NTP has also designed and is carrying out a series of multigenerational assays at the NCTR examining the effects of varying doses of estrogenic agents and one antiandrogen, given in the diet to Sprague-Dawley rats. Agents under study include the soy isoflavone genistein, the detergent byproduct nonylphenol, the drug ethinyl estradiol, and the fungicide vinclozolin. Doses range from those that produce observable adverse reproductive effects, down to levels considered relevant from environmental exposures, or drug use. The study outline is shown in FIGURE 1. End points under study include a range of developmental landmarks particularly focusing on the reproductive tract, and such things as nipple retention in males. Various sexually dimorphic behaviors such as salt and saccharin solution drinking preference are being assessed, as are anatomical brain regions that differ in males and females. Immune function and learning and memory measures were evaluated in range-finding studies, but the findings did not warrant carrying these evaluations into the multigenerational study. Note that the F1 and F3 generations involve 2-year bioassays. For the F1 generation, animals are being carried for 2 years both with and without continuous dosing. Animals in the F4 generation were dosed only through gestational and lactational exposure to their parents. This design is generating data to address questions about the potential for adverse effects, including endocrine-mediated cancers, that result from exposure to endocrine-active agents given at a range of doses to animals during all life stages.

CONTEMPORARY ISSUES IN BIOASSAY DESIGN

The NTP bioassay of today represents the culmination of many design decisions reached over the past 30 years. The assay remains fundamentally able to do the three things that are asked of it: (1) identify substances that cause tumors in rodents; (2) provide information on dose–response for tumors; and (3) characterize other chemical-related toxicities. However, the current design is not the only possible way of doing this. We at the NTP have had the pleasant opportunity of entering into extensive discussions with Drs. Soffritti and Belpoggi of the European Foundation of Oncology and Environmental Sciences (Ramazzini Foundation) under an agreement for scientific collaboration between the Ramazzini and the NIEHS/NTP. This agreement was established by Professor Cesare Maltoni and Dr. Ken Olden in June of 2000. As we have discussed projects of mutual interest, we have also explored the reasons for the differences in basic bioassay designs between the NTP and Ramazzini Foundation studies. As a basis for the following talks and panel discussions, it is instructive to highlight some of these differences and to reflect on how they have come about.

Before doing so, I should caution that there is not necessarily a wrong or a right way to design and perform a bioassay. Each approach has strengths and weaknesses, and this comparison is presented in the hope that the weaknesses can be minimized and the strengths maximized as we go forward.

The NTP exposes male and female F344/N rats and B6C3F1 mice to potential carcinogens for 2 years, employing an aggressive moribund sacrifice program to terminate animals subjectively judged close to death. The Ramazzini Foundation[10] exposes male and female Sprague-Dawley rats to potential carcinogens from weaning for 2 years, and then the animals are allowed to live out their natural lives. In some cases, the pregnant dams are exposed to the agent as well. Dose group numbers are similar in the Ramazzini Foundation and the NTP studies, with 50 animals being typical. The NTP employs randomization of weanlings to dose groups following stratification by weight, with very light or heavy animals excluded from study. The Ramazzini Foundation assigns litters to the same dose group and uses all animals, while keeping track of litter identification information. Dose selection in the NTP studies is keyed to the MTD as judged from pre-chronic studies. Dose selection in the Ramazzini Foundation also seeks to provide a minimally toxic challenge to the animals, but is usually based on information gleaned from the literature. The NTP uses a three-step pathology review employing non-government expert pathologists as needed. The Ramazzini Foundation follows a similar multi-staged review of diagnoses, but uses staff from within the Institute. The NTP has an active disease surveillance program. Data from NTP studies are widely available, largely in electronic form. Efforts are under way to similarly collect pathology data from Ramazzini Foundation studies on the NTP Toxicology Data Management System. The primary statistical method currently employed by the NTP (i.e., the poly K test) is a survival adjusted quantal-response procedure that assigns a weighted risk to each animal in the study based on its survival time and an expected lesion prevalence rate. The primary statistical tests used by the Ramazzini Foundation are the prevalence analysis for non-lethal tumors and the log rank test of Mantel and Cox. The Ramazzini Foundation assigns significance to differences in total benign and malignant tumor-bearing animals and total tumors, while the NTP typically restricts its interpretations to tissue-specific tumor responses.

These design differences stem in part from differences in the expectations for the respective cancer bioassays. The NTP approach places an emphasis on characterization of the carcinogenic properties of the individual chemical. There is a solid and consistent basis on which doses are selected for each individual substance, and quantitative information on toxicokinetics is provided to support interpretation of dose–response findings. Animals are selected for uniformity, randomized to prevent litter effects, and kept free of intercurrent infections. Both rats and mice are used to increase the chances of identifying carcinogens that may be missed if studied only in one species.

The Ramazzini Foundation approach is geared toward the study of what is termed "diffuse carcinogenesis."[10,11] Their studies attempt to identify substances that may influence the cancer prevalence in the population through interactions of humans with their environment. Thus, the study subjects are individual rats, but include the weak, or sickly, as well as healthy subjects, and litters can be followed to identify familial influences on tumor responses. There is less emphasis on using doses geared to an experimentally established MTD, but this may be offset somewhat by allowing the animals to live out their natural life, rather than being terminated at 2 years as with the NTP approach. Interestingly, the current statistical approach used to analyze tumor incidence data by the NTP would seem ideally suited to the lifetime studies of the Ramazzini Foundation.

As stated earlier, there is no single right way to perform a bioassay, and the topics I have raised are just a few of the similarities and differences in the bioassay programs that could influence the study outcomes. Others include different strains of rats, diets, housing conditions, etc. It is likely that strong carcinogens will be equally identified by either bioassay approach. Carcinogens of low intrinsic potency may give different responses in the two bioassays for a variety of reasons. A subsequent talk on chemicals studied by both the Ramazzini Foundation and the NTP emphasizes these consistencies (see paper by James Huff later in this volume).

SCIENTIFIC CONTRIBUTIONS AND CONCLUSIONS

Others in this symposium have commented further on individual bioassay successes or examples of chemicals of public health importance that have been identified as carcinogens through these and other bioassay programs. I would like to simply point out that these bioassay programs, with all their blemishes and imperfections, have identified and predicted a number of human carcinogens, and have stimulated a remarkable number of scientific studies on why tumors have arisen in rodents under the conditions of these assays.[12] This work has led to better, although still very incomplete notions of how carcinogens act, especially in the area of nongenotoxic carcinogenesis. The broad compendium of cancer bioassay results is an invaluable reference set of information for new assay development, as well as for further investigations into mechanisms of action, and will be used in this manner far into the future. These are expensive, complicated, and highly worthwhile studies. It is my belief that the cost in time, effort, manpower and money that has gone into these cancer research programs is small when compared to the potential costs to society of ignoring what they are telling us. The scientific community and the public at large owe a large debt of gratitude to Drs. Maltoni and Rall for their foresight and wisdom in creating and nurturing these bioassay programs.

REFERENCES

1. HASEMAN, J.K. 1984. Statistical issues in the design, analysis, and interpretation of animal carcinogenicity studies. Environ. Health Perspect. **58:** 385–392.

2. BUCHER, J.R. et al. 1996. Workshop overview: National Toxicology Program studies: principles of dose selection and applications to mechanistic based risk assessment. Fundam. Appl. Toxicol. **31:** 1–8.
3. FORAN, J.A., Ed. 1997. Principles for the Selection of Doses in Chronic Rodent Bioassays. ILSI Risk Sciences Institute. ILSI Press. Washington, DC.
4. BUCHER, J.R. 2000. Doses in rodent cancer studies: sorting fact from fiction. Drug Metab. Rev. **32:** 153–164.
5. GAYLOR, D.W. et al. 1985. Experimental design of bioassays for screening and low dose extrapolation. Risk Analysis **5:** 9–16
6. WARD, J.M. et al. 1995. Peer review in toxicologic pathology. Toxicol. Pathol. **23:** 226–234.
7. SILLS, R.C. et al. 1999. High frequency of codon 61 K-ras A → T transversions in lung and Harderian gland neoplasms of B6C3F1 mice exposed to chloroprene (2-chloro-1,3-butadiene) for 2 years, and comparisons with the structurally related chemicals isoprene and 1,3-butadiene, Carcinogenesis **20:** 657–662.
8. DEVEREUX, T.R. et al. 1994. Low frequency of H-ras mutations in hepatocellular adenomas and carcinomas and in hepatoblastomas from B6C3F1 mice exposed to oxazepam in the diet. Carcinogenesis **15:** 1083–1087.
9. VAN BIRGELEN, A.P.J.M. et al. 1997. Design of 2-year bioassays with dioxin-like compounds in female Sprague Dawley rats. Organohalogen Compounds **34:** 154–159.
10. SOFFRITTI, M. et al. 1999. Mega-experiments to identify and assess diffuse carcinogenic risks. Ann. N.Y. Acad. Sci. **895:** 34–55.
11. MALTONI, C. et al. 1999. The scientific and methodological bases of experimental studies for detecting and quantifying carcinogenic risks. Ann. N.Y. Acad. Sci. **895:** 10–26.
12. HUFF, J. 1999. Value, validity, and historical development of carcinogenesis studies for predicting and confirming carcinogenic risks to humans. *In* Carcinogenicity Testing, Predicting, and Interpreting Chemical Effects. K.T. Kitchin, Ed.: **2:** 21–123. Marcel Dekker. New York.

Chemicals Studied and Evaluated in Long-Term Carcinogenesis Bioassays by Both the Ramazzini Foundation and the National Toxicology Program

In Tribute to Cesare Maltoni and David Rall

JAMES HUFF

National Institute of Environmental Health Sciences,
Research Triangle Park, North Carolina 27709, USA

ABSTRACT: The Ramazzini Foundation (RF) in Bentivoglio, Italy and the National Toxicology Program (NTP) in Research Triangle Park, North Carolina have carried out several hundred chemical carcinogenesis bioassays: 200 by RF and 500 by NTP. Of these, 21 have been evaluated by both laboratories. The 14 chemicals for which both laboratories have designed, conducted, and reported bioassay results are: acrylonitrile, benzene, chlorine, diesel fuel, ethylbenzene, methylene chloride (dichloromethane), propylene, styrene, styrene oxide, toluene, trichloroethylene, trichlorofluoromethane, vinylidene chloride, and xylenes. The other seven chemicals (two are fibers) were evaluated by both laboratories, but results have not yet been published. Results of these 14 interlaboratory studies were compared both to explore consistency of carcinogenic responses and to identify possible factors that may reveal reasons for any differences observed. Individual carcinogenesis results from each laboratory were duplicated and complementary. Of the 14 chemicals compared, 11 (80%) were either carcinogenic (9 chemicals) or noncarcinogenic (2 chemicals) in both studies. Eight of the paired chemicals had at least one carcinogenic target site in common. The other three were carcinogenic in one laboratory but not in the other. Possible explanations for these differences include dose, method of administration, duration of follow-up, and whether or not total tumors are counted. The collaboration between these two pioneering bioassay laboratory programs contributes greatly to our understanding of chemical carcinogenesis and results in better protection of workers and the general population from chemical diseases, especially cancers.

KEYWORDS: bioassay; chemical carcinogens; hazard identification; long-term tests; Maltoni; National Toxicology Program; Rall; Ramazzini Foundation

---

Address for correspondence: James Huff, Ph.D., Environmental Carcinogenesis, National Institute of Environmental Health Sciences, P.O. Box 12233, 111 T.W. Alexander Drive, Research Triangle Park, NC 27709. Voice: 919-541-3780; fax: 919-541-5002.
    huff1@niehs.nih.gov

TRIBUTE

Cesare Maltoni and David P. Rall were extraordinary and compassionate physicians as well as steadfast and innovative experimentalists, a grand combination that showed in their unrelenting empathy for workers and their deep understanding of the value of experimental testing and research to better protect and advance public health worldwide. They led their respective programs (the Ramazzini Foundation and the National Toxicology Program) to the forefront of the public and occupational health field, thereby helping to reduce exposure to toxic and carcinogenic chemicals. Their combined strength and uncompromising attitude directed to the betterment of health of the individual stands as a model for all of us. These two giants in the field of public health are sorely missed. This paper is dedicated to my two long-time friends, mentors, colleagues, coworkers, and defenders and champions of public health and primary prevention of diseases, especially cancers.

INTRODUCTION

Long-term chemical carcinogenesis bioassays are the cornerstone for primary prevention and for protection of the worker and the general public from chemically and occupationally associated cancers.[1–73] The two largest, longest existing, and most well-established bioassay programs in the world are the Ramazzini Foundation (RF)[1–19] and the National Toxicology Program (NTP).[20–73] More than 700 chemicals or agents have been tested for carcinogenic activity by these two programs: roughly 200 by RF and 500 by NTP. Twenty-one chemicals have been tested by both laboratories; 14 have been reported by both (TABLE 1). The results of the tests of these 14 chemicals make up the primary aspect of this paper.

TABLE 1. Chemicals evaluated in long-term carcinogenesis bioassays by both RF and NTP[a]

1. Asbestos	12. Styrene oxide[b,113,114]
2. Acrylonitrile[b,74–79]	13. Tetrachloroethylene
3. Benzene[b,80–93]	14. Toluene[b,118–122]
4. Chlorine[b,94–97]	15. 1,1,1-Trichloroethane
5. Diesel fuel[b,98–100]	16. Trichloroethylene[b,123–126]
6. Dichloroethane	17. Trichlorofluoromethane[b,127,128]
7. Ethylbenzene[b,101–103]	18. Vinylidene chloride[b,129–134]
8. Methylene chloride[b,104–106]	19. Vitamin C
9. Nitrilotriacetic acid	20. Wollastonite
10. Propylene[b,107–111]	21. Xylenes[b,136–138]
11. Styrene[b,112,115–117]	

[a]Available published paper on each of these 21 chemicals are listed in the REFERENCES section.
[b]Studies by RF and NTP form the basis for the evaluations and comparisons in this paper. Cited are chemical-specific and other relevant references published by RF and NTP.

CARCINOGENESIS BIOASSAYS

Experimental long-term chemical carcinogenesis bioassays are designed and carried out to identify potential carcinogenic effects for humans.[1,2,5,10,14,15] Carcinogenesis results in rodents, mainly rats and mice, have been shown to be a consistent and reliable indicator and predictor of human cancer risk.[22–24,26–29,35,38–42] All known human carcinogens that have been evaluated adequately in animal bioassays are also carcinogenic in animal bioassay studies. Of the nearly 100 recognized human carcinogens, about one-third were shown first to be carcinogenic in experimental animals.[27,28] Hence, for chemicals discovered to be carcinogenic to laboratory animals, prudent public health policy suggests strongly that eliminating exposures to these carcinogens would reduce or eliminate certain environmentally associated cancers.[2,20,23,24,27–29]

Findings from carcinogenesis bioassays are used for establishing and setting occupational exposure standards; for developing primary prevention strategies by national and international regulatory agencies and other organizations; for carcinogen evaluation organizations like the U.S. Congress–mandated Report on Carcinogens, and among others the California Environmental Protection Agency's PROP 65 program, and the International Agency for Research on Cancer's Monographs Programme; and for formulating and promulgating policy decisions by environmental and occupation research and regulatory agencies.

PERSPECTIVE

For many years, a key to primary cancer prevention was to identify known animal and human carcinogens and to either eliminate or drastically reduce exposures to these carcinogens. In more recent years this long-accepted *prima facie* evidence has been challenged, largely by vested industries and other parties, often using speculative mechanisms of carcinogenesis or modes of action purported to be rodent specific and irrelevant to humans. Many of these arguments have settled on what has been described as "modes of action," rather than "mechanisms," alleging that the "modes" of carcinogenicity in animals are not or will not be the same in humans and that therefore a particularly chemical would *de facto* be safe for human exposure. This results in a lowering of the estimated relative hazards and risks from exposures to rodent carcinogens and thus raises the potential cancer risks of humans exposed to these chemicals.

EXPERIMENTAL DESIGN COMPARISONS

Overall, the designs for long-term bioassays are similar between the RF and NTP laboratories.[1–73] That is, treatment of animals, laboratory characteristics and Good Laboratory Practices, and pathology assessment and reporting (TABLE 2; see Belpoggi *et al.*, Soffritti *et al.*, and Bucher, these proceedings, for more extensive descriptions and details).[1,2,5,27,28,42,47,52] However, there are several important differences between the two laboratories.

TABLE 2. General experimental bioassay designs and conditions typically used by RF and NTP

	RF	NTP
Species/Strain		
Rats	Sprague-Dawley Wistar	Fischer 344/N
Mice	Swiss, RF/J	B6C3F1
Numbers per Group		
	50–100	50–60
Routes and Exposures	MTD + 2–3 doses and controls	MTD + 2–3 doses and controls
Gavage	4 per week [M,T,Th,F]	5 per week [M,T,W,Th,F]
Vehicle	olive oil	corn oil
Inhalation	7 h/d–5 d/wk	6 h/d–5 d/wk
Feed	continuous	continuous
Age at Start		
	6–8 weeks	6–8 weeks
Duration of Exposures and Experiments		
	52–104 weeks exposure, continue with no chemical exposure for natural lifetime	104 weeks
Necropsy/Histopathology		
	complete	complete
Tissue Preservation		
	70% ethyl alcohol	10% formalin

NOTE: For more details about experimental designs and laboratory conditions see the papers in this volume by Belpoggi *et al.*, by Bucher, and by Soffritti *et al.*

Strain of Rat

RF typically uses Sprague-Dawley rats in their bioassays, whereas the NTP uses Fischer 344/N rats. NTP usually uses B6C3F1 mice; RF generally uses rats only (sometimes Wistar instead of, or in addition to, Sprague-Dawley), but occasionally uses mice of Swiss and RF/J strains. Almost without exception, RF and NTP use both sexes of whatever strains are chosen for the long-term studies, and both routinely use 50–60 animals per sex, per group of control and exposed animals.

Exposures

RF exposes animals via inhalation for 7 hours per day and NTP for 6 hours per day; both for 5 days per week. For oral intubation, there are two differences: RF uses virgin olive oil as a vehicle for chemical administration, while the NTP uses corn oil. RF intubates chemical/oil mixtures 4 days a week and NTP does so for 5 days per week.

Duration

This is a key and major difference between these two laboratories. RF exposes animals for 52 to 104 weeks and then allows the animals to live their natural life without any additional exposures. NTP terminates their bioassays at 104 weeks. On

TABLE 3. Routes of exposure for the 14 chemicals evaluated by RF and NTP for carcinogenic activities

Chemical	RF	NTP	S/D
1. Acrylonitrile	inhalation, gavage	inhalation	S
2. Benzene	inhalation, gavage	gavage	S
3. Chlorine	drinking water	drinking water	S
4. Diesel fuel	gavage	skin	D
5. Ethylbenzene	gavage	inhalation	D
6. Methylene chloride	inhalation, gavage	inhalation	S
7. Propylene	inhalation	inhalation	S
8. Styrene	inhalation, gavage, injection	gavage	S
9. Styrene oxide	gavage	gavage	S
10. Toluene	gavage	inhalation	D
11. Trichloroethylene	inhalation	gavage	D
12. Trichlorofluoromethane	inhalation	gavage	D
13. Vinylidene chloride	inhalation	gavage	D
14. Xylenes	gavage	gavage	S

TOTALS: RF for four chemicals used more than one route of exposure; for eight chemicals the two laboratories used the same route, and for six they used different routes; of the 33 routes, 16 were gavage, 13 inhalation, 2 drinking water, 1 skin (dermal), 1 injection.

S/D = Same or Different route of exposure comparisons; S was used if one of the routes for the same chemical was the same.

occasion, each laboratory alters its standard protocol. For example, parent generations are exposed during conception, gestation, and lactation; and exposures continue for the offspring or both parents and offspring.

Routes of Exposure for the 14 Chemicals

For eight chemicals both RF and NTP used the same route of exposure, whereas for another six they used different routes of exposure (TABLE 3). In four cases, RF used multiple routes. Gavage (oral intubation) was the most common route, with inhalation following close behind, and drinking water, skin, and injection used rarely.

Abbreviated Tumor Analysis Results by Chemical

A summary of "plus" or "minus" results for each of the 14 chemicals evaluated and published by both the RF and the NTP are shown in TABLE 4. These simplified designations (+,–) are expanded on in the following chemical-specific text details. For 11 (80%) of the 14 chemicals there is positive or negative concordance. An outline of the comparative organ sites is given in TABLE 5 for each of the 14 chemicals. The most common collective organ sites for carcinogenesis for these 14 chemicals are: total malignant tumors, 10; mammary gland, 8; liver and lung, 6; forestomach and leukemia, 5; head, 4; and kidney, skin, testes, and Zymbal glands, 3.

Generally, NTP does not report total malignant tumors as an indication of carcinogenesis. For this paper, the total malignant tumors "site" is listed for benzene,

TABLE 4. Abbreviated results on chemicals evaluated by RF and NTP

	Chemical	+ or − carcinogenicity results		Concordance
		RF	NTP	
1.	Acrylonitrile	+	+	Y
2.	Benzene	+	+	Y
3.	Chlorine	+	+	Y
4.	Diesel fuel	+	+/−	Y?
5.	Ethylbenzene	+	+	Y
6.	Methylene chloride	+	+	Y
7.	Propylene	−	−	Y
8.	Styrene	+	+/−[a]	Y
9.	Styrene oxide	+	+	Y
10.	Toluene	+	−	N?
11.	Trichloroethylene	+	+	Y
12.	Trichlorofluoromethane	−	−	Y
13.	Vinylidene chloride	+[b]	−[c]	N?
14.	Xylenes	+	−	N?
			Totals	11Y, 3N

NOTE: + = Positive chemical-related carcinogenic response in one or more target organs; − = no evidence of carcinogenic activity related to chemical exposure; Y = correlation of similar positive or no evidence of carcinogenicity responses from both laboratories; N = noncorrelation of carcinogenic responses between both laboratories; ? = questionable correlation or noncorrelation between the two laboratories (explanations given in the text for the particular chemical).
[a]Positive studies reported in literature: Cruzan et al. (see ref. in text).
[b]Embyro exposures from 12 days + 104 weeks.
[c]NTP studies considered basically inadequate because exposure concentrations were low.

whereas the other NTP chemicals did not show significant increases in total tumors compared to controls. Total tumors data (benign, malignant, combined) are listed in each NTP technical report, but no statistical comparisons are made routinely. Comparing target sites for the tested and reported 14 chemicals shows eight having at least one target organ in common or both bioassays exhibiting no carcinogenic responses; six chemicals do not show target organ concordance. Comparative carcinogenic findings by the RF and the NTP for each of the 14 chemicals are summarized alphabetically by chemical.

Acrylonitrile

Acrylonitrile[74–79] induced tumors of the forestomach and harderian glands when administered orally to mice (NTP); neoplasms of the ovary and lung in female mice may have been related to administration of acrylonitrile. Nonneoplastic lesions of the forestomach and harderian gland in males and of the forestomach and ovary in females were associated with exposure to acrylonitrile. Via inhalation (RF), in rats,

TABLE 5. Simplified organ/tissue site carcinogenesis results

Chemical	Organ/tissue site carcinogenesis[a]		Site Concordance[b]
	RF	NTP	
1. Acrylonitrile	Br, MG, Z, L, An, T	FS, HG, Ov?, Lu?	N
2. Benzene	Mult	Mult	Y
3. Chlorine	Leuk	Leuk?	Y
4. Diesel fuel	T, He, Ut/Vg	Skin?	N
5. Ethylbenzene	T, He	K, Te, Lu, L	N
6. Methylene chloride	Lu, T?, MG?	MG, Lu, L	Y
7. Propylene	none	none	Y
8. Styrene	T, MG, Lu?	Lung?	Y
9. Styrene oxide	FS	FS	Y
10. Toluene	T, MG, He, Leuk	none	N
11. Trichloroethylene	Te, K	K, L, Te	Y
12. Trichlorofluoromethane	none	none	Y
13. Vinylidene chloride	T, Leuk	none?	N
14. Xylenes	T, MG, He, Leuk	none	N
		Correlative totals	8Y, 6N

[a]Organs/tissues listed alphabetically, including benzene listed here as multiple (see TABLE 7): An = angiosarcomas, extrahepatic, 1; Br = brain, 1; FS = forestomach, 5; He = tumors of the head, 4 (often includes combined total tumors of Zymbal gland, ear duct, nasal cavities, oral cavity); HG = harderian gland, 2; K = kidney, 3; L = liver, 6; Leuk = leukemia, 5; Lu = lung 6; lymphoma, 3; MG = mammary gland, 8; Mult = multiple organ/tissue sites [TABLE 6]; oral, 2; Ov = ovary, 2; preputial gland, 1; skin, 3; T = total malignant tumors, 10; Te = testis, 3; Ut/Vg = uterus/vagina, 2; Z = Zymbal gland, 3; ? = questionable response.

[b]Concordance: Y = at least one chemical-specific carcinogenic target organ is the same between the two laboratories; N = no chemical-specific target organ is the same.

acrylonitrile caused tumors of the brain, mammary and Zymbal glands, liver, extrahepatic angiosarcomas, and total malignancies (TABLE 6).

Even though there were 10 individual tumor sites, none were overlapping among rats and mice. Conversely, for four tumor sites there was concordance in both sexes within a species.

In contrast to mice, no tumors in rats were observed when acrylonitrile was given by gavage (RF). The use of only a single low dose (5 mg/kg) and a 52-week exposure duration (RF) are in contrast with the use of doses of 2.5, 10.0, and 20 mg/kg and 104 weeks of exposure (NTP study) and likely account for the lack of carcinogenesis found in the RF study.

A structurally related chemical, methacrylonitrile, given by gavage did not induce any tumors in rats or mice (NTP). However, in male and female rats, methacrylonitrile administration caused significant increases in the incidences of nonneoplastic lesions of the nose and liver; these chronic toxic lesions did not lead to tumors.

TABLE 6. Acrylonitrile: organ/tissue site tumors from the RF and NTP in six experiments using one strain of rats and one strain of mice

	Sprague-Dawley (RF) Rat				B6C3F1 (NTP) Mice		Tumor sites
	Inhalation		Gavage		Gavage		
	M	F	M	F	M	F	
Brain	+	+	–	–	–	–	2
Angiosarcoma	+	+	–	–	–	–	2
Forestomach	–	–	–	–	+	+	2
Harderian gland	–	–	–	–	+	+	2
Mammary glands	–	+	–	–	–	–	1
Zymbal gland	+	–	–	–	–	–	1
Liver	+	–	–	–	–	–	1
Lung	–	–	–	–	–	+?	1
Ovary	–	–	–	–	–	+?	1
All malignant tumors	+	+	–	–	–	–	2
Number of tumor sites	5	4	0	0	2	4	15

NOTE: + = positive chemical-related carcinogenic response in those target organs; +? = questionable or marginal positive carcinogenic response; – = no evidence of carcinogenic activity related to chemical exposure.

Benzene

Benzene[80–93] induced multisite and multispecies/strains carcinogenic effects in both sexes in both RF and NTP studies.

There were 10 tumor sites in the RF bioassays and 11 tumors sites in the NTP studies. In both studies, using different strains of rats and mice, there were 8 sites and "total malignant tumors" in common (TABLE 7) and 13 unique sites. The most responsive strains in these experiments were the B6C3F1 mice with 10 sites of carcinogenic activity and Sprague-Dawley rats with 8 sites. Fischer rats had three positive tumor sites. All strains showed increases in total malignant tumors. Tumors of the Zymbal gland were the first and most consistent response from benzene exposure.[88] This tumor site has been criticized because humans do not have an exact replica, but there are modified sebaceous glands of the ear in humans.[90] The second tumor site in prevalence was mammary glands. In three of the experiments benzene induced lung tumors; and, in two others, benzene caused tumors of the skin even though in all these cases benzene was given by gavage.

Chlorine

Chlorine[94–97] showed leukemogenic effects in female Sprague Dawley rats (RF) and in female Fischer rats (NTP) when given by drinking water. No carcinogenic effects were observed in male rats or in mice. Chloraminated drinking water also caused marginal increases in leukemia in female rats (NTP).

TABLE 7. Benzene: organ/tissue site tumors from RF and NTP in seven experiments using three strains of rats and three strains of mice

Strain Species	Sprague-Dawley rats		Wistar rats	Fisher[a] rats	Swiss mice	RF/J mice	B6C3F1[a] mice	
Route	Gavage	Inhalation	Gavage	Gavage	Gavage	Gavage	Gavage	Total
Zymbal gland	+	+	+	+	+	+	−	6
Mammary gland	[+]	[+]	−	−	+	+	+	5
Oral	+	+	+	+	−	−	−	4
Lung	−	−	−	−	+	+	+	3
Nasal cavities	+	[+]	+	−	−	−	−	3
Lymphoma	[+]	−	−	−	−	+	+	3
Liver	+	[+]	−	−	−	−	+	3
Forestomach	+	−	−	−	−	−	[+]	2
Skin	+	−	−	+	−	−	−	2
Uterus	−	−	−	−	−	−	+	1
Ovary	−	−	−	−	−	−	+	1
Harderian	−	−	−	−	−	−	+	1
Preputial gland	−	−	−	−	−	−	+	1
All malignancies	+	+	+	+	+	+	+	7
Total Sites	9	6	4	4	4	4	4	11

[a] = the two strains utilized in the NTP studies. + = a positive carcinogenic response; [+] = marginally increased carcinogenic response; − = no significant carcinogenic activity; sites listed in order of prevalence of responses per organ/tissue.

Diesel Fuel

Diesel fuel,[98–100] or, as in the NTP dermal studies, marine diesel fuel at doses of 250 and 500 mg/kg resulted in dose-related incidences of squamous cell neoplasms of the skin (primarily carcinomas), providing equivocal evidence of carcinogenicity for male and female B6C3F1 mice. The sensitivity of detecting systemic carcinogenicity in female mice dosed with marine diesel fuel was reduced by poor survival. Two-year NTP dermal studies of JP-5 navy fuel at doses of 250 and 500 mg/kg provided no evidence of carcinogenicity for male and female B6C3F1 mice. RF studies included unleaded gasoline, leaded gasoline, gasoil (diesel), kerosene, and several solvents therein; only the first three are mentioned here. All exposures were via gavage, four days per week, at exposures of 0, 500, and 800 mg/kg. Carcinogenic effects of all three of these gasolines included increases in total malignant tumors, mammary gland (except gasoil), head, and uterus/vaginal with unusual neurinosarcomas. Total tumors in the low-dose female mice (but not in male mice) were doubled in the NTP JP-5 navy fuel study. Conversely in the diesel fuel study, female but not male mice controls had a doubling of total tumors over those in exposed animals, but there was poor survival among treated animals.

Ethylbenzene

Ethylbenzene[101–103] induced carcinogenic responses in both sexes of rats (kidney, male and female; and testes) and of mice (lung in males and liver in females) in NTP inhalation studies. Tumors of the head (mainly nasal cavity) and total malignant tumors were increased in rats in the RF gavage studies. Both studies showed carcinogenic activity; the NTP inhalation studies were perhaps more convincing in that cancers appeared in four target organs, and ethylbenzene was carcinogenic in both sexes of rats and mice. This may have been due to the different routes of exposure, strain of rats, or a combination of the two. Nonetheless, taken together these findings show convincing evidence of carcinogenicity for ethylbenzene.

Methylene Chloride (Dichloromethane)

Methylene chloride (dichloromethane)[104–106] exhibited multisite carcinogenesis in both the RF and the NTP studies. Given by gavage (RF), dichloromethane induced lung tumors in male mice, and a marginal increase in total malignant tumors occurred in female rats. By inhalation exposure, dichloromethane induced tumors of the mammary gland in both laboratories, albeit marginally at RF. A nonsignificant increase in total malignant tumors in rats was reported (RF). In the NTP inhalation studies, lung and liver tumors were induced in both sexes of mice. In the NTP studies mammary gland tumors, not a usual occurrence in males, were increased in both male and female rats.

Propylene

Propylene[107–111] exposures by inhalation did not cause any increases in tumors at either laboratory. In the NTP studies, rats and mice were exposed to 0, 5000, and 10,000 ppm, while in the RF studies, rats (104 weeks) and mice (78 weeks) were exposed to 0, 200, 1000, and 5000 ppm. Conversely, the oxide of propylene via inhalation caused a small number of papillary adenomas of the nasal turbinates in male and female rats, and hemangiomas or hemangiosarcomas of the nasal turbinates in male and female mice (NTP).

Styrene

Styrene[112–117] given by inhalation (0, 25, 50, 100, 200, 500 ppm for 52 weeks), by gavage (0, 50, 250 mg/kg for 52 weeks), and by injection (1 sc or 4 times ip at 50 mg) caused only tumors of the mammary gland in rats and total benign/malignant tumors (RF) by inhalation. In addition, by gavage (rats: 0, 500, 1000, 2000 mg/kg; mice: 0, 150, 300 mg/kg), styrene was associated with a marginal increase in tumors of the lung in male mice (NTP).

Styrene Oxide

Styrene oxide[112–117] was studied using the gavage route of administration: RF used 0, 50, 250 mg/kg 4 or 5 times/week for 52 weeks (animals then lived out their life without exposure) and NTP used 0, 275, 550 mg/kg for rats and 0, 375, 750 mg/kg for mice 3 times per week for 104 weeks. The main pathologic findings from both laboratories were high incidences of squamous cell carcinomas or papillomas of the forestomach in both sexes of both rats and mice. Additionally, there was a statistically significant increase in the incidence of hepatocellular neoplasms in male mice receiving 375 mg styrene oxide/kg (NTP). These virtually single target site carcinogens like styrene oxide often cause cancer in the first organ exposed (so-called "application site" carcinogenesis): gavage and forestomach tumors; inhalation and nose or lung tumors; and dermal and skin tumors.

Toluene

Toluene[118–122] did not cause any tumors using the inhalation route with exposures as high as 1200 ppm (NTP). By the gavage route, tumors of the mammary glands, head, total malignant tumors, and leukemias were increased in rats (RF). Tumors of the mammary glands were somewhat increased in the low-dose group (500 mg/kg: 27% versus 14%) but not in the high-dose group (800 mg/kg). The total number of malignant tumors was doubled in the exposed animals compared to controls. Combined tumors of the head were elevated only in the top-dose male rats. Only in the low-dose group of female rats was the incidence of leukemias/lymphomas convincing (2% vs. 17.5%). Thus even though there were four positive responses, the findings overall, while being evidence of carcinogenic activity, were considered less than overwhelming.

Trichloroethylene

Trichloroethylene[123–126] has been evaluated in several laboratories, and in all there is evidence of carcinogenic activity. The two major routes of exposure, oral and inhalation, gave convincing carcinogenesis. Inhalation (RF): Sprague-Dawley rats showed increases in Leydig cell tumors, non-dose-related leukemias, and some rare renal tumors. Swiss mice exhibited lung and liver tumors. B6C3F1 mice had increases in lung, liver (?), and total malignant tumors. Oral gavage (NTP): tumors of the liver in B6C3F1 mice; a few rare tumors of the renal tubular cells in ACI, August, Marshall, Osborne-Mendel, and Fischer rats; interstitial cell tumors of the testis in Marshall rats. Thus, trichloroethylene in these and other studies clearly shows carcinogenic activity in the liver, lung, kidney, lymphoma/leukemia, and testis. Importantly there is evidence in human studies as well: liver and biliary tract, non-Hodgkin's lymphoma, and kidney (NTP report on carcinogens, 2000). These trichloroethylene target tissues/organs in both rodents and humans are particularly consistent.

Trichlorofluoromethane

Trichlorofluoromethane[127,128] exposures by inhalation and by gavage were uniformly negative. Inhalation exposures were 0, 1000, and 5000 ppm to Sprague-Dawley rats and Swiss mice (RF). Oral exposures to Fischer rats were 0, 500, and 1000 mg/kg and to B6C3F1 mice were 0, 2000, and 4000 mg/kg (NTP). The rat gavage experiments were considered inadequate due to high and early chemical-associated mortality.

Vinylidene Chloride

Vinylidene chloride[129–134] was studied using two routes of exposure. The RF inhalation experiments showed increases in leukemias and total malignant tumors in

Sprague-Dawley rats whose exposure began *in utero*. There was a marginal increase in mammary gland tumors in female rats. The NTP gavage experiments (0, 1, 5 mg/kg rats; 0, 2, 10 mg/kg mice), while uniformly negative, were considered less than adequate because the use of a maximum tolerated dose had not been clearly demonstrated. There was a slight increase in the low-dose female mice for lymphoma or leukemia (7/48, 15/49, 7/50).

Xylenes, Mixed

Xylenes, mixed[135–138] were evaluated for carcinogenicity in both laboratories by the gavage route. The NTP studies showed no evidence of carcinogenic activity of mixed xylenes at exposures of 0, 250, or 500 mg/kg for Fischer rats and to B6C3F1 mice at 0, 500, or 1000 mg/kg. The commercial mixture contained 60% *m*-xylene, 14% *p*-, 9% *o*-, and 17% ethylbenzene. The RF bioassay used xylenes composed of 50% *m*-, 27% *o*-, 22% *p*-, and 0.3% toluene at exposures of 0, 500, and 800 mg/kg. Sprague-Dawley rats (RF) showed increases in total malignant tumors, mammary gland tumors, lymphomas/leukemias, and tumors of the head; all were non-dose related.

DISCUSSION

Both the Ramazzini Foundation and the National Toxicology Program are pioneers in the study, design, conduct, evaluation, and interpretation of long-term chemical carcinogenesis bioassays, which serve to identify chemicals that cause cancer in experimental animals and that are most likely to cause cancer in exposed humans.

This paper compares the carcinogenesis results for 14 chemicals studied, evaluated, and reported by both RF and NTP laboratories (TABLE 8). Results between the RF and the NTP for the 14 chemicals studied by both laboratories are remarkably consistent regarding whether a chemical showed a positive or negative carcinogenic effect: that is, 11 of 14 chemicals are concordant. The RF and NTP studies had at least one target organ in common for eight chemicals; for six chemicals, there was no common target organ.

Only three chemicals gave inconsistent results: xylene, vinylidene, and toluene. RF had positive carcinogenicity findings, and the NTP did not. Each of these is discussed in an attempt to ascertain the differences in carcinogenic responses found by the two laboratories.

In the RF studies of xylenes, neither total malignant tumors nor those of the oral cavity were significantly increased until after 112 weeks. Increases in hemolymphorecticular leukemias were not seen until week 144, long after a two-year study would have been terminated. Thus, it appears that the difference in detecting carcinogenicity of xylenes is simply the duration of the experiments. The RF typically exposes animals for 52–104 weeks and then allows the animals to live out the remainder of their natural lives. The NTP bioassay terminates at 104 weeks. The NTP study design was an attempt to mimic an occupational lifetime for workers—that is, from young adulthood through retirement age, the human age that is comparable to 2 years of age for rats and mice. Because of great increases in life span over the past 50 years, it may be beneficial to extend studies in rodents beyond 2 years, particularly because most human tumors occur later in life.

TABLE 8. Summary bioassay findings from RF and NTP

1. Chemicals studied by RF	200
2. Chemicals studied by NTP	500
3. Bioassay designs basically similar, except	
a. RF exposures of 52–104 weeks	
b. NTP exposures 104 weeks	
c. RF duration: natural life span	
d. NTP 104 weeks (2 years)	
e. RF rat species: Sprague-Dawley	
f. NTP rat species: Fischer 344	
4. Same chemicals studied by RF and NTP	21
5. Bioassays evaluated and published	14
6. Routes of exposure	
a. Same for a chemical	8
b. Different for a chemical	6
c. Most used route: gavage	16
d. Next most used route: inhalation	13
7. Carcinogenicity (+,+ or –,–) concordance	11/14
8. Organ/tissue site commonality	8/14
9. Most common tumor sites (per chemical)	
a. Total malignant tumors	10
b. Mammary glands	8
c. Lung or liver	6
d. Forestomach or leukemia	5

Results in the vinylidene chloride bioassays did not agree between the two laboratories. Experiment duration was shorter in the NTP studies, and exposure concentrations were lower. Perhaps more importantly, the RF studies began with *in utero* exposure from 12 days of gestation and continued for two years after birth. The animals then lived out their life span. Tumors were increased in the offspring but not in the breeders. This underscores the importance of evaluating carcinogenicity of chemicals in offspring of women exposed to potentially carcinogenic chemicals before or during a pregnancy.

In the case of vinylidene chloride, one of the tumor categories that showed increases in the RF study was the category of "total malignant tumors," a carcinogenesis category not used by the NTP. The low-dose vinylidene chloride doubling of total malignant tumors (22 versus 9) in female mice (NTP) is similar to the doubling reported by RF (34.1 versus 17.9). The counting of total malignancies also explains the discordance between the RF and NTP for toluene-induced tumors. Thus, it may be prudent to include the total number of tumors caused by a given agent in evaluating its carcinogenicity and potential for causing tumors in humans.

ACKNOWLEDGMENTS

I thank Morando Soffritti and Fiorella Belpoggi for inviting me to this meeting to honor Cesare Maltoni and David P. Rall, and for their long collaboration and friendship. In addition, I thank Myron Mehlman for his efforts in organizing this meeting and clearing the way for me to attend. For their useful and valuable remarks on my slides and on this paper, I thank John Bucher and Ronald Melnick.

REFERENCES

General references for the RF and the NTP are given first. References are then grouped by the chemicals studied in alphabetical order with the most recent year first. For each chemical, references are given first for the Ramazzini Foundation and than for the National Toxicology Program.

Ramazzini Foundation

1. SOFFRITTI, M., F. BELPOGGI, F. MINARDI, et al. 1999. Mega-experiments to identify and assess diffuse carcinogenic risks. Ann. N.Y. Acad. Sci. **895:** 34–55.
2. MALTONI, C., M. SOFFRITTI & F. BELPOGGI. 1999. The scientific and methodological bases of experimental studies for detecting and quantifying carcinogenic risks. Ann. N.Y. Acad. Sci. **895:** 10–26.
3. CASTLEMAN, B., J. DEMENT, A.L. FRANK, et al. 1998. Salud ocupacional. Int. J. Occup. Environ. Health. **4:** 131–133.
4. MALTONI, C. 1997. Biomedical research as a science for development: the case of gasoline. Ramazzini Lecture. Ann. N.Y. Acad. Sci. **837:** 1–14.
5. MALTONI, C. 1995. The contribution of experimental (animal) studies to the control of industrial carcinogenesis. Appl. Occup. Environ. Hyg. **10:** 749–760.
6. MALTONI, C. 1995. The long-lasting legacy of industrial carcinogens: the lesson of asbestos. Irving J. Selikoff Memorial Lecture. Ann. N.Y. Acad. Sci. **837:** 570–586.
7. MALTONI, C., F. MINARDI, M. SOFFRITTI & G. LEFEMINE. 1991. Long-term carcinogenicity bioassays on industrial chemicals and man-made mineral fibers at the Bentivoglio (BT) laboratories of the Bologna Institute of Oncology: premises, programs, and results. Toxicol. Ind. Health. **7:** 63–94.
8. MALTONI, C., P. CARMENTANO & A. PALAZZINI. 1990. Cancer mortality trends analysis for Bologna and province. Programs, methodology, objectives, and early results. Ann. N.Y. Acad. Sci. **609:** 110–130; Discussion, 130–135.
9. SOFFRITTI, M., C. MALTONI, F. MAFFEI & F. BIAGLI. 1989. Formaldehyde: an experimental multipotential carcinogen. Toxicol. Ind. Health. **5:** 699–730.
10. MALTONI, C. 1988. International standards for occupational exposure to toxic agents. Am. J. Ind. Med. **13:** 529–530.
11. PERINO, G., B. CONTI, A. CILIBERTI & C. MALTONI. 1988. Incidence of pancreatic tumors and tumor precursors in Sprague-Dawley rats after administration of olive oil. Ann. N.Y. Acad. Sci. **534:** 604–617.
12. MALTONI, C., F. MINARDI & L. MORISI. 1982. The relevance of the experimental approach in the assessment of the oncogenic risks from fibrous and non-fibrous particles. The oncology project of the Bologna Institute of Oncology. Med. Lav. **73:** 394–407.
13. MALTONI, C., A. CILIBERTI & D. CORRETTI. 1982. Experimental contributions in identifying brain potential carcinogens in the petrochemical industry. Ann. N.Y. Acad. Sci. **381:** 216–249.
14. MALTONI, C. 1978. Predictive carcinogenicity bioassays in industrial oncogenesis. Prog. Biochem. Pharmacol. **14:** 47–56.
15. MALTONI, C. 1976. Occupational carcinogenesis. Predictive value of carcinogenesis bioassays. Ann. N.Y. Acad. Sci. **271:** 431–443.

16. MALTONI, C. 1976. Precursor lesions in exposed populations as indicators of occupational cancer risk. Ann. N.Y. Acad. Sci. **271:** 444–447.
17. MALTONI, C. 1976. Occupational chemical carcinogenesis: new facts, priorities and perspectives. IARC Sci. Publ. **13:** 127–149.
18. MALTONI, C. & G. LEFEMINE. 1975. Carcinogenicity bioassays of vinyl chloride: current results. Ann. N.Y. Acad. Sci. **246:** 195–218.
19. MALTONI, C., G. LEFEMINE, P. CHIECO & D. CORRETTI. 1974. Vinyl chloride carcinogenesis: current results and perspectives. Med. Lav. **65:** 421–444.

National Toxicology Program

20. TOMATIS, L. & J. HUFF. 2002. Evolution of research on cancer etiology. Chapter **9:** 189-201. *In* The Molecular Basis of Human Cancer: Genomic Instability and Molecular Mutation in Neoplastic Transformation. W.B. Coleman & G.J. Tsongalis, Eds. Humana Press Inc. Totowa, NJ.
21. HUFF, J. 2001. Sawmill chemicals and carcinogenesis. Environ. Health Perspect. **109:** 209–212.
22. HASEMAN, J., R. MELNICK, L. TOMATIS & J. HUFF. 2001. Carcinogenesis bioassays: study duration and biological relevance. Food Chem. Toxicol. **39:** 739–744.
23. TOMATIS, L., R.L. MELNICK, J. HASEMAN, *et al.* 2001. Alleged misconceptions distort perceptions of environmental cancer risks. FASEB J. **15:** 195–203.
24. RALL, D.P. 2000. Laboratory animal tests and human cancer. Drug Metab. Rev. **32:** 119–128.
25. HUFF, J. 2000. The Legacy of David Platt Rall. Scientific, environmental, public health, and regulatory contributions. Eur. J. Oncol. **5:** 85–100.
26. BUCHER, J.R. 2000. Doses in rodent cancer studies: sorting fact from fiction. Drug Metab. Rev. **32:** 153–163.
27. HUFF, J. 1999. Long-term chemical carcinogenesis bioassays predict human cancer hazards. Issues, controversies, and uncertainties. Ann. N.Y. Acad. Sci. **895:** 56–79.
28. HUFF, J. 1999. Value, validity, and historical development of carcinogenesis studies for predicting & confirming carcinogenic risks to humans. *In* Carcinogenicity Testing, Predicting & Interpreting Chemical Effects. K.T. Kitchin, Ed.: 21–123. Marcel Dekker. New York.
29. TOMATIS, L., J. HUFF, I. HERTZ-PICCIOTTO, *et al.* 1997. Avoided and avoidable risks of cancer. Carcinogenesis **18:** 97–105.
30. HASEMAN, J.K., G.A. BOORLAND & J. HUFF. 1997. Value of historical control data and other issues related to the evaluation of long-term rodent carcinogenicity studies. Toxicol. Pathol. **25:** 524–527.
31. KARSTADT, M. & J.K. HASEMAN. 1997. Effect of discounting certain tumor types/sites on evaluations of carcinogenicity in laboratory animals. Am. J. Ind. Med. **31:** 485–494.
32. BUCHER, J.R, C.J. PORTIER, J.I. GOODMAN, *et al.* 1996. Workshop overview. National Toxicology Program Studies: principles of dose selection and applications to mechanistic based risk assessment. Fundam. Appl. Toxicol. **31:** 1–8.
33. ABDO, K.M. & F.W. KARI. 1996. The sensitivity of the NTP bioassay for carcinogen hazard evaluation can be modulated by dietary restriction. Exp. Toxicol. Pathol. **48:** 129–137.
34. HASEMAN, J.K. & M.R. ELWELL. 1996. Evaluation of false positive and false negative outcomes in NTP long-term rodent carcinogenicity studies. Risk Anal. **16:** 813–820.
35. RALL, D.P. 1995. Can laboratory animal carcinogenicity studies predict cancer in exposed children? Environ. Health Perspect. **103** Suppl. 6: 173–175.
36. FUNG, V.A., J.C. BARRETT & J. HUFF. 1995. The carcinogenesis bioassay in perspective: Application in identifying human cancer hazards. Environ. Health Perspect. **103:** 680–683.
37. DUNNICK, J.K., M.R. ELWELL, J. HUFF & J.C. BARRETT. 1995. Chemically induced mammary gland cancer in the National Toxicology Program's carcinogenesis bioassay. Carcinogenesis. **16:** 173–179.
38. RALL, D.P. 1994. Shoe-leather epidemiology—the footpads of mice and rats: animal tests in assessment of occupational risks. Mt. Sinai J. Med. **61:** 504–508.

39. FUNG, V.A., J. HUFF, E.K. WEISBURGER & D.G. HOEL. 1993. Predictive strategies for selecting 379 NCI/NTP chemicals evaluated for carcinogenic potential: scientific and public health impact. Fundam. Appl. Toxicol. **20:** 413–436.
40. HUFF, J. & D.P. RALL. 1992. Relevance to humans of carcinogenesis results from laboratory animal toxicology studies. *In* Maxcy-Rosenau-Last's Public Health & Preventive Medicine, 13th edit. J.M. Last & R.B. Wallace, Eds.: 433–440; 453–457. Appleton & Lange. Norwalk, CT.
41. RALL, D.P. 1992. Problems remain to be resolved in the area of quantitative risk assessment. Regul. Toxicol. Pharmacol. **15:** 104–105.
42. HUFF, J., J. HASEMAN & D. RALL. 1991. Scientific concepts, value, and significance of chemical carcinogenesis studies. Ann. Rev. Pharmacol. Toxicol. **31:** 621–652.
43. RALL, D.P. 1991. Carcinogens and human health: Part 2. Science. **251:** 10–13.
44. HUFF, J. & J. HASEMAN. 1991. Long-term chemical carcinogenesis experiments for identifying potential human cancer hazards: collective database of the National Cancer Institute and National Toxicology Program (1976–1991). Environ. Health Perspect. **96:** 23–31.
45. HUFF, J., J. CIRVELLO, J. HASEMAN & J. BUCHER. 1991. Chemicals associated with site-specific neoplasia in 1394 long-term carcinogenesis experiments in laboratory rodents. Environ Health Perspect. **93:** 247–270.
46. RALL, D.P. 1990. Carcinogens in our environment. IARC Sci. Publ. **104:** 233–239.
47. CHHABRA, R.S., J.E. HUFF, B.S. SCHWETZ & J. SELKIRK. 1990. An overview of pre-chronic and chronic toxicity/carcinogenicity experimental study designs and criteria used by the National Toxicology Program. Environ. Health Perspect. **86:** 313–321.
48. HUFF, J.E., S.L. EUSTIS & J.K. HASEMAN. 1989. Occurrence and relevance of chemically induced benign neoplasms in long-term carcinogenicity studies. Cancer Metastasis Rev. **8:** 1–22.
49. HASEMAN, J.K., J.E. HUFF, G.N. RAO & S.L. EUSTIS. 1989. Sources of variability in rodent carcinogenicity studies. Fundam. Appl. Toxicol. **12:** 793–804.
50. RALL, D.P. 1988. Laboratory animal toxicity and carcinogenesis testing. Underlying concepts, advantages and constraints. Ann. N.Y. Acad. Sci. **534:** 78–83.
51. HOEL, D.G., J.K. HASEMAN, M.D. HOGAN, *et al.* 1988. The impact of toxicity on carcinogenicity studies: implications for risk assessment. Carcinogenesis. **11:** 2045–2052.
52. HUFF, J.E., E.E. MCCONNELL, J.K. HASEMAN, *et al.* 1988. Carcinogenesis studies: results of 398 experiments on 104 chemicals from the U.S. National Toxicology Program. Ann. N.Y. Acad. Sci. **534:** 1–30.
53. HASEMAN, J.K. & J.E. HUFF. 1987. Species correlation in long-term carcinogenicity studies. Cancer Lett. **37:** 125–132.
54. RALL, D.P., M.D. HOGAN, J.E. HUFF, *et al.* 1987. Alternatives to using human experience in assessing health risks. Ann. Rev. Public Health. **8:** 355–385.
55. MARONPOT, R.R., J.K. HASEMAN, G.A. BOORMAN, *et al.* 1987. Liver lesions in B6C3F1 mice: the National Toxicology Program, experience and position. Arch. Toxicol. Suppl. **10:** 10–26.
56. HASEMAN, J.K., J.E. HUFF, E. ZEIGLER & E.E. MCCONNELL. 1987. Comparative results of 327 chemical carcinogenicity studies. Environ. Health Perspect. **74:** 229–235.
57. HASEMAN, J.K., E.C. THARRINGTON, J.E. MUFF & E.E. MCCONNELL. 1986. Comparison of site-specific and overall tumor incidence analyses for 81 recent National Toxicology Program carcinogenicity studies. Regul. Toxicol. Pharmacol. **6:** 155–170.
58. HUFF, J.E., E.E. MCCONNELL & J.K. HASEMAN. 1985. On the proportion of positive results in carcinogenicity studies in animals. Environ. Mutagen. **7:** 427–428.
59. HASEMAN, J.K., J.E. HUFF, G.N. BOO, *et al.* 1985. Neoplasms observed in untreated and corn oil gavage control groups of F344/N rats and (C57BL/6N X C3H/HeN)F1 (B6C3F1) mice. J. Natl. Cancer Inst. **75:** 975–984.
60. HASEMAN, J.K., D.D. CRAWFORD, J.E. HUFF, *et al.* 1984. Results from 86 two-year carcinogenicity studies conducted by the National Toxicology Program. J. Toxicol. Environ. Health. **14:** 621–639.
61. HASEMAN, J.K., J.E. HUFF & G.A. BOORMAN. 1984. Use of historical control data in carcinogenicity studies in rodents. Toxicol. Pathol. **12:** 126–135.

62. HUFF, J., J. MOORE & D. RALL. 1984. The National Toxicology Program and preventive oncology. *In* The Cosmetic Industry: Scientific and Regulatory Foundations. N. Estrin, Ed.: 647–676. Marcel Dekker. New York.
63. HART, L.G., J. HUFF, J.E., MOORE & D.P. RALL. 1983. The National Toxicology Program's research and testing activities. *In* Hazard Assessment of Chemicals, Current Developments, Vol. 2. J. Saxena, Ed.: 191–244. Academic Press. New York.
64. HUFF, J. 1982. Carcinogenesis bioassay results from the National Toxicology Program. Environ. Health Perspect. **45:** 185–198.
65. HUFF, J. 1982. Condensations of the Carcinogenesis Bioassay Technical Reports. Environ. Health Perspect. **45:** 199–210.
66. RALL, D.P. 1981. Issues in the determination of acceptable risk. Ann. N.Y. Acad. Sci. **363:** 139–144.
67. RALL, D.P. 1980. Carcinogenicity testing of drugs. JAMA **243:** 1035.
68. RALL, D.P. 1979. Relevance of animal experiments to humans. Environ. Health Perspect. **32:** 297–230.
69. RALL, D.P. 1979. The role of laboratory animal studies in estimating carcinogenic risks for man. IARC Sci. Publ. **25** (1): 79–89.
70. RALL, D.P. 1978. Thresholds? Environ. Health Perspect. **22:** 163–165.
71. RALL, D.P. 1976. Occupational carcinogenesis. Toward an integrated program of government action. Ann. N.Y. Acad. Sci. **271:** 198–199.
72. RALL, D.P. 1973. Risks, research, and reason. Fed. Proc. **32:** 1766–1768.
73. RALL, D.P. 1969. Difficulties in extrapolating the results of toxicity studies in laboratory animals to man. Environ. Res. **2:** 360–367.

Acrylonitrile

Ramazzini Foundation

74. MALTONI, C., A. CILIBERTI, G. COTTI, *et al.* 1988. Long-term carcinogenicity bioassays on acrylonitrile administered by inhalation and by ingestion to Sprague-Dawley rats. Ann. N.Y. Acad. Sci. **534:** 179–202.
75. MALTONI, C., A. CILIBERTI & D. CORRETTI. 1982. Experimental contributions in identifying brain potential carcinogens in the petrochemical industry. Ann. N.Y. Acad. Sci. **381:** 216–249.
76. MALTONI, C., A. CILIBERTI & V. MAIO. 1977. Carcinogenicity bioassays on rats of acrylonitrile administered by inhalation and by ingestion. Med. Lav. **68:** 401–411.

National Toxicology Program

77. GHANAYEM, B.I., A. NYSKA, J.K. HASEMAN, *et al.* 2002. Acrylonitrile is a multisite carcinogen in male and female B6C3F1 mice. Toxicol. Sci. **68:** 59–68.
78. GHANAYEM, B.I. & NTP STAFF. 2001. Toxicology and carcinogenesis studies of acrylonitrile (CAS No. 107-13-1) in B6C3F1 mice (gavage studies). NTP Tech. Rept. Series # TR-506. National Toxicology Program. Research Triangle Park, NC.
79. GHANAYEM, B.I. & NTP STAFF. 2001. Toxicology and carcinogenesis studies of methacrylonitrile (CAS No. 126-98-7) in F344/N rats and B6C3F1 mice (gavage studies). NTP Tech. Rept. Series # TR-497. National Toxicology Program. Research Triangle Park, NC.

Benzene

Ramazzini Foundation

80. MALTONI, C., A. CILIBERTI, C. PINTO, *et al.* 1997. Results of long-term experimental carcinogenicity studies of the effects of gasoline, correlated fuels, and major gasoline aromatics on rats. Ann. N.Y. Acad. Sci. **837:** 15–52.
81. MALTONI, C., A. CILIBERTI, G. COTTI, *et al.* 1989. Benzene, an experimental multipotential carcinogen: results of the long-term bioassays performed at the Bologna Institute of Oncology. Environ. Health Perspect. **82:** 109–124.

82. MALTONI, C., B. CONTI, G. PERINO, et al. 1988. Further evidence of benzene carcinogenicity. Results on Wistar rats and Swiss mice treated by ingestion. Ann. N.Y. Acad. Sci. **534:** 412–426.
83. MALTONI, C., B. CONTI, G. COTTI & F. BELPOGGI. 1985. Experimental studies on benzene carcinogenicity at the Bologna Institute of Oncology: current results and ongoing research. Am. J. Ind. Med. **7:** 415–446.
84. MALTONI, C., B. CONTI & G. COTTI. 1983. Benzene: a multipotential carcinogen. Results of long-term bioassays performed at the Bologna Institute of Oncology. Am. J. Ind. Med. **4:** 589–630.
85. MALTONI, C., G. COTTI, L. VALGIMIGLI & A. MANDRIOLI. 1982. Hepatocarcinomas in Sprague-Dawley rats, following exposure to benzene by inhalation. First experimental demonstration. Med. Lav. **73:** 446–450.
86. MALTONI, C., B. CONTI & C. SCARNATO. 1982. Squamous cell carcinomas of the oral cavity in Sprague-Dawley rats, following exposure to benzene by ingestion. First experimental demonstration. Med. Lav. **73:** 441–445.
87. MALTONI, C. G. COTTI, L. VALGIMIGLI & A. MANDRIOLI. 1982. Zymbal gland carcinomas in rats following exposure to benzene by inhalation. Am. J. Ind. Med. **3:** 11–16.
88. MALTONI, C. & C. SCARNATO. 1979. First experimental demonstration of the carcinogenic effects of benzene; long-term bioassays on Sprague-Dawley rats by oral administration. Med. Lav. **70:** 352–357.

National Toxicology Program

89. TSUTSUI, T., N. HAYASHI, J. HUFF, et al. 1997. Benzene-, catechol-, hydroquinone- and phenol-induced cell transformation, gene mutations, chromosome aberrations, aneuploidy, sister chromatid exchanges and unscheduled DNA synthesis in Syrian hamster embryo cells. Mutat. Res. **373:** 113–123.
90. HUFF, J. 1992. Applicability to humans of rodent-specific sites of chemical carcinogenicity: tumors of the forestomach and of the harderian, preputial, and zymbal glands induced by benzene. J. Occup. Med. Toxicol. **1:** 109–141.
91. HUFF, J.E., J.K. HASEMAN, D.M. DEMARINI, et al. 1989. Multiple-site carcinogenicity of benzene in Fischer 344 rats and B6C3F1 mice. Environ. Health Perspect. **82:** 125–163.
92. HUFF, J.E., W. EASTIN, J. ROYCROFT, et al. 1988. Carcinogenesis studies of benzene, methylbenzene, and dimethyl benzenes. Ann. N.Y. Acad. Sci. **534:** 427–440.
93. HUFF, J. & NTP STAFF. 1986. Toxicology and Carcinogenesis Studies of Benzene (CAS No. 71-43-2) in F344/N Rats and B6C3F1 Mice (Gavage Studies). NTP Tech. Rept. Series # TR-289. National Toxicology Program. Research Triangle Park, NC.

Chlorine

Ramazzini Foundation

94. SOFFRITTI, M., F. BELPOGGI, A. LENZI & C. MALTONI. 1997. Results of long-term carcinogenicity studies of chlorine in rats. Ann. N.Y. Acad. Sci. **837:** 189–208.

National Toxicology Program

95. MELNICK, R.L., M.C. KOHN, J.K. DUNNICK & J.R. LEININGER. 1998. Regenerative hyperplasia is not required for liver tumor induction in female B6C3F1 mice exposed to trihalomethanes. Toxicol. Appl. Pharmacol. **148:** 137–147.
96. DUNNICK, J. & NTP STAFF. 1992. Toxicology and Carcinogenesis Studies of Chlorinated Water (CAS Nos. 7782-50-5 and 7681-52-9) and Chloraminated Water (CAS No. 10599-90-3) (Deionized and Charcoal-Filtered) in F344/N Rats and B6C3F1 Mice (Drinking Water Studies). NTP Tech. Rept. Series # TR-392. National Toxicology Program. Research Triangle Park, NC.
97. DUNNICK, J.K. & R.L. MELNICK. 1993. Assessment of the carcinogenic potential of chlorinated water: experimental studies of chlorine, chloramine, and trihalomethanes. J. Natl. Cancer Inst. **85:** 817–822.

Diesel Fuel

Ramazzini Foundation

98. MALTONI, C., A. CILIBERTI, C. PINTO, *et al.* 1997. Results of long-term experimental carcinogenicity studies of the effects of gasoline, correlated fuels, and major gasoline aromatics on rats. Ann. N.Y. Acad. Sci. **837:** 15–52.
99. MALTONI, C. 1995. The contribution of experimental [animal] studies to the control of industrial carcinogenesis. Appl. Occup. Environ. Hyg. **10:** 749–760.

National Toxicology Program

100. NTP STAFF. 1986. Toxicology and Carcinogenesis Studies of Marine Diesel Fuel (NO CAS) and J.P-5 Navy Fuel (CAS No. 8008-20-6) in B6C3F1 Mice (Dermal Studies). NTP Tech. Rept. Series # TR-306. National Toxicology Program. Research Triangle Park, NC.

Ethylbenzene

Ramazzini Foundation

101. MALTONI, C., A. CILIBERTI, C. PINTO, *et al.* 1997. Results of long-term experimental carcinogenicity studies of the effects of gasoline, correlated fuels, and major gasoline aromatics on rats. Ann. N.Y. Acad. Sci. **837:** 15–52.

National Toxicology Program

102. CHAN, P.C., J.K. HASEMAN, J. MAHLERI & C. ARANYI. 1998. Tumor induction in F344/N rats and B6C3F1 mice following inhalation exposure to ethylbenzene. Toxicol. Lett. **99:** 23–32.
103. CHAN, P.C. & NTP STAFF. 1999. Toxicology and Carcinogenesis Studies of Ethylbenzene (CAS No. 100-41-4) in F344/N Rats and B6C3F1 Mice (Inhalation Studies). NTP Tech. Rept. Series # TR-466. National Toxicology Program. Research Triangle Park, NC.

Methylene Chloride (Dichloromethane)

Ramazzini Foundation

104. MALTONI, C., G. COTTI & G. PERINO. 1988. Long-term carcinogenicity bioassays on methylene chloride administered by ingestion to Sprague-Dawley rats and Swiss mice and by inhalation to Sprague-Dawley rats. Ann. N.Y. Acad. Sci. **534:** 352–366.

National Toxicology Program

105. MENNEAR, J. & NTP STAFF. 1986. Toxicology and Carcinogenesis Studies of Dichloromethane (Methylene Chloride) (CAS No. 75-09-2) in F344/N Rats and B6C3F1 Mice (Inhalation Studies). NTP Tech. Rept. Series # TR-306. National Toxicology Program. Research Triangle Park, NC.
106. MENNEAR, J.H., E.E. MCCONNELL, J.E. HUFF, *et al.* 1988. Inhalation toxicity and carcinogenesis studies of methylene chloride (dichloromethane) in F344/N rats and B6C3F1 mice. Ann. N.Y. Acad. Sci. **534:** 343–351.

Propylene

Ramazzini Foundation

107. CILIBERTI, A., C. MALTONI & G. PERINO. 1988. Long-term carcinogenicity bioassays on propylene administered by inhalation to Sprague-Dawley rats and Swiss mice. Ann. N.Y. Acad. Sci. **534:** 235–245.

National Toxicology Program

108. RENNE, R.A., W.E. GIDDENS, G.A. BOORMAN, et al. 1986. Nasal cavity neoplasia in F344/N rats and (C57BL/6 x C3H)F1 mice inhaling propylene oxide for up to two years. J. Natl. Cancer Inst. **77:** 573–582.
109. QUEST, J.A. & NTP STAFF. 1985. Toxicology and Carcinogenesis Studies of Propylene (CAS No. 115-07-1) in F344/N Rats and B6C3F1 Mice (Inhalation Studies). NTP Tech. Rept. Series # TR-272. National Toxicology Program. Research Triangle Park, NC.
110. BOORMAN, G. & NTP STAFF. 1985. Toxicology and Carcinogenesis Studies of Propylene Oxide (CAS no. 75-56-9) in F344/N Rats and B6C3F1 Mice (Inhalation Studies). NTP Tech. Rept. Series # TR-267. National Toxicology Program. Research Triangle Park, NC.
111. QUEST, J.A., J.E. TOMASZEWKI, J.K. HASEMAN, et al. 1984. Two-year inhalation toxicity study of propylene in F344/N rats and B6C3F1 mice. Toxicol. Appl. Pharmacol. **76:** 288–295.

Styrene/Styrene Oxide

Ramazzini Foundation

112. CONTI, B., C. MALTONI, G. PERINO & A. CILIBERTI. 1988. Long-term carcinogenicity bioassays on styrene administered by inhalation, ingestion and injection and styrene oxide administered by ingestion in Sprague-Dawley rats, and para-methylstyrene administered by ingestion in Sprague-Dawley rats and Swiss mice. Ann. N.Y. Acad. Sci. **534:** 203–234.
113. MALTONI, C., G. FAILLA & G. KASSAPIDIS. 1979. First experimental demonstration of the carcinogenic effects of styrene oxide; long-term bioassays on Sprague-Dawley rats by oral administration. Med. Lav. **70:** 358–362.

National Toxicology Program

114. LIJINSKY, W. 1986. Rat and mouse forestomach tumors induced by chronic oral administration of styrene oxide. J. Natl. Cancer Inst. **77:** 471–476.
115. HUFF, J.E. 1984. Styrene, styrene oxide, polystyrene, and beta-nitrostyrene/styrene carcinogenicity in rodents. Prog. Clin. Biol. Res. **141:** 227–238.
116. NCI STAFF. 1979. Bioassay of a Solution of b-Nitrostyrene and Styrene for Possible Carcinogenicity (CAS No. 102-96-5, CAS No. 100-42-5). NCI Tech. Rept. Series # TR-170. National Cancer Institute. Bethesda, MD.
117. NCI STAFF. 1979. Bioassay of Styrene for Possible Carcinogenicity (CAS No. 100-42-5). NCI Tech. Rept. Series # TR-185. National Cancer Institute. Bethesda, MD.

Toluene

Ramazzini Foundation

118. MALTONI, C., A. CILIBERTI, C. PINTO, et al. 1997. Results of long-term experimental carcinogenicity studies of the effects of gasoline, correlated fuels, and major gasoline aromatics on rats. Ann. N.Y. Acad. Sci. **837:** 15–52.
119. MALTONI, C., B. CONTI, G. COTTI & F. BELPOGGI. 1985. Experimental studies on benzene carcinogenicity at the Bologna Institute of Oncology: current results and ongoing research. Am. J. Ind. Med. **7:** 415–446.

National Toxicology Program

120. HUFF, J. Absence of toluene carcinogenicity in rodents following long-term inhalation exposure. Intl. J. Occup. Environ. Health. In press.

121. HUFF, J. & NTP STAFF. 1990. Toxicology and Carcinogenesis Studies of Toluene (CAS No. 108-88-3) in F344/N Rats and B6C3F1 Mice (Inhalation Studies). NTP Tech. Rept. Series # TR-371. National Toxicology Program. Research Triangle Park, NC.
122. HUFF, J.E., W. EASTIN, J. ROYCROFT, et al. 1988. Carcinogenesis studies of benzene, methylbenzene, and dimethyl benzenes. Ann. N Y Acad. Sci. **534:** 427–440.

Trichloroethylene

Ramazzini Foundation

123. MALTONI, C., G. LEFEMINE, G. COTTI & G. PERINO. 1988. Long-term carcinogenicity bioassays on trichloroethylene administered by inhalation to Sprague-Dawley rats and Swiss and B6C3F1 mice. Ann. N.Y. Acad. Sci. **534:** 316–342.

National Toxicology Program

124. NCI STAFF. 1976. Carcinogenesis Bioassay of Trichloroethylene (CAS No. 79-01-6). NCI Tech. Rept. Series # TR-2. National Cancer Institute. Bethesda, MD.
125. NTP STAFF. 1988. Toxicology and Carcinogenesis Studies of Trichloroethylene (CAS No. 79-01-6) in Four Strains of Rats (ACI, August, Marshall, Osborne-Mendel) (Gavage Studies) NTP Tech. Rept. Series # TR-273. National Toxicology Program. Research Triangle Park, NC.
126. NTP STAFF. 1990. Carcinogenesis Studies of Trichloroethylene (without Epichlorohydrin) (CAS No. 79-01-6) in F344/N Rats and B6C3F1 Mice (Gavage Studies) NTP Tech. Rept. Series # TR-243. National Toxicology Program. Research Triangle Park, NC.

Trichlorofluoromethane

Ramazzini Foundation

127. MALTONI, C., G. LEFEMINE, D. TOVOLI & G. PERINO. 1988. Long-term carcinogenicity bioassays on three chlorofluorocarbons (trichlorofluoromethane, FC11; dichlorodifluoromethane, FC12; chlorodifluoromethane, FC22) administered by inhalation to Sprague-Dawley rats and Swiss mice. Ann. N.Y. Acad. Sci. **534:** 261–282.

National Toxicology Program

128. NCI STAFF. 1978. Bioassay of Trichlorofluoromethane for Possible Carcinogenicity (CAS No. 75-69-4). NCI Tech. Rept. Series # TR-106. National Cancer Institute. Bethesda, MD.

Vinylidene Chloride

Ramazzini Foundation

129. COTTI, G., C. MALTONI & G. LEFEMINE. 1988. Long-term carcinogenicity bioassay on vinylidene chloride administered by inhalation to Sprague-Dawley rats. New results. Ann. N.Y. Acad. Sci. **534:** 160–168.
130. MALTONI, C. & D. TOVOLI. 1979. First experimental evidence of the carcinogenic effects of vinylidene fluoride; long-term bioassays on Sprague-Dawley rats by oral administration. Med. Lav. **70:** 363–368.
131. MALTONI, C. 1977. Recent findings on the carcinogenicity of chlorinated olefins. Environ. Health Perspect. **21:** 1–5.
132. MALTONI, C., G. COTTI, L. MORISI & P. CHIECO. 1977. Carcinogenicity bioassays of vinylidene chloride. Research plan and early results. Med. Lav. **68:** 241–262.
133. MALTONI, C. 1976. Occupational chemical carcinogenesis: new facts, priorities and perspectives. IARC Sci. Publ. **13:** 127–149.

National Toxicology Program

134. CHABRA, R. & NTP STAFF. 1982. Carcinogenesis Bioassay of Vinylidene Chloride (CAS No. 75-35-4) in F344 Rats and B6C3F1 Mice (Gavage Study). NTP Tech. Rept. Series # TR-228. National Toxicology Program, Research Triangle Park, NC.

Xylene

Ramazzini Foundation

135. MALTONI, C., A. CILIBERTI, C. PINTO, *et al.* 1997. Results of long-term experimental carcinogenicity studies of the effects of gasoline, correlated fuels, and major gasoline aromatics on rats. Ann. N.Y. Acad. Sci. **837:** 15–52.
136. MALTONI, C., B. CONTI, G. COTTI & F. BELPOGGI. 1985. Experimental studies on benzene carcinogenicity at the Bologna Institute of Oncology: current results and ongoing research. Am. J. Ind. Med. **7:** 415–446.

National Toxicology Program

137. HUFF, J., W. EASTIN, J. ROYCROFT, *et al.* 1988. Carcinogenesis studies of benzene, methylbenzene, and dimethyl benzenes. Ann. N.Y. Acad. Sci. **534:** 427–440.
138. EASTIN, W. & NTP STAFF. 1986. Toxicology and Carcinogenesis Studies of Xylenes (Mixed) (60% m-Xylene, 14% p-Xylene, 9% o-Xylene, and 17% Ethylbenzene) (CAS No. 1330-20-7) in F344/N Rats and B6C3F1 Mice (Gavage Studies). NTP Tech. Rept. Series # TR-327. National Toxicology Program. Research Triangle Park, NC.

Index of Contributors

Belpoggi, F., 26–45, 46–69, 70–86, 87–105, 106–122, 123–136
Bua, L., 70–86
Bucher, J.R., 198–207

Cattin, E., 70–86, 106–122
Cevolani, D., 46–69, 123–136
Ciliberti, A., 106–122

Guarino, M., 46–69, 123–136

Huff, J., 208–229

Lambertini, L., 87–105, 123–136
Lauriola, M., 87–105, 106–122

Maltoni, C., 26–45, 46–69, 70–86, 87–105, 106–122, 123–136
Mehlman, M.A., ix–xii, 1–25, 137–148, 149–159
Melnick, R.L., 177–189
Minardi, F., 26–45, 70–86, 106–122

Padovani, M., 46–69, 87–105

Soffritti, M., 26–45, 46–69, 70–86, 87–105, 106–122, 123–136
Suzuki, Y., 160–176

Tomatis, L., 190–197

Yuen, S.R., 160–176

OHIO UNIVERSITY LIBRARY

Please return this book as soon as you have finished with it. In order to avoid a fine it must be returned by the latest date stamped below. All books are subject to recall after two weeks or immediately if needed for reserve.

CF